Before Nature

Before Nature

Cuneiform Knowledge and the History of Science

FRANCESCA ROCHBERG

The University of Chicago Press

CHICAGO AND LONDON

The University of Chicago Press, Chicago 60637
The University of Chicago Press, Ltd., London

Published 2016
Paperback edition 2020
Printed in the United States of America

29 28 27 26 25 24 23 22 21 20   1 2 3 4 5

ISBN-13: 978-0-226-40613-8 (cloth)
ISBN-13: 978-0-226-75958-6 (paper)
ISBN-13: 978-0-226-40627-5 (e-book)
DOI: https://doi.org/10.7208/chicago/9780226406275.001.0001

Library of Congress Cataloging-in-Publication Data

Names: Rochberg, Francesca, 1952– author.
Title: Before nature : cuneiform knowledge and the history of science / Francesca Rochberg.
Description: Chicago : The University of Chicago Press, 2016. | Includes bibliographical references and index.
Identifiers: LCCN 2016019567 | IISBN 9780226406138 (cloth : alk. paper) | ISBN 9780226406275 (e-book)
Subjects: LCSH: Assyro-Babylonian literature—History and criticism. | Science—Assyria. | Philosophy of nature—Assyria. | Learning and scholarship—Assyria. | Astronomy, Assyro-Babylonian.
Classification: LCC Q125 .R716 2016 | DDC 509.35—dc23
LC record available at https://lccn.loc.gov/2016019567

♾ This paper meets the requirements of ANSI/NISO Z39.48-1992 (Permanence of Paper).

Dedicated to the memory of my father,
George Rochberg
1918–2005

Molto affettuoso

Contents

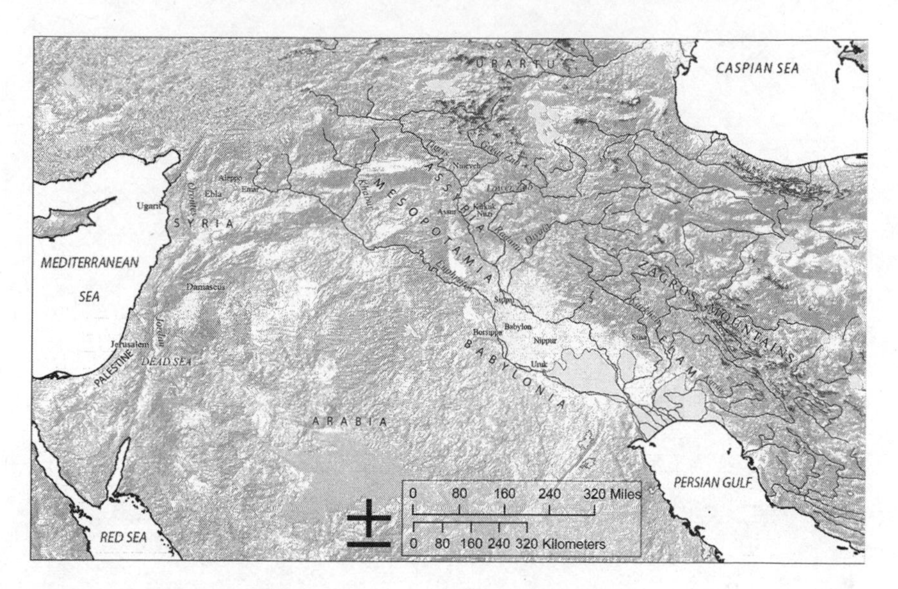

Map of the Ancient Near East

INTRODUCTION

The Ancient Near East, Science, and Nature

The aim of this book is to raise and explore questions about observing and interpreting, theorizing and calculating what we think of as natural phenomena in a world in which there was no articulated sense of nature in our terms, no reference or word for it. This seems to be the case in the cuneiform world. What I mean by the cuneiform world is nothing more (or less) than that which is represented by the cuneiform corpus of the Babylonian and Assyrian *literati*, the scholarly specialists in bodies of knowledge relating to the phenomena.

The sources in question come from a variety of textual corpora produced by elite Assyro-Babylonian scribes from the last two millennia BCE. While the texts of these diverse corpora have their distinct content, form, purpose, orthographic style, intellectual goals and attitudes, the phrase *the cuneiform world* can serve to evoke the context of cuneiform scholarship in contrast to that of other corpora from antiquity. The purpose of raising questions of knowing, predicting, explaining, and interpreting the phenomena is to reflect on the nature and aim of knowledge in cuneiform sources and the relationship between that complex body of knowledge and the broad scope of the history of science.

The project of coming to some comprehension of an intellectual culture in which the conception of nature did not occupy the enormous place in the organization of the world as it does today, or has since early in the history of Western natural philosophy, may not seem to intersect with or impinge upon the history and philosophy of natural science virtually by definition. If we are to pursue the history of science beyond the boundaries of the

Western classical tradition into the ancient Near East, obviously the category *nature* was not native in these parts, and our histories of the engagement with natural phenomena in the ancient Near East need some further explication of the framework in which such phenomena were understood.

Geoffrey Lloyd argued that the idea of nature as a domain inclusive of phenomena occurring in the nonhuman-made environment was an invention of the Greeks.[1] Edward Grant viewed nature as a given, not an invention.[2] But however given the phenomena of the physical environment might seem to some of us to be, the ways in which such phenomena are seen, understood, and described are a function of a specific intellectual and experiential orientation toward the world around. As such, nature is not a static concept, and the content and form of science in history is vivid testament to that fact. Still, one would like to understand what it means to say that Assyro-Babylonian inquiry into the phenomena belongs to science, when nature, in any way comparable to other historical conceptions or usages of that category, was, in the terms of cuneiform texts, not the object of inquiry.

The basic historical question, therefore, is: If science is understood as the inquiry about nature and its phenomena, how do cuneiform texts relate to the larger framework of the history of science? A related philosophical question is: What kind of science is it that does not have nature as its conscious object of inquiry? And finally, from the point of view of the cultural values we attach to science and scientific knowledge, we might ask in what way can knowledge that is not consciously directed at nature be usefully counted as science?

There is much existing scholarship on the relation between natural philosophy and science, but less has been said about these questions. It seems to me that the cuneiform corpus presents a different reference point from any other body of systematized knowledge and should be considered on its own with respect to the usual historical sources for telling the story of science. Although posed from three vantage points, the questions raised here can be reduced to one fundamental one—namely, what is the relationship between cuneiform knowledge—and knowledge practices—and what we call science? How to understand cuneiform knowledge in relation to science without recourse to later ideas of nature is the leitmotif of this book.

The scribes interested in the phenomena of their particular surrounding environment as signs—phenomena such as eclipses, animal and bird behavior, the features of the human form, or the appearance of the liver of a sacrificed sheep—were engaged in the world in a way that immediately resonates with any number of historical sciences in their use of observation, analysis of regularities and anomalies, calculation, and prediction. It is this particular orientation to the world expressed in and through the textual output of the scribal elite of Babylonia and Assyria, the *literati* of those societies, that concerns the following chapters.

If there is a place for the cuneiform world in the history of science, it is not to be defined only in terms of astronomy, for which the Babylonians were famed in antiquity and admired in modern times. The clay tablets containing the work of scholars specialized in the calculation of lunar and planetary phenomena belong to a wider tradition of cuneiform scribal learning focused on a vast array of phenomena, not only those of the heavens. Indeed, the cuneiform scholars' inquiries produced diverse bodies of knowledge for the study of signs, for medicine, magic, and astrology, and employed all the varieties of reasoning that we now associate with rationality and with science, that is, empirical, inductive, deductive, and analogical.

The field of the history of science has already come to terms with major change in the idiom of science and its objects of knowledge over the *longue durée*. The meaning or understanding of technical terms of scientific discourse, such as *mass*, *lunar latitude*, or *eclipse*, change over time, and some, such as *phlogiston*, *celestial spheres*, or the *luminiferous ether* fall away entirely. The meanings of *episteme* for Aristotle, or *scientia* for Augustine, or for the Scholastics, are not applicable to other contexts of scientific discourse outside their own. Some room in this epistemic plurality, therefore, has to be made for the language of the Assyrian and Babylonian scholars if the terms *nēmequ* or *ṭupšarrūtu*, roughly "knowledge (of skills, of bodies of scholarly texts)" and "scribal learning, or scholarship," respectively, can even be considered in relation to other historical ideas of knowledge.

In setting the cuneiform corpus and its native epistemic discourse alongside those of other premodern scientific cultures, a certain kinship between knowledge and methods of knowing developed in the cuneiform world can be identified with those of later periods. To describe relation in terms of kinship implies more than borrowings and parallels, which have long been identified between Babylonian and Greek, or even later antique and medieval tradition. It implies what Marshall Sahlins called "mutuality of being," that is, in his words: "kinfolk are persons who participate intrinsically in each other's existence; they are members of one another."[3] What I argue for is membership for cuneiform knowledge within the family of knowledge systems, ways of knowing and engaging with the world of phenomena, that make up the history of what we call science.

Kinship, however, does not imply sameness. In reflecting on the nature of knowledge in cuneiform sources and the related question of its relationship to the broader scope of the history of science, an important dissimilarity or discontinuity to be recognized is that the object of cuneiform knowledge was not conceived of as nature. The fact that what we consider to be the realm of nature with all its phenomena did not always provide the framework for scientific knowledge is one idea that so far has not been taken up as a serious historical question. To more fully depict and elucidate the nature of cuneiform knowledge in terms of its own objects of inquiry, however, this question must be considered.

The concept of nature has a long and complex history and has represented the framework for the world construed in various ways, including or not including gods, or God, taken as self-moving or as materialistic. Clarence Glacken said "the word 'nature,' as everyone knows, has many meanings in Greek and Latin and in modern languages. With all of its failings it is a grand old word."[4] He went on to say: "When Huxley wrote *Evidence as to Man's Place in Nature* (1863), he discussed man's place in the evolutionary scale of being. When Marsh wrote *Man and Nature* in 1864 he described the earth as modified by human action. Sometimes the word is synonymous with the physical or natural environment; sometimes it has a more philosophical, religious, theological aura than these more matter-of-fact terms express. Occasionally it attains grandeur as in Buffon's reference to it as 'le trône extérieur de la magnificence Divine.'"[5] R. G. Collingwood prepared his lectures that eventually became *The Idea of Nature* (1944) over the course of the 1930s, leaving the manuscript unfinished at his death. His book opens with the statement: "In the history of European thought there have been three periods of constructive cosmological thinking; three periods, that is to say, when the idea of nature has come into the focus of thought, become the subject of intense and protracted reflection, and consequently acquired new characteristics which in their turn have given a new aspect to the detailed science of nature that has been based upon it."[6] The assumption that science is the result of the focus of thought upon the idea of nature poses a problem for the astronomy/astrology, medicine, magic, and divination from the two millennia prior to Collingwood's three periods of European cosmological thinking. In the context of the cuneiform world, I propose, the prehistory of the conception of nature is not found to be the equivalent of the prehistory of science.

In Raymond Williams's *Key Words: A Vocabulary of Culture and Society*, the words *nature* and *culture* were both singled out for their semantic and historical complexity. Culture, he deemed "one of the two or three most complicated words in the English language,"[7] but nature—

> "Nature" is perhaps the most complex word in the language. It is relatively easy to distinguish three areas of meaning: (i) the essential quality and character *of* something; (ii) the inherent force which directs either the world or human beings or both; (iii) the material world itself, taken as including or not including human beings. Yet it is evident that within (ii) and (iii), though the area of reference is broadly clear, precise meanings are variable and at times even opposed. The historical development of the word through these three senses, and the main variations and alternatives within the two most difficult of them, are still active and widespread in contemporary usage.[8]

Parsing the meanings and usages of the word *nature* is complex for all the reasons given by Williams. But when the conception of nature becomes a tool for understanding systems of thought that do not, or did not, make use of the term, yet another degree of complexity is added to the mix. It is *our* term, *our* conception, *our* tool of understanding. Of course we are free to use whatever intellectual tools we have at our disposal, but, to quote Williams again, "The complexity of the word ['nature'] is hardly surprising, given the fundamental importance of the processes to which it refers. But since 'nature' is a word which carries, over a very long period, many of the major variations of human thought—often, in any particular use, only implicitly yet with powerful effect on the character of the argument—it is necessary to be especially aware of its difficulty."[9] In pursuing a line of inquiry into the history of science as it might relate to cuneiform knowledge and cuneiform intellectual practices, it seems to me particularly necessary to be especially aware of the difficulty of assuming that, like most sciences we know of, cuneiform knowledge had any stakes in knowing nature.

If anything seems fixed in our minds, here in the modern scientific world, it is nature, and with it, the dichotomy of nature and culture. We seem to need one to define the other. The appeal of and to this dichotomy culminated in the famous expression "nature versus nurture" of eugenics creator Francis Galton in the late nineteenth century.[10] It remains a deeply rooted given in some quarters until today. Many tropes and debates about human behavior still point to a division into internals and externals predicated on the idea of the material realities of nature inhabiting our bodies (genetic material, bacteria, amino acids, chemical reactions) while we create and inhabit social worlds. Other ways of seeing how nature and culture interact or intersect in the human being have long been central to philosophy on the question of where nature as matter falls in relation to the human mind. The duality of nature and culture once thoroughly underpinned anthropological discourse, epitomized as Lévi-Strauss's "fundamental structures of the human mind." Terence Turner saw that "Lévi-Strauss conceived the nature-culture relation ambiguously as both external and internal: externally as a boundary between human culture and the world of nature beyond the village; and internally as the psychological divide between the mental processes of perception and association and the consciousness of the cultural subject."[11] Nature and culture, as Turner made clear, remained separate and distinct ideas for structuralism.

Geoffrey Lloyd's recent reassessment of the dichotomy, or what he called *some* dichotomy between nature and culture, raised the question of how universal to human thought and experience the dichotomy really is:

> On one line of interpretation it would be the nature/culture division itself that would be an underlying assumption common to all societies, though then it becomes important to consider whether or how far there are culture-specific differences in the apprehension of that dichotomy. The opposing line of argument would have it that that dichotomy is a typical example of the imposition of Western categories on other people's world-views, though that then raises the problem of whether those views are incommensurable with our own. Either way, a first observation would be that we seem to have to make room for a variety of view-points on the question for two distinct reasons. First, there are . . . societies where there is no explicit concept of 'nature' as such at all, though that should not be confused with a denial that they have an implicit grasp of that domain. Second, how 'culture' is interpreted also varies remarkably, even before we take into consideration the fact that primatologists and others claim that certain species of animals possess culture at least in a restricted sense.[12]

Lloyd's suggestion that room be made for other viewpoints on how the world is understood to be constituted is of the essence for explicating texts from ancient Assyria and Babylonia that have to do with the observation, interpretation, and calculation of many phenomena, and thus, I would argue, for the analysis of the practice of science in the ancient Near East.

The inception of science was once thought to begin with the invention and study of nature in the Greek philosophical milieu. The remarkable achievements of the late Babylonian astronomers in the Achaemenid and Hellenistic periods were a primary mitigating factor in the revision of that narrative. But in the course of the four-thousand-year-long civilization of the ancient Near East, for two thousand years of that prodigious period, diverse efforts to describe, understand, schematize, and calculate aspects of observable, possible, and conceivable phenomena are attested in forms of Assyrian and/or Babylonian divination, medicine, and astronomy/astrology. In their own terms, these efforts, however, did not explicitly focus on nature, nor did they require that their objects of inquiry be explained in terms of that separate domain, or of what we would consider to be natural causes. A science not focused upon nature so as to understand nature's independent existence and functioning is not something we are used to considering, but cuneiform texts force this consideration upon us.

If the epistemic goal of science is to understand, consciously, the workings of nature, where are cuneiform texts to be placed, and how are they to be understood within science's history? The problem lies not only with the term *nature*, but also with the term *science*. The ways we conceive of the meanings of both of these terms are insufficient for encompassing cuneiform intellectual culture with its particular forms and practices of knowledge, reasoning, and values. Just as naturalism provides one way to understand the world, there was another way attested in the cuneiform Near East. What that way was is expressed by the interests of scholarly cuneiform texts dealing with phenomena.

This book does not set out to translate nature into a cuneiform idiom; that is, it does not seek to find the counterpart to the conception of nature in cuneiform texts. It does not take up the history of the conception of nature in order to demarcate just exactly when it made its first or clearest appearance. It does not try to modify a definition of nature to accommodate cuneiform textual evidence. It does attempt to discover another way of understanding cuneiform scholars' interest in their phenomenal world, the things that were known as the result of the particular aims of their knowledge, and why the study of the ancient cuneiform scholars' interest in phenomena can repay us in a broader and deeper understanding of the history of science and the scientific imagination.

The title *Before Nature* requires a few further words of clarification beyond what has been said. In the middle of the last century, as the commitment to sociocultural evolutionism waned in anthropology, but scientism still soared, a collection of essays was published by Henri Frankfort, H. A. Groenewegen-Frankfort, John A. Wilson, Thorkild Jacobsen, and William A. Irwin titled *The Intellectual Adventure of Ancient Man: An Essay of Speculative Thought in the Ancient Near East* (1946). The volume was reissued in 1949 by Penguin under the more evocative title *Before Philosophy: The Intellectual Adventure of Ancient Man*. In the paperback edition, *Before Philosophy* was widely disseminated and became authoritative as a depiction of a distinctive cognitive-intellectual character for ancient civilizations—namely, Mesopotamia (geographically equivalent for the most part with Iraq), Egypt, and Israel, in contrast to that of the better known Greeks and Romans. The thesis of a special ancient Near Eastern cognitive attitude (called primitive in accordance with the terms of the day) was formulated in terms of what and how various groups of people (designated by the quasi-national divisions, "Mesopotamia," "Egypt," and "Israel") thought about humankind's place in the world of nature and in relation to the divine. It was not, however, portrayed as an intellectual history of the ancient Near East inasmuch as philosophy and science were not regarded as having taken root there, hence, *Before Philosophy*.

Based on an evolutionary approach to the relationship between culture and mind, what *Before Philosophy* aimed to describe and analyze was rather an essential form of ancient Near Eastern civilization in terms of its stage of cognitive development. Despite its particular cognitive-historical reasons for a separation of ancients from moderns—evidenced principally in the use of the word *primitive* to designate ancient thought compared to *modern*—*Before Philosophy*, written by distinguished archaeologists and philologists, was nonetheless an attempt to reach an understanding of this ancient thought-world not in terms of Western models or constructs (though how successful that was can be debated) but from within the various worlds of its sources. The book's claim that scientific thought was developmentally excluded in these cultures was based, however, on an analysis of mythological and religious texts, as well as on the use of mid-twentieth-century standards in defining what counted as scientific thought. According to the cognitive history of *Before Philosophy*, thought in the ancient Near East lacked "cogency of reasoning," was "wrapped in imagination," and "tainted with fantasy." What the Frankforts found lacking was, in particular, the capacity for abstraction. Ancient thought was classified as "prescientific," the result of a failure to separate nature from culture, two domains that in their view and according to their time were clear and distinct.

Times have changed. For historians, sociocultural evolutionism and scientism are long out of fashion. The highly cultural and particularly Western nature of nature has been acknowledged. From a historical viewpoint it becomes increasingly difficult to see a direct and simple connection between science as knowing subject and a pure realm of nature as its object of knowledge. Nonetheless, the very idea of science still seems to imply or even depend on a concept of nature and on inquiry aimed at knowing it.

The question of the nature and aim of the cuneiform scribes' inquiry into phenomena before the name *nature* was given to any separate and discrete order of being or knowing is therefore central to this book.

The present study did not set out to be in dialogue with the Frankforts, or with Thorkild Jacobsen, Sumerologist and mythographer, who wrote the essay on Mesopotamia in that collection. But I soon saw that engagement with the Frankforts would be especially useful as a point of departure for the present discussion of science and nature with respect to ancient Mesopotamia. What is particularly problematic about *Before Philosophy* is that it purported to offer a synthesis of ancient Near Eastern ideas concerning human beings' relation to (what we distinguish as) the physical and the metaphysical, and how various social groups of Near Eastern antiquity interpreted their experience, yet no evidence for a direct engagement with physical phenomena, such as astral or other omens, or astronomy itself, little known to nonspecialists (even less so than it is today) at the time of *The Intellectual Adventure*'s writing, was brought to bear. The discussion focused on mythological texts as evidence for a lack of "scientific thought." There is, however, much more textual evidence that the ancient cuneiform scribes were fully engaged in the study of many more phenomena than can be represented in their mythological texts. Even in view of this greater scope of material, the Assyrian and Babylonian conceptual world cannot be so easily mapped onto our own, nor is our desire to classify things into natural phenomena or natural kinds useful for explaining what interested the cuneiform scribes.

The sources for the present study are wholly distinct from those of Thorkild Jacobsen's essay. My discussion focuses on Akkadian cuneiform texts that deal with a variety of phenomena (physical and nonphysical, that is, hypothetical, imagined, possible, and conceivable phenomena) as objects of inquiry, and which, among other things, concerned prediction from and prediction of those phenomena. Such texts include a broad spectrum of omens, both celestial and terrestrial, and together constitute the quintessential form and also the bulk of scholarly learning in cuneiform culture. Also included are the texts we classify as astronomical, or astronomical/astrological, magical, and medical. These were all part of the vast literature the scribes of the first millennium termed *ṭupšarrūtu*, "the art of the scribe," or "scholarship," the intellectual province of the master scribe, the *ummânu*, that is, "expert," or "scholar." The history of *ṭupšarrūtu* itself exhibits shifts and changes in the treatment and understanding of phenomena over its long history.

The relation between cuneiform knowledge and science has both historical and philosophical dimensions. The philosophical dimension comes from contemporary concerns with understanding the nature of science, but contemporary concerns have deep roots in ancient Greek philosophy that were tied to the conceptual realm of nature. The historical dimension is supplied by clay tablets that anywhere from four to two thousand years ago were written by scribes inhabiting the cities of what is now mostly Iraq. There is also the later filtration of some of these sources in our Western intellectual heritage from Late Antiquity through the Middle Ages and into the Renaissance. These two dimensions, the historical and the philosophical, in my view, interestingly intersect in this context.

On the historical side the intellectual heritage of the West begins in the ancient Near East, and so there is already reason to look for ways in which the cuneiform inquiry into the phenomena of the world on one hand and of later science on the other might be, in Sahlins's terms, "members of one another."[13] The Hellenistic reception of Babylonian astrological and astronomical traditions not only perpetuated aspects of cuneiform knowledge but also created new traditions anchored in but not wholly defined by such knowledge.[14] The idea of Babylon as the source of astronomical science is a common thread. One of the latest vestiges is found in the thirty folios of the *Liber Nimrod*, a work in which the fictive astronomer Nimrod, perhaps the latest exemplar of the topos of the wise Babylonian astronomer, teaches, mostly in mythological terms, his disciple, Ioanton, aspects of astronomy and cosmology.[15] Although Nimrod is derived ultimately from the Bible (Gen. 10:8, also 1 Chron. 1:10 and Mic. 5:6), later tradition identified him with Ninus, a fictive king and founder of Nineveh, and established spurious connections for him with the history of ancient Mesopotamia. The transformation of Nimrod into an astronomer from Babylonia is an example of the impact of cuneiform knowledge upon the Western imagination.

Legitimate traces of the original Babylonian astronomy and astrology, the celestial divination and prognostication, observational records and ephemerides written on cuneiform tablets, entered Europe too, as evidenced in Ptolemy's *Tetrabiblos*, or Geminus's *Introduction to the Phenomena*.[16] Later, however, despite occasional ascriptions of certain traditions to Chaldeans, genuine Babylonian ideas were largely unknown to all but the rarest and most astute of textual critics, such as the classicist and chronographer Scaliger. We may note with a bit of irony that not many years after Scaliger's death, the celebrated seventeenth-century traveler Pietro della Valle brought back to Rome the first cuneiform texts ever seen in Europe.

On the philosophical side the cuneiform source material can be profitably analyzed in terms of themes familiar to the philosophical investigation of science, such as Empiricism, rationality, causality, and model-making. Such analysis, it is hoped, can add to ongoing debates about epistemology in the history of science more generally, particularly those concerned with local versus universal knowledge, or pluralism versus monism about scientific knowledge.

To respond both to historical and philosophical dimensions of the relationship of cuneiform knowledge to science, this book is divided into four parts: Part I is historiographical; part II is epistemological; part III has to do with reasoning, causality and law, and part IV with the phenomenological framework within which the cuneiform scholarly imagination took shape and developed through celestial observation, prediction, and explanation.

As the historiographical part, part I sets out to bring cuneiform sources to the table in a consideration of the historiography of science. Chapter 1 establishes a conceptual reference point for such a history of science that includes cuneiform knowledge as a representation of other ways in which the world was studied, organized, and understood beyond the borders of the notional territory of nature. Chapter 2 returns to Henri Frankfort and colleagues' *Before Philosophy: The Intellectual Adventure of Ancient Man* as a historiographical point of departure and aims to put the present work into its historical context and to mount a detailed critique of its approach.

Critiques of the validity of the Western category of science for non-Western or ancient sources have been aired by others elsewhere, making the case that the theological component of ancient knowledge systems makes it difficult to separate science from religion in antiquity (if not up until the early modern period as well). The anachronistic separation of ancient science and religion is not the battle engaged in the chapters of part II. Depending on what aspects of cuneiform texts and their historical and social contexts we seek to analyze and elucidate, the conceptual apparatus of either science or religion or both can come into play. Although the constructs *science* and *religion* share the problem that each uses a present category to understand a past reality, this book is not aimed at a resolution of that particular category problem. In the evidence presented in the chapters of part II, the gods are ever present, but their presence does not mitigate the approach to understanding cuneiform knowledge systems in terms of science rather than religion any more than the presence of God in late antique, medieval, and early modern sources limits the history of science in those periods.

Part II is interested in what was known in the cuneiform scholarly world, and in what context things were objects of knowledge. These are the two aspects central to chapter 3, which looks at ominous signs as a principal, though not the only, element for understanding how knowledge of the heavens, of plants, animals, and stones was developed in the cuneiform world. Understanding that cuneiform knowledge took shape because certain social structures and values both required and produced it, this chapter looks at the major bodies of knowledge produced by scholars of the royal court and divine temple best known from the activities of the scholars in the palaces of Neo-Assyrian monarchs as well as in later developments from Neo-Babylonian and Hellenistic periods. It focuses on written bodies of knowledge of the heavens, of plants, animals, and stones. An exploration of cuneiform knowledge could just as easily focus on what is known as the lexicographical tradition, which offers a wholly different and equally significant dimension to the history of cuneiform epistemology. I leave this complex investigation to specialists in that extensive domain of sources.[17]

The concerns of cuneiform divinatory and astronomical texts with norms against which regularity and irregularity, the normal and the anomalous, were defined, is the subject of chapter 4. It is through this exploration of the systematic consideration of norms, ideals, malformations, and anomalies that a cuneiform modality of order can be established without dependence on a conception of nature. The search for and understanding of order and anomaly also constitutes one of the principal aspects of the kinship relation between science and cuneiform knowledge.

Part III considers how to approach the fact that the ancient cuneiform scribes engaged with the entities of the world from a perspective different from that which lies at the basis of science in its Western form, that is, in the form presumptive of naturalism, together with its distinct methodological, metaphysical, ontological, and epistemological implications. How and what phenomena were known from such a perspective, as examined in part II, is central to the question of the scribes' engagement with their world. No less important is the question of how we are to understand the use of empirical, deductive, and analogical reasoning within the framework of their various inquiries.

Chapter 5 takes up the cognitive-historical matter of the rational, both in contemporary scholarship on cuneiform divination and astronomy and in the evidence for forms of reasoning in various cuneiform knowledge systems. Rationality in cuneiform knowledge is here found to stand on grounds other than those established by Greek philosophical criteria. In the face of the evidence of cuneiform texts, such criteria, once presumed to be universal for defining scientific rationality, can be questioned. Further, because of the absence of nature as a meaningful category in the cuneiform world, the problem of "apparently irrational beliefs," usually associated with explanation by recourse to supernatural agency, must be addressed differently. Indeed, to invoke supernatural agency presupposes a conception of the order of nature, and to base causation upon a supposed category of the supernatural distorts the evidence of cuneiform magical texts that do not operate on the basis of such ideas. Instead, analogical and associative reasoning is considered in the context of cuneiform knowledge, specifically as it functioned independently of a framework of nature and natural phenomena.

Chapter 6 explores causality and law, famously interrelated aspects in the philosophy of science, in relation particularly to Assyro-Babylonian divination, taking up questions about how the scribe-scholars viewed connections between things, and how an idea of cosmic order akin to the laws of nature was expressed before the conception of nature was formulated.

Part IV focuses on celestial divination, astronomy, and astrology, separable more in terms of text types and less as a matter of intellectual domain, to examine how observation, prediction, and explanation of heavenly phenomena were fundamental to those scholarly pursuits, contributing to the scribes' conception of what it meant to understand the phenomena of the heavens.

Chapter 7 looks at the role of observation in the cuneiform knowledge of the heavens. Two corpora shed considerable light on this subject, together spanning close to a thousand years of observational recording. Combined, they provide a legitimate site for the study of the earliest selective attention to heavenly phenomena, meaningful within a context of interdependent knowledge and practices—namely, divination, astronomy, and astrology.

One of the obvious and principal goals of cuneiform knowledge of celestial phenomena was that of prediction. Whether to predict from the signs, or to predict the signs themselves, prediction relates to the empirical and rational dimensions of cuneiform knowledge. Chapter 8 considers two classes of prediction represented by divinatory and astronomical practice, as well as two forms of explanation that divide similarly along the lines of divination and astronomy. A logical symmetry between prediction and explanation was long ago canonical in the philosophy of science. This position has since been abandoned, but more recent discussion of other ways to see interconnection casts an interesting light on the landscape of the scribes' use of prediction and explanation, which do not attest to a logical symmetry but can be seen to have a certain interdependence.

We might classify many of the phenomena of interest in the cuneiform world as natural phenomena, though the ancient scholar did not. What difference does it make? Is the difference merely terminological? I submit that it is not primarily a terminological problem but is fundamentally conceptual, and as such makes a very great difference. Without a reference to nature, and therefore no sense of it comparable to that held by Greek, medieval, or early modern natural philosophers, certainly not of modern scientists, our understanding of the particular orientation of cuneiform scholars to their contemporary environment needs to establish itself on different grounds.

Indeed, our construal of cuneiform bodies of knowledge and their associated activities, as well as every aspect of the philosophy of such knowledge—epistemology, reasoning, causality, observation, explanation, and prediction—must be sensitive to its native conceptual grounds, the universe of cuneiform observation, interpretation, and prediction. In this regard cuneiform knowledge tests our sensitivity to the limit. With this in mind, the concluding chapter addresses what I hold to be the legitimate place of cuneiform scholarly knowledge in the history of science, and the unique framework it established within which to exercise its particular scientific imagination.

PART I

Historiography

CHAPTER ONE

Science and Nature

pirišti ilāni rabûti "secret knowledge of the great gods"

> Modern universalism flows directly from naturalist ontology, based as it is on the principle that beyond the muddle of particularisms endlessly churned out by humans, there exists a field of truths reassuringly regular, knowable via tried and trusted methods, and reducible to immanent laws the exactness of which is beyond blight from their discovery process. In short, cultural relativism is only tolerable, indeed interesting to study, in that it stands against the overwhelming background of a natural universalism where truth seekers can seek refuge and solace. Mores, customs, ethos vary but the mechanisms of carbon chemistry, gravitation and DNA are identical for all.
>
> PHILIPPE DESCOLA, "Who Owns Nature?"

> Where the social world of another culture is at issue, we have learned, against our own deep-seated ethnocentric resistance, to take shock for granted. *We can, and in my view must, learn to do the same for their natural worlds.*
>
> THOMAS KUHN, "The Natural and the Human Sciences," in *The Road since Structure: Thomas S. Kuhn*

THE CUNEIFORM WORLD AND THE NATURAL WORLD

There is nothing self-evident about nature. Neither the phenomena of nature nor the environment as such have always been understood in the way we now understand them, and yet we tend to assume that what we call the natural world with its myriad forms, regularities, and irregularities confronts us all and always has. Haven't we always stood and still stand "before nature," observing, inquiring, and seeking to understand it as a whole, or in its parts? There is a serious question of reference that fails to justify such an assumption. Neither the sense nor the reference of the word is absolute across cultures and history. What we make of "the natural world" is a direct function of our particular moment in history, our particular cultural idiom and imagination. R. G. Collingwood concluded his extended historical essay on the idea of nature with the observation that

> nature, though it is a thing that really exists, is not a thing that exists in itself or in its own right, but a thing which depends for its existence upon something else . . . that natural science is not a tissue of fancies or fabrications, mythology or tautology, but is a search for truth . . . but that natural science is not, as the positivists imagined, the only department or form of human thought about which this can be said, and is not even a self-contained and self-sufficient form of thought, but depends for its very existence upon some other form of thought which is different from it and cannot be reduced to it.[1]

At issue in the following chapters is not the ontological problem of whether there is a mind-independent universal state of physical reality that we call nature, a realm apart from human interaction with it, operating in accordance with its own immutable and universal laws. Nor is it the historical problem of the development of the conception of that universal nature. Neither are the various naturalisms relevant to the study of cuneiform knowledge and the world to which it refers—methodological naturalism, where the supernatural has no place in scientific explanation; metaphysical naturalism, where the supernatural does not exist and all supervenes on nature; epistemological naturalism, where knowledge is attached to things such as natural kinds; or ontological naturalism, where only science ascertains what exists (sometimes but not always equated with physicalism). None of these positions will help us gain purchase on the cuneiform world of inquiry about phenomena.

The effort to understand the world of our perception and experience of phenomena might be a basic way to define the impetus for science, even though quite different ways of construing objects of inquiry and different methods of knowing have been historically encompassed by what we call science. There have been any number of claims as to when and where science emerged, from the beginnings of European modernity, to the European Middle Ages, to the classical Greek period, and even in ancient Babylonia. A claim that parts of the ancient cuneiform corpus contain evidence of the thinking and doing of science may well be an affirmation that Theseus's ship retained at least something of its identity after having had all its planks replaced. At the same time, and whether one agrees or not that the "ship of science" was launched from ancient Mesopotamia to be utterly changed by the modern period, it seems plain that the ancient Assyrian and Babylonian scribes of the cuneiform world had no stakes in knowing nature as such.

From a modern Western scientific perspective such a world is difficult to imagine. But as Philippe Descola stated the matter: "Seen from the point of view of a hypothetical Jivaro or Chinese historian of science, Aristotle,

Descartes, and Newton would not appear so much as the revealers of the distinctive objectivity of nonhumans and the laws that govern them, rather, they would seem the architects of a naturalistic cosmology altogether exotic in comparison with the choices made by the rest of humanity in order to classify the entities of the world and establish hierarchies and discontinuities among them."[2] So too did Assyrian and Babylonian scribes engage with the entities of the world from a perspective different from that which lies at the basis of science in its later Western forms. The scribes' engagement with the world, as preserved in learned cuneiform texts, was not presumptive of nature, consequently not open to the possibilities for methodological, epistemological, ontological, and metaphysical naturalisms. In that sense, "before nature" is meant to represent the perspective before the conception of nature was formulated, a perspective wholly open to reconstruction and interpretation. What is offered here is but one such reconstruction and interpretation.

In the cuneiform world, ominous phenomena in general, and astral phenomena in particular, were observed, analyzed, calculated, and predicted by learned and specialized scribes. Entities of the external world were of interest, for example, to diviners, and to medical practitioners, who knew about the efficacy of plants used in administering the sick by mouth, aromatic woods for his or her fumigation, and stones used in making amulets to aid in conjuring evil spirits viewed as responsible for disease. As well, lists and descriptions of objects in the physical environment are attested in extensive bilingual Sumerian and Akkadian lexical lists of the writings of words for trees, birds, domestic and wild animals (from the series *Ura* Tablet XIII, for example),[3] or, indeed, in the simile-filled descriptions of landscape and terrain traversed by the Assyrian army as well as the flora and fauna encountered and described in the highly literary accounts of Neo-Assyrian annals. All of these texts are evidence of the interest in, awareness of, and observance of all kinds of things that exist in the topographies of heaven and earth. But as Niek Veldhuis said in reference to the Sumerian lexical corpus, "the aim of this scholarship was *not to understand nature* or geography but to understand Sumerian and Sumerian writing."[4] To better represent the interests and aims of cuneiform scribal culture vis-à-vis the physical environment, it seems that we ought to divest ourselves of nature as a heuristic category.

Functioning as a posit for the discussion to follow, therefore, is that a sense of nature in our or any other historical terms, such as in Greek or Roman antiquity—in which already there were multiple meanings of nature, including that which encompassed the divine—was not a factor in cuneiform sources bearing on the study of phenomena. And yet a great many phenomena that we would classify as natural exerted a strong pull on the intellectuals of cuneiform society, seen most sharply in the areas of what we designate as astronomy, celestial divination, natal astrology, as well as other kinds of divination, where many different phenomena were taken as signs from the gods.

Cuneiform texts seeking to know about the phenomena were not alonc in antiquity in their interest in the connection between divinity and the world. Ancient Greek astronomy and cosmology exhibit a strong commitment to the idea of the heavens as divine.[5] Even among the philosophers of the Old Academy entities such as demons and spirits, somewhere between human beings and gods had to be fitted into the world as a whole. As P. Merlan put it, "We should call them supernatural, but for a Platonist, as for many other Greeks, the concept of nature was much wider than for us and simply included such entities. We must not forget that in the philosophy of Democritus and Epicurus even gods become 'natural' entities and are simply nature's products."[6] It is, therefore, not simply a matter of the gods versus nature. It is the question of how to reimagine a framework for phenomena that does not involve all-encompassing nature, and how methods of knowing can function within such a framework through the use of empirical, deductive, associative, and analogical reasoning. The relationship between what we think of as natural phenomena and what the Assyro-Babylonian scribes thought of as signs will be a principal point of entry into these questions.

SIGNS AND WONDERS

Signs, or omens, in cuneiform texts are to be distinguished from anomalies, prodigies, or what later might have been seen as marvels or wonders. Unlike the miraculous, signs in the cuneiform omen corpus are not limited to anomalous occurrences, although certainly anomalies were an important part of the vast divinatory enterprise. An entire omen series was devoted to anomalous births (*Šumma izbu*).[7] Anomalies were, however, not defined against nature, or as preternatural, but against certain patterns within which phenomena were observed to occur in an ordered world.

A phenomenon such as a solar eclipse, before its regularity and periodicity were known, must assuredly have been a wonder, and a terrifying one at that. But by the time such signs were recorded and systematized into textual series, Babylonian and Assyrian treatment of solar eclipses as signs is no longer explainable wholly in terms of wonders. Given that by the middle of the first millennium BCE, the Saros period governing eclipse possibilities, both lunar and solar, was understood by Babylonian astronomers,[8] the response to a solar eclipse, for those who could calculate one, can hardly be explained in terms of wonders. At the same time, however, to count the Saros cycle texts as testimony to an understanding of, or even interest in, the laws of nature would be an anachronistic way of viewing them.

The biblical passages making reference to the "signs and wonders" (Deut. 26:8 and Neh. 9:10) that accompanied the Israelites' exodus from Egypt refer to divine miracles sent by God against the regular order of nature, the plagues that affected only part of the population, people speaking in "tongues." When Joshua commanded the sun to stand still over Gibeon, and the moon to do likewise over the Valley of Aijalon, and Yahweh stopped the celestial bodies in their tracks "in the middle of the sky" for an entire day (Josh. 10:12–14), these too were divine miracles, decisive for the Israelite army against the Amorites. Each miracle demonstrated the power of God to disrupt cosmic order at will and also to drive the narrative about the Israelites' successes against their enemies.[9]

The demonstration of divine power by manipulation of celestial bodies was not unknown in the cuneiform literary tradition. In the Akkadian poem "When Above" (*Enūma Eliš* IV 19–26), authorization of Marduk's supremacy among the Babylonian pantheon is based precisely upon such a demonstration:

> They set up among them a certain constellation,
> To Marduk their firstborn they said (these words),
> "Your destiny, O Lord, shall be foremost of the gods,"
> "Command destruction or creation, they shall take place.
> "At your word the constellation shall be destroyed,
> "Command again, the constellation shall be intact."
> He commanded and at his word the constellation was destroyed,
> He commanded again and the constellation was created anew.[10]

And it was upon this demonstration of limitless power that Marduk was entitled to make the world (from the body of the slain Tiamat) and establish order in the celestial universe, maintaining the courses of the stars by shepherding them like sheep (*Enūma Eliš* VII 130–31).[11]

The examples of "wondrous signs" in the biblical passages, as well as the destruction and creation of the constellation in the Akkadian poem, were one-off events in literary contexts, reflecting not empirical but metaphysical value, which is to say, not facts but truths. They had a didactic, if not a political, agenda, focusing on the power of God/Marduk. As such the literary texts are to be distinguished from the cuneiform omen texts. The literary texts tell of wonders in a didactic (or politically informed) narrative framework about God, or a god, whereas the omen texts compile the phenomena classified as ominous for the purpose of interpreting both regularly and anomalously occurring phenomena in an observable world.

The understanding of signs as compiled in omen texts was intrinsically related to the notion of divine action in the world. An ominous phenomenon seems to have been understood as part of a divine plan, a design (*uṣurtu*), for human events, not as a demonstration of divine disruption of world order. The relation between the miraculous in the literary context of *Enūma Eliš* and the signs in omen texts does not indicate discrepant worldviews but different aims of different texts and textual categories (religio-historical, literary-scholarly, divinatory-scholarly). Having pragmatic rather than literary or didactic value, omen texts provided a systematic scholarly reference compilation of the various ways to read the meaning of phenomena in heaven and on earth for the events of human life.

The attention to signs and wonders in the later history of science is the subject of Lorraine Daston and Katharine Park's *Wonders and the Order of Nature, 1150–1750*, where they noted that "wonders as objects marked the outermost limits of the natural,"[12] and they read a history of the orders of nature in the "history of wonders as objects of natural inquiry." Their book focused on wonders, not miracles, that is, on what they called preternatural wonders, not the supernatural, employing a distinction theologians and philosophers of medieval Christian culture make between two ontological orders, one in which the laws of nature can be suspended by the Almighty, the other, as Daston and Park described it, "suspended between the mundane and the miraculous."[13] These early views on the order of nature precede the formation of naturalism and were firmly rooted in antiquity. What may functionally resemble an "order of nature" in the cuneiform tradition, the "designs of heaven and earth" (*uṣurāt šamê u erṣeti*), however, is not to be so readily assimilated to its medieval and early modern counterparts.

Even when an anomaly in the world of cuneiform divinatory signs was construed in relation to a notion of right order, laid down by the gods as a part of the "designs of heaven and earth," there seems to be a qualitative difference between that notion of a divinely determined right or ideal order and the Greek conception of anomaly with respect to the essential character of something (its "nature," as expressed in the term *phusis*). It is because of the absence of the background idea of nature that in the cuneiform world the category "sign" (Sumerian GISKIM, Akkadian *giskimmu, ittu*) had a different sense from the wonders observed and studied by the medieval Christians and Renaissance natural philosophers that form the subjects of Daston and Park's *Wonders*.

Concerning extispicy, the inspection of the entrails of the sacrificed animal, the Assyrian scribes addressed the sun-god Shamash, saying "in the exta of the sheep you (Shamash) write omens,"[14] and the liver is referred to as the "tablet of the gods."[15] In reference to celestial signs, the final paragraph of *Enūma Anu Enlil* Tablet 22 refers to the three great cosmic deities, Anu, Enlil, and Ea, drawing the constellations as though on a road map of heaven, and describes that "When Anu, Enlil, and Ea created heaven and earth, they made known the ominous signs (*giskimmu*), they set up the stations (*nanzazu*), secured the positions (*gisgallu*) of the gods of the night . . . divided the (celestial) roads of the stars their (the gods') likenesses, drew the constellations."[16] The great gods' drawing of the constellations (expressed with the verb *eṣēru* "to draw") upon the heavens and the establishment of the cosmic "designs of heaven and earth" (expressed with the derived noun *uṣurtu/uṣurātu*) are conceptually related.

The divine production of signs through writing and drawing can be seen as a mode of expression unifying liver and celestial divination. But the writing of signs upon the liver and entrails of the ritually sacrificed animal by the sun-god Shamash, and sometimes also by the storm-god Adad, differed operationally from the production of signs in heaven. From the evidence of the celestial omens themselves, Shamash produced solar signs and Adad weather signs by their own physical manifestations, just as Sin produced lunar signs or Ishtar signs from the planet Venus. A native distinction, therefore, within divination techniques may be inferred between signs occurring through divinely manifested phenomena on one hand, and on the other, signs resulting from the diviner's asking the gods for a sign, whether in the entrails of sacrificed sheep, in the patterns exhibited by dropping oil into water, releasing smoke from a censor, or sprinkling flour. The objects of interest to celestial and terrestrial divination were the phenomena that could be observed or imagined in the world without such entreaty.

The temptation to compare these different procedures to the Ciceronian categories of divination brought out in *De Divinatione* is great, particularly in the use of the terms *oblativa* ("unsolicited") and *impetrativa* ("solicited") to describe Babylonian omens that were, so to speak simply "given" by the gods, as opposed to those requested by the diviner. The inspection of the liver and exta was indeed preceded by a ritual request for an answer to a specific question. An example from an extispicy query written by the scribes Šumâ and (probably) Nabû-ušallim on behalf of King Esarhaddon reads (note that passages in square brackets are restored securely from similar passages in other texts): "Shamash, great lord, [give me a firm positive ans] wer to what I am asking you! If Urtaku, King of Elam, has se[nt this proposal for making peace[17]] to Esarhaddon, King of Assyria, [has he honestly sent] true, sincere words of re[conciliation to Esarhaddon, King of Assyria]? Be present in this ram, [place (in it) a firm positive answer, favorable designs], favorable, propitious omens [by the oracular command of your great divinity, and may I see them.]"[18] A question was asked and a liver was inspected for the answer, positive or negative. This is procedurally different from what the celestial diviners did, which was to keep a watch (*maṣṣartu*) of the heavens for any signs produced there by the gods, who were themselves made manifest in the phenomena. The diviner, in that case, was required to be skilled in the written repertoire of omens in various domains of phenomena to be able to recognize them as such.

The comparison with Cicero's classification is apt to mislead if what we are after is an understanding of what the idea of signs has to say about the diviner's perception and orientation to the external world, the world of his understanding. Cicero's treatise is predicated on a skepticism about the whole business of signs sent by or requested of gods. In its opening passage he called divination "the foresight and knowledge of future events. A really splendid and helpful thing it is—if only such a faculty exists."[19] When he comes to relate his conversation with his brother, Quintus, the question arises, from Quintus's assumption that divination, if indeed it is legitimate, proves the existence of gods, "and, conversely, if there are gods then there are men who have the power of divination."[20] Cicero's well-known response is that "you are defending the very citadel of the Stoics in asserting the interdependence of these two propositions: 'if there is divination there are gods,' and, 'if there are gods there is divination.' But neither is granted as readily as you think. For it is possible that nature gives signs of future events without the intervention of a god, and it may be that there are gods without their having conferred any power of divination upon men."[21] Quintus is made to reply: "There are two kinds of divination: the first is dependent on art, the other on nature."[22] The Ciceronian analysis of divination, therefore, stems from a distinct cosmological as well as an ontological scheme of things, in which the gods can be differentiated from nature. This places the discussion on a basis that cannot be extended to cuneiform texts.

Neither can the ancient Babylonian and Assyrian omens be classified with miracles. Distinctions such as we make between natural phenomena, natural signs or wonders, and miracles cannot be drawn in the cuneiform world. Regularities and irregularities were certainly a part of the understanding of phenomena as signs, gauged in terms of the periodic nature of some signs. Other ways of finding norms and deviations from norms were not the result of understanding how things are, but rather what they might mean. Put another way, the understanding of how or what things are as the product of an interdependence of observation and theory about visible phenomena, was never given epistemic priority over what things meant as the product of divinatory speculation about those phenomena.

CUNEIFORM KNOWLEDGE VERSUS NATURAL KNOWLEDGE

There are many ways to define science, by its process and methods, its practice, and its principal and most general product—knowledge. Yet from our point of view, the multiple ways we can define science seem to involve the conception of nature and the relation of science to it. We would be inclined to say that the principal product of science was knowledge of nature. In antiquity that relation took shape in the world of Greek philosophy, giving rise to a complex history of key importance to this day. The word *nature* (*phusis*) had a variety of senses in Greek philosophical works, from Plato and the Platonists to Aristotle and the Hellenistic philosophers (Cynics, Stoics, Epicureans). Robert M. Grant pointed out that the word has four meanings in Aristotle's *Physics*, seven meanings in the *Metaphysics*.[23] Cicero gives four.[24] But before the concept of nature took shape, and changed shape, across the long history of European and Islamic natural philosophy and science, within which God entered the picture and assumed centrality until modernity, for an equally long period beginning roughly in the early second millennium BCE a learned cuneiform world in the ancient Near East engaged in activities manifestly kindred with science in some of the ways it observed and understood phenomena, yet did not seek to ground its understanding in a physically constituted framework.

In his review of Pierre Hadot's *The Veil of Isis: An Essay on the History of the Idea of Nature*, Ian Hacking began with the following from Robert Boyle's *A Free Inquiry into the Vulgarly Received Notion of Nature*, 1686: "The word 'nature' is encountered everywhere, notably in the writing and talk of poets, scientists, ecologists and even politicians. But though they frequently employ the word, they seem not to have much considered what notion ought to be framed of the thing, which they suppose and admire, and upon occasion celebrate, but do not call in question or discuss." Hacking commented: "Boyle found eight meanings for the word, and pretty much suggested we scrap the lot. No one paid him any heed. Nature is too deeply entrenched in our awareness of the world. Nature is awesome. Nature is gentle tranquillity itself. Nature is terrifying. Nature is the Lake District. Nature is female. Nature is how things ought to be. Nature is crueller even than Man, so that from its very beginning, the human race has had to shield itself from the forces of nature. Nature, above all, is other than us—except that we are part of nature."[25] Hacking conveyed a nature that is many things. But Raymond Williams had said that it was "a singular abstraction," taking it to be "a major advance in consciousness,"[26] adding that, "I think we have got so used to it, in a nominal continuity over more than two millennia, that we may not always realize quite all that it commits us to."[27] Williams, while certainly not denying the multiplicity of meaning conveyed by the term, held the significance of the idea of nature, in its post-Christian form, to be its singularity: "From many early cultures we have records of what we would now call nature spirits or nature gods: beings believed to embody or direct the wind or the sea or the forest or the moon. Under the weight of Christian interpretation we are accustomed to calling these gods or spirits pagan: diverse and variable manifestations before the revelation of the one true God. But just as in religion the moment of monotheism is a critical development, so in human responses to the physical world, is the moment of a singular Nature."[28] The transition to a singular uniform nature was equally significant in Amos Funkenstein's study of changes in attitudes in the history of knowledge. He placed the transition in the Renaissance, when "Aristotle's hierarchy of 'natures' became truly *one* nature."[29]

Singular or multiple, divine and animate or not, nature functions as a category whose boundaries and contents vary cross-culturally and historically. As Hacking said in the above quoted remarks, it is "deeply entrenched in our awareness of the world." But however much we might grant a relativity or a plurality to the conception of nature, recognizing, with Bruno Latour, the "seamless fabric" of nature and culture,[30] the whole complex of thought seems to stop at the borders of the ancient Near East, roughly taken to represent the region east to west, from Iran to the Mediterranean, and north to south, from Turkey to Egypt. Indeed, the particular synthesis of experience that is the modern Western view stops at the borders of the many other cultures not directly descended from the classical Greeks.[31]

At this juncture it should be noted that the area from which the sources in the present study come, sometimes called the "Fertile Crescent," is nearly coextensive with the area now called the Middle East. Taking the entirety of the area that saw the invention of the earliest scripts, cuneiform (in Mesopotamia) and hieroglyphs (in Egypt), we have defined the area typically known as the Cradle of Civilization, that is, of Western civilization. The Tigris and Euphrates River valley states, where cuneiform was first used, have a complex cultural relationship to the West. The identification of an ancient Mesopotamian world as Western originally came as a result of the recognition of precursors to political or religious ideas, artistic motifs, and social practices known from biblical or classical literature and archaeology. Numerous examples of the continuity from the ancient Near East to the West have been documented. Recent and more regular inclusion of cuneiform astronomical texts within the scope of the Western history of science, and along with it the recognition of characteristic empirical and mathematical approaches to lunar and planetary phenomena, somehow highlights a cultural identification between Babylonian knowledge of the heavens and the West.

There is no question that the complex of Babylonian celestial divination, astronomy, and astrology were important foundation stones for the *astrologia*, *astronomia*, and *mathematike* of the Hellenistic period known from Greek and Latin texts.[32] The mathematical astronomical texts from late Babylonia (from circa 500 BCE onward) are now considered an indisputable part of Western astronomy, and the astrological tradition of the Babylonians, which was also clearly known to Greco-Roman writers, is a significant part of that science. In addition to the circulation and reception of astronomical and divinatory methods outside the area of Mesopotamia proper, there is the well-attested "Orientalizing" period in Greek art, the astralization of the divine in Canaanite and biblical religion as an import from Babylonia, and possibly even the later influence of such astralization on Greek religious philosophies where the divinity of the heavenly bodies is common in works of Plato, Aristotle, and the Stoics. In Jean Bottéro's words, the land between the rivers Tigris and Euphrates was "the homeland of the first discernible fathers of our Western world."[33] A number of traditions of knowledge represented in cuneiform texts can be seen as ancestral to both Western (geographically west of the "Near East" and including the western Mediterranean regions and Europe) and Eastern (geographically from the Hellespont and the eastern coast of the Mediterranean to the Indus Valley) sciences, therefore including Byzantine, Islamic, and Indian sciences.[34]

Despite the evidence of continuity from the ancient Near East to the later West, to the biblical and the classical worlds, ancient Mesopotamia is not reducible to a Western framework. In the preface to his "Ancient Mesopotamia: Portrait of a Dead Civilization," A. L. Oppenheim drew attention not to kinship with the West but rather to the unique and alien nature of the civilization reflected in cuneiform sources.[35] Other Assyriologists have made similar remarks, such as Simo Parpola's comment on "deep cultural differences" in the introduction to his edition of the Neo-Assyrian royal letters. "No translation," he said, echoing Benno Landsberger's notion of conceptual autonomy, "no matter how good it is, can make these texts familiar or immediately understandable to a modern non-technical reader. One is bound to admit the existence of a cultural barrier which can be—even partially—removed only through a more thorough acquaintance of the texts themselves or related documents from the same period."[36]

Against the grain of those who emphasize continuity and familiarity over discontinuity and strangeness, some cuneiform evidence was included in Helaine Selin's *Encyclopedia of the History of Science, Technology, and Medicine in Non-Western Cultures.*[37] There are entries on "Astrology in Babylonia," "Medicine in Ancient Mesopotamia," "Mathematics in Mesopotamia," and "Geometry in the Near and Middle East." Coverage of the ancient Near East in the volume is uneven, and the authors of the entries, with the exception of Jens Høyrup ("Geometry in the Near and Middle East"), are not specialists in cuneiform texts. Selin, however, understood this material as appropriate within a non-Western framework, and the inclusion of those entries, as well as Eleanor Robson's "The Uses of Mathematics in Ancient Iraq, 6000–600 BC" in the other of Selin's edited reference works, *Mathematics across Cultures: The History of Non-Western Mathematics,*[38] is indicative of a shift in thinking about the cultural identity of ancient Iraq, or Mesopotamia. Among the writers on ancient Iraq in Selin's encyclopedias of non-Western science and mathematics only Robson addressed the question of the ambiguity of Mesopotamia's cultural identity, saying, with respect to mathematical texts, "since its discovery in the early twentieth century AD, this mathematics has been treated implicitly as part of the 'Western' tradition; even now one finds 'Mesopotamian' mathematics categorized as 'Early Western mathematics,' while Iraqi mathematics in Arabic, some of which is directly related to its compatriot precursors, appears under 'other traditions' (e.g., Cooke 1997)."[39]

The Mesopotamian prehistory of many elements of Western culture, statecraft, religion, and science, elements that may legitimately warrant the epithet "Cradle of Western Civilization," does not exclude the fact that some aspects of Mesopotamian culture are not characteristically Western. The dramatic discoveries of relation, parallels, and heritage do not define the identity of a civilization. The flood story of Noah from the biblical book of Genesis may be found foreshadowed in the Akkadian literary works such as *The Epic of Gilgamesh* and *Atra-ḫasis*, but the presence of the motif of immortality gained through divine favor and survival of an epic flood, and even finer details of its narrative, such as the construction of a large wooden boat in which to weather the storm, are not sufficient to classify *The Epic of Gilgamesh* or *Atra-ḫasis* as exemplars of Western literature, much less all cuneiform literature as Western. As Piotr Michalowski observed, the modern designation "Epic of Creation" or the "Babylonian Genesis" for the Akkadian poem "When Above" (*Enūma Eliš*) reflects Western interests in etiological stories of creation, and the metaphysics of beginning and end times, whereas the Babylonian work, despite its inclusion of a creation account, one among diverse others, is really a paean to the Babylonian god Marduk.[40] *Enūma Eliš* is a creation account in the service of an instauration. A more finely tuned analysis of literary form, poesy, and aesthetics will find more features that differ from rather than duplicate those known from Western literature, a lack of metrics in poetry, for example, and instead, the favoring of repetition or structural features such as chiasmus.[41]

The same applies to the cuneiform contexts within which phenomena were described, that is, fundamental differences in focus and interest when it comes to inquiry into the phenomena. It seems to me that the world-picture in which the cuneiform scribe operated was constructed of other structures, other scaffolding, to use a metaphor of Wittgenstein's, than that which we call nature.[42] The *scaffolding* of the cuneiform scribes' world might be described in terms of divine design, as in the expressions "the designs of heaven and earth" (*uṣurāt šamê u erṣeti*) or "the enduring designs of the gods" (*uṣurāt ilī kīnāti*), but these expressions do not convey notions of a material essence or an independent rationality apart from divine will.

If we are inextricably to link science and nature for all times past and present, then cuneiform texts will be defined out of the picture of the history of science. But recent thinking on both science and nature has revealed the many layers and often contested meanings of each of these terms. The history and philosophy of science since the 1960s and 1970s has subjected the category *science* to a steady critical evaluation and refiguration, yet a sustained discussion of key historiographical and philosophical issues for defining science has not yet been offered with respect to cuneiform evidence.

My claim is that the various efforts to know, understand, predict, and explain phenomena in cuneiform sources are of interest for an increasingly historicized understanding of science, with and without an accompanying conception of nature. I maintain that there was a long-lived learned Assyro-Babylonian tradition of knowledge about phenomena that bears relation to the history of science regardless of the absence of a conscious category of nature around which to focus its epistemic tradition. The chapters of this book cannot possibly address all the aspects of this complex of relationships, but make up an exploration of some of the issues.

PLACING CUNEIFORM KNOWLEDGE IN THE HISTORY OF SCIENCE

How knowledge in the cuneiform scholarly world relates to the history of scientific knowledge is a question as important for the study of ancient Assyria and Babylonia as it is for the history of science. It addresses both the nature of knowledge in the oldest literate culture and the historical reach of what we call science. Twentieth-century epistemologists devoted considerable efforts to establishing a philosophical definition of what it is to know something, most often focusing on propositional knowledge and criteria for the truth and justification of such propositions. That analysis of knowledge, however, never provided the necessary and sufficient conditions for scientific knowledge that would unify science over the course of its history.

Transhistorical epistemic criteria for science, such as were promoted by logical Empiricists and other modernists, therefore gave way as inadequate bases for assessing earlier ways of knowing. Because the form, content, and purpose of knowledge change as science and its practices change, the evaluation of historical ways of knowing has led almost inevitably to a historicism and pluralism about scientific epistemology. Historicism requires a commitment to the situation of knowledge systems within their own time and place. Pluralism raises an array of philosophical questions concerning the nature of our knowledge of nature. I would only emphasize that, in Hasok Chang's words, "pluralism can deliver its benefits without a paralyzing relativism."[43] Although Chang underscores the plurality of science as a function of the complexity of nature, the study of the interpretive framework for cuneiform knowledge can also contribute a significant dimension to the idea of pluralism in science by virtue of its engagement with the world of phenomena independent of the interpretive framework of nature.

The idea of historical epistemologies is the fruit of the integration of the social/historical and the cognitive approaches to knowledge. Going back to 1980, Mary Hesse said, "It is now a platitude to hold that the two approaches to the history of science labeled respectively 'internal' and 'rational,' or 'external' and 'social' are complementary and not contradictory."[44] Helen Longino put it that "the social and contextual embeddedness of scientists and of their reasoning practices does not banish reason from science, nor does it mean that human rationality is not a key element of science."[45] The so-called internalist, or cognitive, and externalist, or social, approaches to the history of science have in some quarters joined forces. If, as Longino and others have argued, knowledge exists in the social, then separating the rational from the social would render science "an epistemology of the ideal,"[46] abstract, inherently ahistorical, detached from actual practice. Longino found the point of view that maintains a dichotomous understanding of the rational and the social to be unproductive in the face of "the growing recognition of the social character of scientific inquiry and the increasing acknowledgment of explanatory plurality in various scientific fields."[47]

Chang too, more recently, spoke of the "false dichotomies" of the social and intellectual, the cultural and intellectual, the social/cultural and rational, of theory and practice, and the internal and external.[48] He suggested a redefinition of the internal that would rehabilitate the distinction consistent with and building upon that which Dudley Shapere proposed in an early formulation of this issue.[49] Accordingly, what is internal to science is not reducible to its empirical or theoretical content, but develops historically in a process of "learning how to learn about nature."[50] Shapere asked, "How did we, in seeking knowledge about nature, come to learn how to go about determining what needs to be studied, what is relevant to that study, the methods of going about that study, and the character of explanations which would be the goals of those studies?"[51] Thus redefined, the internal aspect of science would consist in the set of assumptions about the world that served as criteria for the success of science.[52] From the perspective of the historical evidence of Roman science Daryn Lehoux expressed the same idea, arguing for a pluralistic approach to the predictive success of scientific theory in historical contexts.[53]

It is of interest here that these arguments for localism and pluralism in the philosophy of science have all tied science to the conscious study of and learning about nature. In the face of cuneiform evidence, it seems to me, localism and pluralism need to take an additional step in defining the practice of science, as suggested by Jouni-Matti Kuukkanen, who said, "Rather than referring to some universally shared internal logic of science and invoking the existence of 'nature' and neutral empirical data derived from it as non-problematic explanatory notions, science is now explained by reference to the factors that can be found, and shown to be found, in particular localities."[54] Here that locality was the one that produced cuneiform investigations of phenomena. Within a fundamentally different framework, the Babylonian and Assyrian scribes determined what phenomena they needed to study, what were the relevant aspects of that study, the methods for carrying it out, and the appropriate kinds of explanations to serve the goals of their study.[55] Predictive success within the Assyro-Babylonian system of knowledge was defined by the particular framework (or frameworks) for observation, modeling, and interpreting the phenomena we reconstruct from cuneiform texts.

Our reconstruction of knowledge in the cuneiform world is particularly dependent upon how textual evidence is tied to place, purpose, and other elements of context. It was the scribal institutions found in royal palaces and/or temple complexes that produced diverse fields of knowing for purposes specific to the roles played by such learned scribes, whether devoted to divination from ominous signs, magical ritual response to signs of misfortune, incantations for use in such rituals, medical knowledge, or astronomical knowledge and the ability to make computations. Not all such texts were housed in major palaces or temples. Some were found in private houses, where scholars, such as the seventh-century Assyrian Kiṣir-Aššur, or the fourth-century Babylonian Iqīša, kept collections of their own.[56]

A more fine-grained perspective on the social contexts of cuneiform knowledge is beset by problems in the politicized world of archaeology and in the acquisition of cuneiform sources. This dimension will not be taken up here as it has been dealt with ably by Eleanor Robson[57] and others, as well as by the Cambridge team on the website Geography of Knowledge in Assyria and Babylonia, 700–200 BC: A Diachronic Comparison of Four Scholarly Libraries.[58] In essence, during the Neo-Assyrian period, from which the richest evidence comes, the scholars were advisors to kings on the basis of knowledge mastered in fields of divination, medicine, ritual performance, and astronomy. For other periods, both earlier and later, the social role of the scholars is less vividly revealed. Less is known of their interactions with kings and how the Late Babylonian temples in Uruk and Babylon functioned as preserves of ancient cuneiform learning without royal support, or without the royal support we are more familiar with from the Sargonid period.

In addition to the textual evidence of the divinatory and related concerns of elite scribes is the corpus of the curricular repertoire developed for the training of scribes. This school material, largely Sumerian and bilingual Sumerian and Akkadian materials, reflects the essentials of scribal training in cuneiform writing from the Old Babylonian to Hellenistic periods.[59] As the basis for the teaching of cuneiform literacy, the bulk of this enormous lexical corpus belongs to a different strand of the scribal knowledge industry,[60] and it therefore lies outside the domain of the present study. Obviously, there are relationships between the two corpora that have to do with the nature of reading signs, both in the world and on clay tablets, manifesting in both areas some essential ways in which reading and interpreting belonged to a fundamental level of the value and function of knowledge in the cuneiform context.

Apart from the basic work to read and translate the many extant cuneiform tablets that relate to the field of ancient Mesopotamian knowledge, one of the chief questions that will no doubt be at the center of much Assyriological research for some time to come is the question of how those texts refer to the world of the scribes' perception, experience, and knowing. Even more basically, the issue of what we mean when we say *knowledge*, or *knowing*, has a long history of philosophical investigation behind it. In order to get at the meaning of what is known, how knowledge refers to a "world," and how to account for "other minds" that engage with the world in a way different from ours, the history of philosophy is replete with imaginary scenarios about Antipodeans,[61] thinking bats,[62] brains in vats,[63] and such. Here, on the other hand, we have ready-made historically existent other minds and a historically existent world that was not understood, described, or investigated in a way that now, at this distance, can be easily mapped onto our own. Assyriologists are not alone in this, as any anthropologist, or for that matter all historians, will readily affirm.

Although we can try to define a conception or conceptions of knowledge in the ancient Near East, no equivalent classification in cuneiform languages serves to unify particular forms of knowledge, reasoning style, or social practice in quite the same way as does our classifier *scientific*. Even granting how variable the meaning of that term is across the historical span, reasons for including cuneiform knowledge in that history are less intuitive for us than, say, Greco-Roman, medieval, or even Islamic science. I submit that the reason for this has to do with the underlying disjunction between the interests of the cuneiform scribes in their investigations of the world on one hand, and those of what we call natural science, or natural knowledge, on the other. That the scribes were not directed toward a description of nature puts the entirety of the cuneiform knowledge culture into a different epistemic framework, even the work of the Late Babylonian mathematical astronomers.

Hub Zwart referred to science and nature as "the two basic poles of the knowledge process, the object and the subject pole."[64] He described the intimate relation between nature and science in terms of nature being "primordial" and science "derivative," saying: "In the beginning there is nature, presenting itself to us, and science is a 'representation.' But if we look at their relationship more closely . . . nature is not simply there, she must be made discernable. Without science, our knowledge of nature would be rather limited and superficial; our experience of nature would no doubt be impoverished and less precise. . . . Without nature, science would make no sense, but the opposite is also true: no nature without science."[65] Zwart's book on comparative epistemology "starts from the conviction . . . that there are *other* ways of knowing about nature besides science."[66] This book, on the other hand, raises the question of whether there have been other ways of doing science besides that which aims to know nature. Zwart said that science makes no sense without nature, a fair statement from a Western point of view. Is it fair for cuneiform science? The hypothesis here is that it is not.

Critics may say that cuneiform knowledge *included* knowledge about nature, even though the ancients did not conceive of their inquiries in this way. What is argued here is not that natural phenomena were not known. Obviously many of the scribes' objects of knowledge, for example, eclipses, the appearance of stones, or the effects of medicinal plants, are classifiable as natural phenomena according to our manner of categorization, and, no doubt, some of what we would view as the physical properties of these things exerted certain constraints on what and how the Assyro-Babylonian scribes knew of their world. But if, as Ernest Gellner put it, "metaphorically speaking, categories . . . are shadows which the uses of language cast on the world,"[67] my hypothesis is that the cuneiform world remained unshaded by categories emerging from and dependent upon the conception of nature. One such category is the supernatural. Another is the notion of a natural kind. In the absence of the categories that stem from the idea of nature, so firmly implanted in our own understanding of things, is it even possible to reconstitute what served to structure the various bodies of cuneiform knowledge with respect to the world in which they were conceived?

Textual knowledge was highly valued in the cuneiform world, a prestigious commodity both in terms of its intellectual as well as its material properties.[68] Tablet colophons indicate as much, from recording the attention to checking and collating originals to injunctions against scribes who were not "knowledgeable" (*mudû*), as in the statement, "Let the knowledgeable show the knowledgeable (*mudû mudâ likallim*). The one who is not knowledgeable (*la mudû*) may not see (the tablet). (It is) a taboo of (such-and-such) god." In his translation of this formula, Rykle Borger implied the sense of "the one who is inducted," or "*in* the know," in his use of the German "der Eingeweihte."[69] Markham Geller and Klaus Geus used the expressions "insiders" for the learned *mudû* and "outsiders" for the not learned *la mudû*.[70] The tablets containing such injunctions are largely those containing specialized learning of the highest order, reserved for the most proficient, that is, those termed *ummânū* "masters."

With respect to the Late Babylonian corpus, Paul-Alain Beaulieu observed that "the highest statistical incidence of these formulas is found in expository texts: learned compendia explaining parts of rituals, lists of gods with their sacred attributes, star catalogues, astrological explanatory lists, descriptions of gods and other mythical beings. In short, texts exposing the theological and (pre)philosophical speculations of Babylonian scholars."[71] Eckart Frahm, too, noted that the "secrecy formulae . . . are attested in various groups of cuneiform texts, but hardly anywhere else more often than in explanatory texts."[72] As well as expository or explanatory texts, ritual and astronomical texts were also claimed by the *ummânū mudû* "learned masters."

Even before the late period, during Neo-Assyrian times (seventh century BCE), already an exclusivity attended the specialization in divinatory materials as well as the materials themselves. Nils Heeßel pointed out that "the knowledge of extispicy . . . was available only to a small group of scholars in possession of the exclusive prerogative of interpretation, who guarded the dissemination of this knowledge,"[73] citing the following passage from a colophon: "These are the extraneous omens which not every diviner knows. The diviner as a father will preserve them for his favourite son, swear him and teach him."[74] A well-known text concerning the ritual qualifications of the extispicy expert, the *bārû*, also states that "the learned master (*ummânu mudû*), who guards the secrets of the great gods, will bind by oath before Shamash and Adad the son whom he loves,"[75] underscoring the importance of familial lineage among elite scribes.

In reference to the *bārû* specified as an expert in oil divination (*apkal šamni*), exclusivity derived both from his lineage, that he was a "well-born" member of a family from Nippur, Sippar, or Babylon, considered a descendant of the mythical antediluvian sage, Enmeduranki, and also from the perfection and purity of his physical person: "When a diviner, an expert in oil, of abiding descent, offshoot of Enmeduranki, king of Sippar, who sets up the holy bowl, hold the cedar, benediction priest of the king, long-haired priest of Shamash, a creature of Ninhursag, begotten by a reverend of pure descent, he himself, being without defect in body and limbs, may approach the presence of Shamash and Adad where (liver) inspection and oracle (take place)."[76] For one who was not of the required lineage, or having squinting eyes, chipped teeth, a finger cut off, a ruptured(?) testicle, or suffering from leprosy to approach the place of the gods for an oracle was interdicted, an offense (*ikkibu*/NÍG.GIG) to the gods, as was the attempt of a diviner "who has not mastered [his] training" to perform divination.[77] Insistence upon the cultic correctness of the diviner is consistent with the required purity of the bull to be sacrificed for making the ritual kettledrum head in a Hellenistic ritual, which instructs that, "When y[ou] want [to cover] the kettledrum (proceed as follows). A knowledgeable expert (LÚUM.ME.A *mu-du-ú / ummânu mudû*) will carefully inspect an ungelded black bull, whose hooves and horns are intact, from its head to the tip of its tail; if its body is black as pitch, it will be taken for the rites and rituals. If it is spotted with 7 white tufts (which look) like stars, (or if) it has been struck with a stick, (or) touched with a whip, it will not be taken for the rites and rituals."[78]

Exclusivity and high value were equally placed on the repertoire of the expert scribes. Their texts were designated a secret of the great gods (AD.HAL DINGIR.MEŠ GAL.MEŠ/*pirišti ilāni rabûti*), secret of the sage (*niṣirti apkalli*), secret of divination (*niṣirti bārûti*), secret of extispicy (MUNUS.ÙRU NAM.AZU/*niṣirti bārûti*), of the exorcists (*niṣirti āšipu*/lúMAŠ.MAŠ), or of the scholar (MUNUS.ÙRU UM.ME.A or AD.ḪAL UM.ME.A/*niṣirti* or *pirišti ummânī*), indeed, of heaven and earth (AD.ḪAL AN u KI/*pirišti šamê u erṣeti*), or even of kingship (*niṣirti* LUGAL-*ti*).[79] The high value of cuneiform knowledge was not only tied to textual scholarship. It was also tied to mathematics, as numerical schemes to generate the dates or durations of visibility of astral and lunar phenomena were already an early part of the history of cuneiform knowledge. These mathematical methods underwent significant modification and were applied to increasingly many phenomena over the course of the long development of cuneiform astronomy, culminating in the astronomical tables and procedure texts of Seleucid Babylonia.[80]

What it was to have knowledge in ancient Mesopotamia may only be partly translatable to later Western scientific values and practice, but it is in part translatable. Threads of continuity picked up from the perspective of history looking backward, such as one finds in theoretical astronomy, are different from those that show themselves by looking forward, such as in the pervasive interest in prognostication and signs. The persistent importance of the theological in the study of nature in later periods of the history of science also bears relation to the cuneiform scribes' attribution of ultimate agency and world design to the gods. Legitimate connections made from the cuneiform world to later sciences can be established from both points of view, that is, the importance of signs and of the divine in the world. And both are necessary for a more complete account of the earliest history of science.[81]

Much like Western scientific knowledge (*scientia*), cuneiform knowledge underwent transformations in its cultural locus and in its implementation of various cognitive strategies throughout its history while maintaining a number of common elements, such as the focus on signs in heaven and on earth, the prediction of the phenomena valued as signs, observation of cyclical celestial phenomena, response to untoward omens, and increasing attention to interpretation of words used in divinatory contexts. That this scribal activity was widespread throughout Babylonia long after the last of the Babylonian kings (Nabonidus, conquered by Cyrus the Persian in 539 BCE), and that it supported substantial innovations in astronomy, astrology, and medicine well into the Seleucid period (after 312 BCE), is a testament not only to the cultural fertility and adaptability of cuneiform knowledge, within Mesopotamia proper as well as across its borders into other areas, but also, and consequently, its epistemic power.

CHAPTER TWO

Old Ideas about Myth and Science

kīma labīri "as in former times"

> Why do men study ancient history, acquire a knowledge of dead languages, and decipher illegible inscriptions? What inspires them with an interest not only in the literature of Greece and Rome, but of ancient India and Persia, of Egypt and Babylonia? Why do the puerile and often repulsive legends of savage tribes rivet their attention and engage their thoughts?
>
> FRIEDRICH MAX MÜLLER, *Chips from a German Workshop*, vol. 2

HISTORICIZING "THE INTELLECTUAL ADVENTURE OF ANCIENT MAN"

Henri Frankfort, H. A. Groenewegen-Frankfort, John A. Wilson, Thorkild Jacobsen, and William A. Irwin's *The Intellectual Adventure of Ancient Man: An Essay of Speculative Thought in the Ancient Near East* (University of Chicago Press) appeared in 1946 in the postwar atmosphere of social evolutionism and scientism. Its objective was to depict the distinctive intellectual character of the best known ancient Near Eastern civilizations—namely, Mesopotamia, Egypt, and Israel. The authors were interested in what and how ancient social groups thought, especially about humankind's place in the world of nature and in relation to the divine. These are questions at the heart of the history of science as well as of religion and natural philosophy. But *The Intellectual Adventure* was not a work in the history of science inasmuch as the authors did not see science as having taken root in the ancient Near East. Instead, accepting and perpetuating a nineteenth-century model of unilinear social evolution, the book aimed to describe and analyze ancient Near Eastern civilization in terms of a stage of thought in relation to nature in particular.

The term used to describe this evolutionary stage was *mythopoeic*, literally "myth-making," which had long before been taken into Victorian scientific and historical discourse and embedded into modernism, according to Shanyn Fiske,[1] from George Grote's *History of Greece* (1846–56). It was Friedrich Max Müller, however, who put a fine point on the mythopoeic, saying,

> before the appearance of the first traces of any national literature, there is a period, represented everywhere by the same characteristic features,—a kind of Eocene period, commonly called the Mythological or Mythopoeic Age. It is a period in the history of the human mind, perhaps the most difficult to understand, and the most likely to shake our faith in the regular progress of the human intellect. We can form a tolerably clear idea of the origin of language, of the gradual formation of grammar, and the unavoidable divergence of dialects and languages. We can understand, again, the earliest concentrations of political societies, the establishment of laws and customs, and the first beginnings of religion and poetry. But between the two there is a gulf which it seems impossible for any philosophy to bridge over. We call it the Mythic Period. . . . Although later poets may have given to some of these fables a charm of beauty, and led us to accept them as imaginative compositions, it is impossible to conceal the fact that, taken by themselves, and in their literal meaning, most of these ancient myths are absurd and irrational, and frequently opposed to the principles of thought, religion, and morality which guided the Greeks as soon as they appear to us in the twilight of traditional history.[2]

Despite the Frankforts' assumptions about what separated ancients from moderns cognitively, evidenced principally in the use of the word *primitive* to designate ancient thought as compared to *modern*, the power of *The Intellectual Adventure* was the result of its use of many original sources from Sumerian, Akkadian, and Egyptian literature. On balance, however, this concise and readable portrait of ancient Mesopotamian, Egyptian, and Israelite thought focused on mythological and religious texts as evidence, taking that material as indicative of a certain stage of cognition and leaving out of the equation the many other sources in which phenomena were objects of observation, systematization, schematization, and prediction within diverse technical literatures of divination, medicine, and astronomy.

The Intellectual Adventure enjoyed widespread reception in the years following its appearance. At the same time Assyriology was making huge strides, notably in the publication and study of Babylonian astronomical texts. The Frankforts' mythopoeic Mesopotamians did not fit into the new history of the exact sciences that was taking shape in midcentury based on the edition of the cuneiform astronomical ephemerides by Otto Neugebauer in *Astronomical Cuneiform Texts* (1955). The difficulty with mythopoeicism would only increase in light of further studies of Babylonian astronomy in the ensuing years, culminating with Neugebauer's Book II of *A History of Ancient Mathematical Astronomy* of 1975.[3]

Since the mid-1970s, coincident with and influenced by the many "turns" in the history of science, the view of science in cuneiform texts has expanded to encompass, in addition to the mathematical astronomical texts of the late period (after circa 500 BCE), texts that do not fit into the classification *exact sciences*, but nonetheless offer evidence of a variety of investigations and resulting systematic compendia of knowledge about other sorts of (observational and nonobservational) phenomena. The bulk of such sources for the "nonexact" sciences takes the form of omen texts, both celestial and terrestrial, including extispicy (especially liver omens). Also extant are medical diagnoses, therapeutic texts, and the so-called nonmathematical astronomical texts. Further discussion of these and other sources, in particular how through them we can say what knowledge was for cuneiform scribes, is found in chapter 3.

The Intellectual Adventure rejected the possibility of science in the cuneiform world because, according to its thesis, mythopoeic logic lacked the necessary qualifications for its development. The point was that this kind of reasoning did not operate with a "detachment which a purely intellectual attitude implies,"[4] detachment being code for objectivity. Thus, according to the philosophical standards of the day, mythopoeic reasoning was inherently unscientific. As a consequence, the book was not primarily interested in science, but it *was* interested in the relationship between the ancients and the natural world.

The Frankforts did not find in cuneiform sources a modern Western conception of nature as an independent physical realm of material phenomena subject to its own laws, one that could be considered with "detachment," but, on the basis of Sumerian and Akkadian mythological texts, found that "the gods were in nature."[5] They concluded that "the mainspring of the acts, thoughts, and feelings of early man was the conviction that the divine was immanent in nature, and nature intimately connected with society,"[6] and further, that "the assumption of an essential correlation between nature and man provided us with a basis for the understanding of mythopoeic thought."[7]

Raising the question of whether "the determination of nature as pure materiality—absent gods, incarnate spirits, or any such nonhuman persons—is a unique Western invention,"[8] Marshall Sahlins cited the Frankforts' observation that it was first in the biblical text that nature was made material—that "the ancient bond between man and nature was destroyed."[9] The differentiation between natural and supernatural resulting from the fact that "worldly things could represent or be signs of God, but they are not God," however, was not, in Sahlins view, "the same as the nature-culture distinctions widely practiced around the world."[10] Rather, "that nature is only *res extensa*, made of nothing, lacking subjectivity. The idea, moreover, becomes the ontological counterpart of an equally singular epistemology, insofar as knowledge of nature cannot be achieved by communication and the other ways subjects understand subjects."[11] This "ontological counterpart" and the "singular epistemology" that goes with it is what is of interest to the chapters of the present book, and why it pays to revisit the Frankforts' analysis.

Mythopoeic thought reflected not only divine immanence in nature, but, as the Frankforts put it, "its logic, its peculiar structure, was seen to derive from an unceasing awareness of a live relationship between man and the phenomenal world."[12] Mythopoeic thought was to scientific thought as ancient Near Eastern civilization was to modern, that is, remote and foreign. The description of ancient Near Eastern thought as so utterly at odds with our own was compelling for its mid-twentieth-century modernist audience. Scholarship in Babylonian astronomy notwithstanding, the idea that the ancient cuneiform writers possessed an alien mind that made myth (*mythos* + *poiein* "to make myth") instead of science became virtually iconic for an entire generation.

What is somewhat surprising, given how differently we think about the subject today, is that nothing has replaced *The Intellectual Adventure* as far as a comprehensive reassessment of the particular cultural orientation of the ancient Near East to what we think of as nature. A. L. Oppenheim's 1978 entry in the *Dictionary of Scientific Biography*, "Man and Nature in Mesopotamian Civilization," already saw the man nature relationship differently from that of the *Intellectual Adventure*. Although the Frankforts and Oppenheim alike took the nature-culture dichotomy as a starting point, where the Frankforts found no nature in Mesopotamia, Oppenheim did.[13]

Interestingly, Oppenheim chose to use as his first exemplary passages the two versions of creation in the biblical book of Genesis. In the first passage (Gen 1:26) God made humankind in order that he have rulership over wild and domesticated animals, birds, and fish. This theme of human control ("rulership") over nature is primary for the Western dichotomy of nature and culture. In the second passage (Gen 2:19) the beasts and the birds are made and brought to Adam to be named, another form of control. The Akkadian expression *šikin* (or *šiknat*) *napišti* "all living things" or "all creatures," including humankind, is not alone evidence for a separate conception of a nonhuman nature, nor for the conceptual dichotomy of nature and culture.[14] In addition, in a cuneiform cultural sense, naming assumes importance in a somewhat different way from that in the biblical text. Although in the Babylonian poem "When Above" (*Enūma Eliš*) it is expressed that only once things had names could they properly exist, there was no primal man who named. As Jean Bottéro put it, "The ancient people were convinced that the name has its source, not in the person who names, but in the object that is named; that it is an inseparable emanation from the object, like a projected shadow, a copy, or a translation of its nature. They believed this to such an extent that in their eyes 'to receive a name' and to exist (evidently according to the qualities and the representation as put forward in the name) was one and the same."[15] *Enūma Eliš* emphasizes the creative and existential power of names and naming, from the first line "when above the heavens were not (yet) named, nor below was the earth (yet) named" (*enūma eliš la nabû šamāmu šapliš ammatu šuma la zakrat* EnEl I 1) to the last tablet containing the fifty names of Marduk (from which comes the line *ina zikri ḫanšā ilū rabûtu ḫanšā šumēšu imbû ušātiru alkassu* "with the name 'Fifty' the great gods had called his fifty names and made his way pre-eminent." EnEl VII 144). In the Bible, however, the separation of the man, Adam, from the rest of the creatures as ruler and designator of all is not something characteristically expressed in cuneiform texts.

But Oppenheim analyzed the relationship between "being named" and "being" as a form of control over nature, thus supporting a nature-culture dichotomy. He said, "in a civilization that knew writing, from such an attitude there might have developed the desire to enumerate such names so as to demonstrate both erudition and the power of the human intellect in confronting nature, the world around the scribe."[16] Following a brief description of the Sumerian, and the bilingual Sumerian to Akkadian, lexical lists (the twenty-four tablets of *Ura*), he added that, in addition to the importance of those lists for the education of the scribes, they were also "for organizing, classifying and defining the phenomena of nature in an established order. . . . It cannot and should not be claimed, of course, that the word lists containing, for example, the names of plants, animals, or stones constitute the beginnings of botany, zoology, or mineralogy in Mesopotamia. They are not a scientific (not even a prescientific) achievement; rather, they result from a peculiar interaction of a genuine interest in philology (or, at any rate, lexicography) and a traditional Near Eastern concern for giving names to all things surrounding the scribe, thus linking nature to man."[17] It is clear that Oppenheim extrapolated what we call nature from the word lists of ancient Mesopotamia. He imputed the notion of nature to the scribes, assuming that they considered and sought to order and organize nature in the form not only of word lists but also in poetic, divinatory, and medical texts.

Appearing in 2011, the preface to a collection of papers titled *The Empirical*

Dimension of Ancient Near Eastern Studies / Die empirische Dimension altorientalischer Forschung[18] reflected the desire for the gap since *The Intellectual Adventure* to be filled, making reference to the need for an updated overall description of the Mespotamian weltanschauung, its particular epistemology and "intellectual orientation in the world."[19] This project will require the expertise of many cuneiformists. The present study offers just one point of view. Toward a reassessment of cuneiform sources in terms of what they tell us about how phenomena were understood in the cuneiform world, the Frankforts' essays are a useful, even necessary, point of departure. For all the deficits of its theoretical framework from a contemporary point of view, *The Intellectual Adventure* raised the fundamental question of worldview—namely, how what we call natural phenomena, and the external world in general, was construed by the literate elite in Near Eastern antiquity.

On the subject of nature and the physical environment in the conception of diverse sources from the ancient Near East, the book was a first, but the very problem of bridging our own attitudes and assumptions to those of a remote time has been an explicit concern ever since Assyriologist Benno Landsberger's inaugural lecture at Leipzig in 1924, titled "Die Eigenbegrifflichkeit der babylonischen Welt" ("The Conceptual Autonomy of the Babylonian World"). There Landsberger asked the question: "To what extent is it possible to reconstruct vividly and faithfully an ancient, alien civilization by philological means, without the help of a tradition continuing down to the present day?"[20] Landsberger's discussion was not oriented toward nature or science, but nevertheless was concerned with what is now routinely referred to as the problem of "otherness," the emic/etic distinction (a red herring),[21] and actor versus observer categories.

The question of how to reconstruct vividly and faithfully the ways a premodern non-Western society engaged with the world, whether we call it science or something else, indeed poses a number of challenges. First there is the question of the classification of Mesopotamian (literate and scribal) culture vis-à-vis the West, where the use of the term *nature* (*phusis*) is first attested in Greek texts. Then there is the question of the "multiplicity of faces" displayed by nature "since the speculations of the *physiologoi*," as Donald R. Kelley put it,[22] and the consequent necessity to grant a kind of relativity about nature insofar as it is embedded in cultural ways of knowing, describing, and explaining. As far as explicit discourse on nature is concerned, its history does indeed begin with the famous *physiologoi*, and what Kelley called "the recognizable canon of physical science."[23] Kelley's study of the idea of law (*nomos*), paralleling Collingwood's *Idea of Nature*, recognized the intertwining of views about law and society with those about nature, and in this context he noted, "As Henri Frankfort has written, 'The realm of nature and the realm of man were not distinguished in ancient Near Eastern speculative thought.'"[24] Thus while the separation of *nomos* from *phusis* became axiomatic in Western thought, each with its own respective norms, neither had independent status in cuneiform texts, which then remained outside the bounds of those histories.

It is true that there is no lexical counterpart to nature in cuneiform languages, nor, consequently, was there a conceptual counterpart, no realm of knowing or of being with its own internal forces or classification of certain things. Classification of animals in terms of domesticated and wild, however, was known, as is clear in the fact that an entire tablet of *Ura* is devoted to wild animals. But how "domesticated" is defined in relation to "wild" does not necessarily depend upon cognizance of a realm equivalent to nature, or imply the notion of the nature-culture dichotomy. Marilyn Strathern addressed the problematic issue of assuming a correspondence between the classificatory dichotomy of wild and domesticated to that of nature and culture in the context of the Hagen of Papua New Guinea,[25] and in general for other non-Western cultures. She cited Jack Goody as having characterized the nature-culture dichotomy as "'a highly abstract and rather eighteenth-century' piece of western intellectualism"[26] alien to non-Western, and I would add, ancient Near Eastern, cultures.

In the *Epic of Gilgamesh*, when the alter-hero Enkidu is first created by the goddess Aruru in the place called "the steppe," we translate the Akkadian *ina* EDIN/*ṣēri*, literally, "the steppe," as "the wild."[27] It is generally taken that the seduction and subsequent "taming" of Enkidu by the woman Shamhat is both symbolic of the transformation of the wild to the domesticated and an act of control by culture over nature. The English word "wild," like *ṣēru*, denotes a place, but "wild" also connotes a characteristic, or many characteristics, of the separate realm of nature, where different norms from those of human culture are sustained without interference from human beings.

Gary Snyder spoke of "the wild" as "the process and essence of nature" and "an ordering of impermanence," that it runs by its own rules, "often associated with unruliness, disorder, and violence."[28] He pointed out that while the word *nature* can connote "the physical universe and all its properties," wild has been defined more by negation. He enumerated:

> The Oxford English Dictionary had it this way:
> Of animals—not tame, undomesticated, unruly.
> Of plants-not cultivated.

Of land-uninhabited, uncultivated.
Of foodcrops-produced or yielded without cultivation.
Of societies-uncivilized, rude, resisting constitutional government.
Of individuals-unrestrained, insubordinate, licentious, dissolute, loose. "Wild and wanton widowes"—1614.
Of behavior-artless, free, spontaneous. "Warble his native wood-notes wild"—John Milton.[29]

It is no wonder that Enkidu would be classed as "wild," from his birthplace to his behavior as one living among the herds (though if we are looking at the OED, unrestrained, licentious, and loose behavior is more Gilgamesh's calling card before Enkidu ever arrived on the scene). Still, the translation of *ṣēru* as "wild" in Gilgamesh I: 102–3 is an extended usage coming from the principle meaning of the word *ṣēru* as "back," thence "back country, hinterland." "Wild" does not appear in the meanings of *ṣēru* given in *The Assyrian Dictionary*, with good reason, as "wild" carries implications that "back country," "open country," or even "steppe" do not.

The juxtaposition of Gilgamesh, king of Uruk and lord over the city Uruk, and Enkidu, who becomes Gilgamesh's friend and near brother, creates the impression that we are dealing with a spatial and even conceptual duality between city as culture and wild as nature. Gilgamesh is a king, whereas Enkidu takes nourishment from the milk of herd animals and slakes his thirst at their watering holes. But as Andrew George pointed out, Enkidu's transformation has to do with loss of innocence; it is part of the poem's exploration of what it is to be human. Enkidu's meetings with human beings, first the hunter and then the seductress Shamhat, ultimately cost him his innocence and strength, and set him up for his ignoble death before his time. At Enkidu's death, Gilgamesh, as described by George,

> initiates the rites of mourning with a great lament, apostrophizing his friend in words that recall Enkidu's early life among the herds of gazelles and wild donkeys (1–6) and calling on those who knew him and witnessed his life to grieve for him. . . . The mourners are to include inanimate parts of the natural world, the 'paths of the Cedar Forest,' by which is meant the tracks trodden by the untamed hero when still in the wild and retraced in the journey with Gilgameš (7–8), and the mountains, meadows, and rivers that also figured in his heroic career (11–13, 18–20).[30]

George adduces a parallel from a Neo-Assyrian text concerning a son's mourning his father, in which the river canals, trees, and fruit participate in wailing and weeping for the dead.[31] These passages seem to reflect more of a connection between the human being and the environment than a separation of nature from culture. Strathern's observation "that a non-western wild-domestic dichotomy triggers off an interpretation in terms of 'nature-culture' in the presence of environmental control"[32] may well be relevant here.

It was because of the lack of a distinction between the "realm of nature" and the "realm of man" that the Frankforts found an incapacity for scientific activity or for producing scientific knowledge among the writers of cuneiform texts. Because the "mind"—logical, scientific, and rational—that would be required to do these things would have to have both made that all-important distinction and made nature an object of inquiry, the scribes could not have and did not produce scientific knowledge. *The Intellectual Adventure* proceeded as though the making of myth, for Sumerians, Akkadians, Assyrians, and Babylonians, was the only tool available for explaining natural phenomena and the only means available for thinking about what we call nature. But the notion of the mythopoeic mind and mythical thinking, put to rest long ago, was perhaps best summarized by G. S. Kirk, when he called it "the unnatural offspring of a psychological anachronism, an epistemological confusion and a historical red herring."[33] Myth, that is to say, myths, were indeed one of the ways in which what we call the natural was framed in the cuneiform textual record.[34] It was not, however, the only way.

A critique of *The Intellectual Adventure* can serve as a starting point from which a new consideration of the different ways phenomena of what we think of as the natural world were treated in cuneiform texts. A full critique, that is of the whole of Frankfort and colleagues' *Intellectual Adventure/Before Philosophy*, would go beyond the scope of the present project, but in the next section, the ideas of the Frankforts' introductory and concluding chapters will be put in context before moving on to consider different cuneiform contexts that reflect a multiplicity of ways in which learned scribes came to terms with a variety of phenomena.

A CRITIQUE OF MYTHOPOEIC MESOPOTAMIA

The theoretical framework for the chapters on Mesopotamia, Egypt, and Israel was set out in the introduction and conclusion to *The Intellectual Adventure of Ancient Man*, coauthored by Henri Frankfort and Henriette Groenewegen-Frankfort, called "Myth and Reality" and "The Emancipation of Thought from Myth," respectively. The publisher's description, surely penned by the authors, shows the scientism of the day (note how myth was to be "overcome"):

> To the people in ancient times the phenomenal world was teeming with life; the thunderclap, the sudden shadow, the unknown and eerie clearing in the wood, all were living things. This unabridged edition traces the fascinating history of thought from the pre-scientific, personal concept of a "humanized" world to the achievement of detached intellectual reasoning. The authors describe and analyze the spiritual life of three ancient civilizations: the Egyptians, whose thinking was profoundly influenced by the daily rebirth of the sun and the annual rebirth of the Nile; the Mesopotamians, who believed the stars, moon, and stones were all citizens of a cosmic state; and the Hebrews, who transcended prevailing mythopoeic thought with their cosmogony of the will of God. In the concluding chapter the Frankforts show that the Greeks, with their intellectual courage, were the first culture to discover a realm of speculative thought in which myth was overcome.

The first sentence set the tone and point of view for the volume as a whole. It reads: "If we look for 'speculative thought' in the documents of the ancients, we shall be forced to admit that there is very little indeed in our written records which deserves the name of 'thought' in the strict sense of that term."[35] As the only attempt to offer a synthesis of ancient Near Eastern ideas concerning the physical and the metaphysical, and how various social groups of Near Eastern antiquity interpreted their experience, the book had tremendous impact on others engaged in research on non-Western cultures. David Wengrow cited Henri Frankfort's influence upon Godfrey Lienhardt's 1961 ethnographic study *Divinity and Experience* as "a unique instance of a major anthropologist drawing significantly upon the thought of an archaeologist."[36] The historian of precolonial Africa, Basil Davidson, credited the book for emboldening him to attempt for African sociocultural history a "new synthesis of cultural patterns and values."[37] He confessed a certain despair about ever reaching such a synthesis, particularly for the history of African cultures where available written records, apart from the many ancient Egyptian inscriptions, were scant and systematic research was limited. "Then," he said, "I chanced in 1962 on the Frankforts' *Intellectual Adventure of Ancient Man* where some vivid contrasts are drawn between the mood and temper of the ancient Egyptian and Mesopotamian civilizations. . . . These civilizations could, it appeared, be given a character."[38]

The book left an imprint on many readers, lay and scholarly alike, interested in ancient thought and in gaining a general grasp of the civilizations in question. A sense of responsibility to present such synthetic treatments of ancient Near Eastern civilization, whose existence had at that time been for less than one hundred years known at all from original sources, was felt in the fields of Near Eastern archaeology, Assyriology, and Egyptology. Not too long after the appearance of *The Intellectual Adventure*, a collection of papers edited by Robert C. Dentan, *The Idea of History in the Ancient Near East*, was published by Yale University Press (1955), which similarly tried to present a concise portrait of the historical attitudes of the various civilizations of the ancient Near East, in this case not only Mesopotamia, Egypt, and Israel, but also Persia, the Hellenistic and Early Christian Mediterranean, as well as early Islam, though without a unifying theoretical framework.

While not expressly a work of cognitive anthropology, *The Intellectual Adventure* was certainly interested in cognition, how the ancients perceived and understood their world as diagnostic of their thought. Using religious and mythological texts as evidence, the aim was to reconstruct the ancients' perceptions and understandings of their physical environment. A cognitive-historical interpretation of ancient Near Eastern civilizations was therefore offered on the premise that, in the absence of science, myth was the only mode of understanding available. But such an approach seems to be based upon a confusion, or conflation, of the cognitive and the imaginative, as Leonid Zhmud remarked: "The intellectualization of a myth and its transference into the sphere of cognitive processes led to a situation where thinking and faith in mythological fictions were equated with one another. Strictly speaking, *the whole complex of ideas connected with mythological thinking has been based on this confusion up to the present*."[39] It was precisely the invention of "mythological thinking" as a particularly "primitive" mode of thought that allowed *The Intellectual Adventure* to adapt the nineteenth-century Comtean model of a progression from religion to science to its own purpose.

The Intellectual Adventure was a work of no small allegiance to progressivism, according to which cultural, intellectual, and cognitive history were all equally viewed as inexorably advancing with time, with the corollary that culture (the arts, technology, and science) established progressive control over, knowledge of, and an ever greater distance from nature over the course of history. Accordingly, in the Ur-time of human history, culture had not established sufficient distance from nature to have legitimate (that is, scientific) knowledge of it. As the Frankforts put it, "the life of man and the function of the state are for mythopoeic thought imbedded in nature, and the natural processes are affected by the acts of man no less than man's life depends on his harmonious integration with nature."[40]

William Y. Adams identified progressivism as "the single most powerful influence in Western historical thought," and as "the tap root of anthropology."[41] Relevant for placing *The Intellectual Adventure* within its own intellectual context, is Adams's "Idealist Progressivism," which sees progress as a function of the human mind maturing over time, as opposed to a materialist version in which the human condition improves as a result of better ways of handling the concrete realities of social and political life. The Idealist Progressivism underpinning *The Intellectual Adventure* determined that history manifests an intellectual evolution from myth through religion to science, signaling a certain and demonstrable maturity of mind, capable of abstract thought, rationality, and the intellectual rigors of logic and theorization.

The question of rationality, particularly whether it was a universal defining feature of human cognition, has long been at stake in anthropological theory. The logical unity of humankind, and with it "the very notion of humanity," as Rodney Needham said, was a question driving much anthropological research over the course of the first half of the twentieth century, and, to paraphrase Needham, the question of whether human reasoning was the same everywhere or not, and, if not, whether differences were a consequence of genetics or culture, was a matter for comparative ethnography.[42] But writing in the early 1970s Clifford Geertz found a reluctance on the part of anthropologists even to accept differences based in culture as fundamental in "defining man." He said, in characteristically vivid terms, that "anthropologists have shied away from cultural particularities when it came to a question of defining man and have taken refuge instead in bloodless universals . . . faced as they are with the enormous variation in human behavior, they are haunted by a fear of historicism, of becoming lost in a whirl so convulsive as to deprive them of any fixed bearings at all."[43]

Those wanting "fixed bearings" lined up on one side of an intellectual fence Richard Shweder described as the dividing line between theorists with commitments to an enlightenment idea of progress and the universal sway of rationality on one side and the romantic rebels, the pluralists and symbolists, on the other. Subscribers to "the enlightenment" were thereby unwilling to relinquish their commitment to the universality of reason and other ideas intimately connected to and projected from it, such as, in Shweder's terms, "the idea of natural law, the concept of deep structure, the notion of progress or development, and the image of the history of ideas as a struggle between reason and unreason, science and superstition."[44] His enlightenment anthropologists, James Frazer, E. B. Tylor, and Claude Lévi-Strauss (among others), saw the human mind as a unity, that is, the capacity for rationality and the ability to think logically was common to all human beings. Here the gap between ancient/"primitive" and modern was a function of evolution. Ancients and primitives, while having the capacity for reason, were however insufficiently developed and were as a consequence prone to logical fallacies, false causality, and magical thinking.

On the other side were Shweder's "romantics," Lucian Lévy-Bruhl, Marshall Sahlins, Clifford Geertz, and Paul Feyerabend (among others), who saw "that ideas and practices have their foundation in neither logic nor empirical science, that ideas and practices fall beyond the scope of deductive and inductive reason, that ideas and practices are neither rational nor irrational but rather *non*rational."[45] This gave rise, Shweder said, to an interpretation of cultures as arbitrary and pluralist, committed to local criteria for truth. The gap between ancient/"primitive" and modern is therefore erased for romantics, who view the history of ideas as "a sequence of entrenched ideational fashions"[46] and culture as "a self-contained 'framework' for understanding experience."[47] A good example, and relevant here because of its impact on *The Intellectual Adventure,* are Lévy-Bruhl's ideas on "how natives think," that is, in a manner not dependent upon the logic of temporal or spatial realities and the "law of non-contradiction," but rather in accordance with another logic, the prelogical, and another law, the law of participation.[48] Lévy-Bruhl's conception of the prelogical, however, was not meant as a commentary on cognitive evolution, or even on the capacity of people in traditional societies to understand logical relationships. In his words, "By *prelogical* we do not mean to assert that such a mentality constitutes a kind of antecedent stage, in point of time, to the birth of logical thought."[49] To further clarify, he said, "It is not *antilogical*; it is not *alogical* either."[50] He remarked that this mode of thought was true for collective representations only, that individuals operated in the world with the same practical reason we would (seeking shelter, escaping from danger, and such), but "as far as it [mental activity] is collective, it has laws which are peculiar to itself, and the first and most universal of these is the law of participation."[51]

The Frankforts traversed the intellectual territories mapped by Shweder, being both progressivist and romantic. Or, perhaps, one could say they embedded their cultural relativist leanings within a progressive enlightenment framework. Nonetheless, the theoretical framework of *The Intellectual Adventure,* most clearly set out in the Frankforts' two chapters, owes its principal philosophical debt to "romantics" Ernst Cassirer and Lucian Lévy-Bruhl, as well as to Martin Buber. Although it was not the Frankforts' intent to trace the history or argue for the theory of a primitive mentality (à la Lévy-Bruhl), of mythopoeic thought (à la Cassirer[52]), or the I-Thou relationship (à la Buber[53]), they took up these ideas without engaging explicitly or critically with the original works in which these ideas were found.[54] Intellectual kinship with these philosophers centered on the interest in myth. Like them, the Frankforts subscribed to the idea that myth was the special product of the human mind unfettered by logic or physics and, in accordance with the progressivism of their time, saw this invention of the mind as primitive compared to the intellectual achievement of science.

Because of the Frankforts' conviction that no philosophy or science akin to anything known in the Western tradition was evident in the ancient Near East, Lévy-Bruhl's separation of myth from philosophy and science must have struck a chord, not only from the cognitive evolutionary standpoint (which Lévy-Bruhl would repudiate[55]) but also in relation to the idea that myth expressed another form of thought about the world, one that did not obey the same logic of so-called ordinary reality, but was instead the consequence of the participation of individuals in a mystical reality, obeying the "law of participation."[56] Here too is where Cassirer's differentiation between the cognitive domain of myth and that of logical analysis and abstraction seems to have been found relevant. Like Cassirer, the Frankforts' assessment of myth was that it reflected a mode of thought unlike that of religion and science. Indeed, Cassirer found myth to be a form that "defies and challenges our fundamental categories of thought. Its logic—if there is any logic—is incommensurate with all our conceptions of empirical or scientific truth."[57]

Yet more influential was Cassirer's take on the perception of nature in myth—namely, that an empirical sense of nature governed by strictly material causes was not yet formed.[58] Myth conceived of nature rather as "a dramatic world—a world of actions, of forces, of conflicting powers,"[59] lending myth its emotional dimension. Cassirer went even further in this argument to say something that one sees again in *The Intellectual Adventure*, that in the mythic perception of the realm of physical phenomena, "things" are not, in Cassirer's words, "dead or indifferent stuff. All objects are benignant or malignant, friendly or inimical, familiar or uncanny, alluring and fascinating or repellant and threatening"[60]—in other words, personalized. Cassirer's juxtaposition of the mythopoeic experience of a personalized emotional world with "the ideal of truth that is introduced by science,"[61] was echoed in *The Intellectual Adventure*. Cassirer, in dialogue here with Lévy-Bruhl, agreed that myth manifests its own "mystic" causality but disagreed with the idea that this mode of thought was generally characteristic of traditional cultures, existing even outside of the holy.[62] This, however, was the very thing Lévy-Bruhl would later retract.[63]

Martin Buber, on the other hand, did not separate his postulated two modes of being in the world as though they were different mentalities. Rather, it was a matter of presence (or "the present") and object (*Gegenwart und Gegenstand*). What is present is what lives and is lived and is not epistemological. I-It is epistemological as it allows us to describe, analyze, and classify. I-You, in contrast, exists as "the mystery of reciprocity."[64] This is indeed what the Frankforts were drawing on when they said, "'Thou' is a live presence, whose qualities and potentialities can be made somewhat articulate—not as a result of active inquiry but because 'Thou,' as a presence, reveals itself."[65] They, however, took the awareness or experience of this "presence" to be a form of cognition all its own. In their words, "The fundamental difference between the attitudes of modern and ancient man as regards the surrounding world is this: for the modern, scientific man the phenomenal world is primarily an 'It'; for ancient—and also for primitive—man it is a 'Thou.'"[66]

The further clarification of this idea in the ensuing pages was clearly and heavily dependent upon Buber. Interestingly, they chose to quote not from Buber but from Alfred Ernest Crawley, most known for his ethnology of marriage, and whose claim was that to the "primitive" the world is "personal."[67] The Frankforts found this to resonate with the attitude in Near Eastern antiquity, not that the world was animistic or that natural phenomena were personified, rather that the world was simply and thoroughly animate, that is, "personal." The relation the Frankforts described between the individual and a phenomenon, experienced as a mysterious and reciprocal relationship without the detachment that typifies our experience of "things," however, came directly out of Buber's *I and Thou*.[68]

Myth in *The Intellectual Adventure* not only belonged to a different and incompatible kind of thinking from that employed in philosophy or science, it was also chronologically prior and prelude to them, more primitive as well as more ancient. This was the fundamental premise of these two chapters, hence the revised title for the second edition, *Before Philosophy*. The Frankforts' cognitive-historical view was that whatever products of intellection were evident in the ancient Near East, they represented earlier stages of thought than philosophical or scientific reason. To argue more persuasively for this view, the book's conclusion compared the mythological thought-world of the ancient Near East against intellectual activity in ancient Greece, where, it was argued, philosophy and science were first born with the pre-Socratics.

The principal question driving *The Intellectual Adventure* was, therefore, how the ancient civilizations of Mesopotamia (Iraq), Egypt, and Israel, as well as Greece, for contrast's sake, understood and explained humankind's place in relation to the world of nature. Differences were drawn along linguistic, geographical, and cultural lines. In contrast to what developed in "The West" (that is, Greece), the ancient Near East (less so Israel) was portrayed as a realm apart. Interestingly, not long after the appearance of *The Intellectual Adventure*, E. R. Dodds's *The Greeks and the Irrational* (University of California Press, 1951) focused attention on areas of Greek thought that did not conform to stereotypes of rationalist Ionian philosophers such as were portrayed by the Frankforts in chapter 8, "The Emancipation of Thought from Myth." Dodds showed that to feature rationality in ancient Greece to the exclusion of other aspects of its literature and even philosophy was an oversimplification. That rationality, logic, and scientific explanation supplied new answers to questions about nature was shown to be only part of a complex history in which religion and the putatively darker regions of the human psyche—supposed to be contrasted with the light of rationality—still had a place. About the Greeks and their not unmitigated focus on the rational, Dodds said of Socrates, for example, that "we should not forget that he took both dreams and oracles very seriously, and that he habitually heard and obeyed an inner voice which knew more than he did (if we can believe Xenophon, he called it, quite simply, 'the voice of God')."[69] Dodds did not challenge the rational/irrational dichotomy as an interpretive tool in cultural analysis, though he did offer a more complex account of Greek intellectual culture.

With reference to ideas about natural phenomena, the principal problem with *The Intellectual Adventure* was in its assumption that myth was all there was. The opening chapter, "Myth and Reality," made brief reference to "the 'science' of omens," saying it was aimed at consolidating "a harmony with nature, a coordination of natural and social forces" and meant "to facilitate humankind's chances of success in its undertakings."[70] Divination is surely "meant to facilitate humankind's chances of success in its undertakings," but nowhere in *The Intellectual Adventure* was there an analysis of the contents of omen texts for evidence of other dimensions of thought, such as the empirical, rational, or predictive, relative to the phenomenal world. Omen texts were simply taken as consistent with other products of mythopoeic thought.

Modern science, according to the Frankforts, with its knowledge of the universal laws of nature, answers questions about nature with precision and truth, while the ancients, using myths, answered similar questions with narratives about gods. The Frankforts opposed "scientific" to "speculative" thought as a way to distinguish between modern and ancient ways of explaining, unifying, and ordering experience. When the Penguin edition came out, however, Henri Frankfort took pains to point out that, contrary to the criticism of some reviewers, he did not "sing the praise of rationalism or equate religion with superstition."[71] While he may have wanted to distance himself from a Humean evaluation of the difference between the kind of thought (abstract reasoning) characteristic of science and that reserved for religious or metaphysical knowledge (sophistry and illusion),[72] he most definitely accepted the same divide. And despite his not seeing the ancient patterns of thinking as inferior to modern, rather as an alternative mode—another mentality altogether (his romanticism)—as a man of his time he accepted that a developmental progress in the history of ideas led the way from myth and religion to reason and science (his progressivism).

By now, as a result of changed attitudes about the nature of science and how to define it, divinatory texts in the ancient Near East have been accepted within the purview of the history of science. What has not been discussed, however, is the central theme around which *The Intellectual Adventure* was developed—namely, the conception of nature in the ancient Near East, or perhaps, the lack of a conception of nature as defined by later Western cultures. Although nature functions as a historically variable category, it seems particularly endemic to Western cultures descended from the classical world. In ancient Mesopotamia, as in any number of premodern and non-Western cultures, engagement with the world did not give rise to a term for nature, nor to a conception of nature as an alleged objective reality. But this peculiar conception of the objective reality of nature was a comparatively late development in the West. For the ancient Near East, then, it is not a matter of reconstructing a notional counterpart to nature, so to speak before the beginnings of the conceptual history of nature, but of defining the various ways in which phenomena that we think of as belonging to nature were of interest in cuneiform texts. A certain relativity of nature, that is to say, a culturally relative approach that might account for the pluralities and changes in the conception of nature(s) cross-culturally and historically, does not seem to be enough for an understanding of cuneiform sources that reflect on what we classify as natural phenomena.

NATURE HISTORICIZED

The word *nature*, in the phrase "the nature of," connotes the essential and defining characteristic of a thing, and by extension, as Raymond Williams put it, nature came to represent "the essential constitution of the world."[73]

He saw the formation of the idea of a singular and abstract nature as stemming from idealist thought. Thus: "Many of the earliest speculations about nature seem to have been in this sense physical, but with the underlying assumption that in the course of the physical inquiries one was discovering the essential, inherent and indeed immutable laws of the world. . . . A singular name for the real multiplicity of things and living processes may be held, with an effort, to be neutral, but I am sure it is very often the case that it offers, from the beginning, a dominant kind of interpretation: idealist, metaphysical, or religious."[74] Williams captured the historical changes in the idea of nature's relation to humankind and to God in a number of well-chosen metaphorical epithets, "God's minister and deputy," "the absolute monarch," "the constitutional lawyer."[75] Accordingly, nature's laws went from being handed down from on high in the Middle Ages to being "but an accumulation and classification of cases" from the seventeenth century onward.[76] In broad outline, he described nature as evolving from a personification to an object. And with respect to the place of humanity in nature's order, he distinguished a medieval religious idea of humankind as belonging to the order of nature, as against a modern secular idea of the human being separate from that order, and thereby in a position to "intervene."[77] All in all, Williams's objective was to historicize nature in order to bring out its varied and plural "natures."

The idea that perceptions of the natural world are a direct function of our cultural ways of knowing and our linguistic ways of describing is by now normative in the history of science. Latour famously offered the coinage "nature-culture" to underscore the idea.[78] Furthermore, in David Delany's sense, nature is "an extremely versatile cultural artifact."[79] It is, as Delany said, "a conceptual tool that is used for making sense, for making things meaningful. . . . People *do* things with 'nature.' Things, places, events are *made* meaningful."[80] Whereas the two domains, nature and culture, have been said to be cleaved from each other since modernity, neither their status as absolutes nor their dichotomous relationship are historically or anthropologically substantiated.[81] Stated perhaps most directly by Strathern: "There is no such thing as nature or culture. Each is a highly relativized concept whose ultimate significance must be derived from its place within a specific metaphysics. No single meaning can in fact be given to nature or culture in western thought; there is no consistent dichotomy, only a matrix of contrasts."[82]

Eduardo Viveiros de Castro, in an investigation of what he termed perspectivism in Amerindian culture, noted that, "as many anthropologists have already concluded (albeit for other reasons), the classic distinction between Nature and Culture cannot be used to describe domains internal to Non-Western cosmologies without first undergoing a rigorous ethnographic critique."[83] Speaking in terms of a "reshuffling of our conceptual scheme," he suggested the expression "'multi-naturalism,' to designate one of the contrastive features of Amerindian thought in relation to Western 'multiculturalist' cosmologies," and explained, "where the latter are founded on the mutual implication of the unity of nature and the plurality of cultures . . . the Amerindian conception would suppose a spiritual unity and a corporeal diversity. Here culture or the subject would be the form of the universal, whilst nature or the object would be the form of the particular."[84] This statement about nature is important here not because ancient Near Eastern cosmological mythology and the Amerindian are comparable, but because it points to the insufficiency of the idea of the unity of nature amid a plurality of cultures.

Clearly the perception of the phenomenal world reflected in cuneiform texts is neither reducible to that of any modern ethnographic societies nor to that of the modern West. The study of it, therefore, requires a different methodological approach. The problematic use of terms like *ethnomedicine* and *ethnophysics* to underscore the differences between indigenous systems of medicine or physical sciences and their modern Western counterparts will be of no use for the ancient Near East.[85] These terms have come to have a marginalizing force, implying, perhaps unwittingly, that the cultures that practice "ethnosciences" do so because they have not had the opportunity to assimilate or be assimilated into the modern scientific mainstream.

From the point of view of contemporary constructivism about science, there is a kind of irony about *Before Philosophy: The Intellectual Adventure of Ancient Man* in its settling on the nature-culture relationship as the key to comparing ancients against moderns, or "Mesopotamian" against scientific thought. One cannot argue with the Frankforts that there was no clearly defined boundary for conceiving of nature as separate from either humankind or the divine. One can, however, argue with the fact that the Frankforts differentiated the ancient Near East from Israel and Greece by reference to a conception of nature that only emerged fairly late in Western natural philosophy and science, indeed, not until the early modern period, when nature came to be, in Raymond Williams's words "an object, even at times a machine . . . a fixed state."[86] The Frankforts employed the universal and all-encompassing conception of nature commensurate with their modernist and scientistic views. But Latour challenged even this point of view, that we only purport to know nature independently of culture. With characteristic irony he said: "We do not mobilize an image or a symbolic representation of Nature, the way the other societies do, but Nature as it is, or at least as it is known to the sciences—which remain in the background, unstudied, unstudiable, miraculously conflated with Nature itself."[87] He argued that making knowledge and doing science always involves an interplay between culture and what we call nature, even in modern science, and called into question the claim that an undifferentiated nature-culture belongs only to the remote time before modernity.[88]

Recognizing the difference between the study of contemporary ethnographic and ancient cultures on the one hand, and the importance of granting an integrity and independence of indigenous knowledge separate from the history of Western science on the other, there is a cultural ambiguity to be defined about "ancient Mesopotamia" that makes the relation of cuneiform knowledge particularly interesting for the history of science. "Ancient Mesopotamia" exists in our cognitive geography both inside and outside the boundaries of Western civilization. Neither wholly Western nor non-Western, Babylonian astronomy and astrology, medicine and other forms of knowledge cannot be legitimately consigned to the ethnosciences. Its astral scientific tradition is too much implicated in what came later, in Greece, Rome, and Europe, to be non-Western. This no doubt accounts for the appearance of only selected entries on the ancient Near East in the *Encyclopedia of the History of Science, Technology, and Medicine in Non-Western Cultures*. There is, for example, no entry for Babylonian astronomy.[89]

For understanding the way knowledge took shape and developed in the ancient Near East, and how we might relate it to later sciences, it seems to me useful to bear in mind Latour's observation that the "seamless web," as he called it, comprising nature and culture, still exists. Modernity may have removed the divine from nature, but not nature from culture. Latour's point was that the imagined gulf between Us and the putative Them is really just so much classificatory rhetoric. We do not have direct access to nature either, he said, and he proposed looking at other cultures as well as our own, comparing "natures-cultures," as he put it, and laying aside the "moderns' victory cry" of our absolute cognitive difference.[90] For interpreting cuneiform sources, the solution is not in distinguishing between knowledge that represents nature (science) versus that which (merely) reflects culture (nonscience, magic, religion, ritual), but, in dismantling the dichotomy, as far as is possible for our way of expressing ourselves, so as, at least, not to take the categories themselves as absolutes.[91]

We have come a long way since Henri Frankfort and Henriette Groenewegen-Frankort formulated their interpretations of ancient Near Eastern thought about nature. Of course the absence of conceptual boundaries around nature and culture was not a function of an archaic stage of cognitive development. Nor was myth the sole manner in which phenomena, our natural phenomena, were dealt with in cuneiform texts. In fact, on balance, when divinatory, astronomical, and medical texts are taken into account, literary (hymnic, mythological) contexts for defining or explaining physical phenomena are proportionately small. In historicizing the Frankforts, it is clear not only that the very idea of a mythopoeic stage of thought was a holdover from a unilinear evolutionary social anthropology, but, more interestingly, the idea of nature that underpinned the Frankforts' reconstruction of "speculative thought" in the ancient Near East was too singular, too absolute, and too anachronistic to be useful for that historical analysis. A new historiography aimed at integrating cuneiform texts into the history of science can recognize a broader corpus of relevant evidence than that which underpinned *The Intellectual Adventure*, and will consider as an important dimension the various ways in which that corpus sought to fit phenomena, physical and otherwise, into its own sense of an order of things.

PART II

Cuneiform Knowledge and Its Interpretive Framework

CHAPTER THREE

On Knowledge among Cuneiform Scholars

mudē UZU *irrī* "Knowledgeable in Intestines"

Science itself is what we know, and philosophers as much as anyone else should proceed on the basis of what we know. It is mistaken as well as self-defeating for philosophers to discuss science from a vantage point independent of or prior to their actual historical situation, replete with the science of their day.

BAS VAN FRAASSEN, "The False Hopes of Traditional Epistemology"

[*mudē* UZU *ir*]*rī* "(the diviner) knowledgeable in (the interpretation of the) intestines"

THE KNOWLEDGEABLE AND THE KNOWN

The literal translation of the phrase quoted in the epigraph above, *mudē* UZU *irrī* as "knowledgeable in intestines,"[1] or the slightly more informative "(the diviner) expert in (interpretation of) the intestines" draws us, as modern readers, into a culture of knowledge not immediately reconcilable with our own. The adjective *mudû* generally qualified an expert in a scholarly field, and the phrase "knowledgeable in intestines" takes its meaning from the idea that possessing knowledge was equivalent to the mastery not of the anatomy of the internal organs of animals, though anatomical understanding of what was seen was assuredly a part of it, but of the meaning of a feature of the exta in a divinatory context. The meaning associated with various features of the exta was derived from a written corpus of omens, constituting a sizable repertoire for the extispicy specialist.

If the present discussion is to contribute in some way to the conversation about the nature of science in history, particularly of its epistemic content, a descriptive outline of what knowledge was in the cuneiform scribal-scholarly context should first be offered.[2] This chapter presents only some features on the relief map of that terrain, as the evidence for scholarly knowledge in Akkadian texts is too vast, and the aim is not to be comprehensive or exhaustive. Specialists will no doubt find the mention of certain important sources or scribal names and families missing. The object is not to provide intellectual biographies of well-known scribes, such as Sin-leqe-unnīnī or Esagil-kīn-apli, although that would be a worthy project elsewhere. My focus is rather on selected parts of cuneiform scholarship having to do with the reckoning of signs in diverse phenomena and in related fields of knowing that supported scholarly divination, such as astronomical knowledge. By focusing thus on the contents of selected texts, and considering what the scribes, who were in some periods identified as "knowledgeable" (*mudû*), held to be objects of knowledge, the goal is to limn out some elements of the cuneiform intellectual landscape. The cognitive historical dimension of that landscape that speaks to reasoning and rationality, complementary to and embodied in the social, is further taken up in chapter 5.

An abstract Akkadian noun "knowledge" from the verb "to know" *idû* occurs only in lexical lists (**idūtu/edūtu* = giš.á.zu.zu[3]) and nowhere in actual use in Akkadian context, while the abstract *mudûtu* "knowledge" from the same verb is used to refer to knowledge as information in a more restricted sense than the sum of what is known. Illustrative of this restricted sense in the Akkadian usage is a passage from the *Code of Hammurabi* that says, "the witnesses who know the lost property will tell what they know (*mudūssunu*, literally, "their knowledge") before the deity."[4] The adjective *mudû* meant "knowing" or "knowledgeable," as in "expert in a specific craft, etc., wise, competent, learned, knowledgeable, expert,"[5] and was said of gods, kings, and artisans.

Although there does not seem to be an Akkadian term for the sum of things known, mastery of words and of the textual repertoire of divination had a term, *lē'ûtu* "knowledge," meaning ability or skill in scholarly literature, derived from the verb "to be able" (*le'û*). A passage in one of the royal letters from the Neo-Assyrian corpus says in reference to the king's own rhetoric, "is not this (skillful rhetoric) the epitome of scholarship/knowledge?"[6] The qualifier *lē'û* "skilled, competent," is found in contexts describing those who mastered skills in the crafts, in divination specifically, and knowledge in general,[7] as when King Assurbanipal boasted of discussing the text, "If the liver is a mirror of heaven," "with the knowledgeable experts (*itti apkallī lē'ûti*)."[8]

Knowledge expressed with the word *lē'ûtu* also had the connotation of understanding, as in this variant on the sentiment that the divine "mind" was inscrutable: "The mind (literally "inside") of the divine is remote like the inner reaches (literally "inside") of heaven. Knowledge (*lē'ūtu*) of it is (so) difficult that humankind cannot understand (it)."[9] The unknowable nature of the divine is found elsewhere in Akkadian literature, perhaps best exemplified by the question: "Where have any of the numerous (human beings) ever (really) understood the way of the gods?"[10]

Acquisition of knowledge was expressed with the verb *aḫāzu*, in its meaning "to learn" or "to understand."[11] Thus another term for things known was the derived noun *iḫzu*, written with a logogram whose Sumerian meaning is literally "thing known" (NÍG.ZU). Its usage suggests a range of meaning from knowledge to teachings or instruction.[12] The term is used in the colophon of the late second millennium scribe and *ummânu* Esagil-kīn-apli, who established the text of the medical diagnostic omen series SA.GIG "for knowledge" (*ana* NÍG.ZU), warning: "Do not neglect your knowledge! He who does not attain(?) knowledge must not speak aloud the SA.GIG omens!"[13]

In the realm of elite scribal learning, to be knowledgeable meant being skilled in the textual repertoire, knowing the contents of the texts containing "wisdom" (*nēmequ*).[14] The scribe's skill in the repertoire included the technologies of cuneiform orthography used in writing or copying the repertoire itself. The term "wisdom," or "knowledge," *nēmequ*, from the same root as the adjective *emqu* "skilled, wise, educated," is found in the colophons to scholarly texts, associating this knowledge of texts and of writing with the god Ea, god of wisdom and esoteric knowledge par excellence (*bēl nēmeqi* "lord of wisdom"), and also with Nabû, god of the scribal craft (*nēmeq Nabû tikip santakki* "the wisdom of Nabû, the cuneiform signs").[15] Specific mention of the nature of these esoteric texts includes the divinatory corpus of extispicy (*bārûtu* "inspection [of the liver]" and *nēmeq Šamaš u Adad* "wisdom of Shamash and Adad [lords of divination]") and of celestial divination (*enūma* ᵈ*Anum* ᵈEN.LÍL.LÁ "When and") and of the craft of the ritual purifier, who dispelled the bad influences of gods, or demons (*kakugallūtu*).[16] The "wise and the knowledgeable" (*enqu mudû*) described the scribes who would understand and explain the fifty names of Marduk, a tour de force of cuneiform scribal learning as set down in Tablet VII of *Enūma Eliš* (lines 145–46 and 157–58). Wisdom and knowledge, therefore, connoted education, expertise, and skill of the highest order in cuneiform writing as in its literary and scholarly repertoire.

Knowledge in the form of divination and its supporting intellectual activities (including various ritual activities and astronomy) remained a central part of cuneiform scholarship for nearly two thousand years. Extant source material stems from the following periods: (1) the Old Babylonian period from roughly 2000 to 1600 BCE;[17] (2) the Middle Babylonian and Middle Assyrian periods, from roughly 1600 BCE to the end of the second millennium;[18] (3) the Neo-Assyrian period, circa 900 to 600 BCE; (4) the Neo-Babylonian period, from circa 620 to 540 BCE; and (5) the so-called Late Babylonian period, very approximately referring to texts from the second half of the first millennium, concentrated in the Seleucid period from circa 330 to 125 BCE, but continued in the Arsacid, or Parthian, period (third century BCE to first century CE) to the end of cuneiform writing itself.

Particularly in the Neo-Assyrian period, during the seventh century BCE, cuneiform scribal scholarship was referred to as *ṭupšarrūtu*, an abstract noun formation from the word for scribe (*ṭupšarru*, a loan word from Sumerian DUB.SAR, literally "to write a tablet"). The corpus of texts belonging to *ṭupsarrūtu* contained the "wisdom of the gods" *nēmeq ilī*; the "wisdom of Ea" *nēmeq Ea*; the "word of the Apsû" *amat apsî*; and the "wisdom of the highest rank" *nēmeq anūti*. The scholars who produced and maintained *ṭupšarrūtu* focused primarily on divinatory disciplines, such as celestial omens (from the series *Enūma Anu Enlil*) and the practice of extispicy, literally "the practice of inspection" (*bārûtu*),[19] meaning of the liver and other exta of the sacrificed sheep. Also included were the practices of conjuration by spells and incantations (*ašipūtu*), medicine (*asûtu*), and the chanting of lamentations (*kalûtu*).

Scribal specialists had professional designations, such as the diviners who interpreted the heavens (*ṭupšar Enūma Anu Enlil* "scribe of *Enūma Anu Enlil*"), the diviners who inspected the exta (*bārû* "diviner," literally, "who makes an inspection [of the exta]"), the conjurer (*āšipu*, from the verb "to cast a spell," also written as *mašmaššu*), whose chief role was in dealing with illness, the chanter (of lamentations) (*kalû*, a Sumerian loanword from GALA "singer"), and the physician (*asû*). A document listing these personnel in the Neo-Assyrian court, most likely of Esarhaddon, tallies seven *ṭupšar Enūma Anu Enlil*, nine *āšipu*s, five *bārû*s, nine *asû*s, six *kalû*s, as well as observers of birds (*dāgil iṣṣūrē*)[20] and foreign diviners, the Egyptian *ḫarṭibi* "chief lector priest," here possibly "dream interpreters,"[21] as well as "Egyptian scribes (A.BA.MEŠ *Muṣuraya*).[22] Each field had its *ummânū*, meaning "savants," "experts," or "masters (of the discipline)."[23]

The picture that emerges in the Neo-Assyrian texts shows that scholars were trained for the purpose of serving the king by acting as interpreters of divine wisdom. Some scholars were particularly expert in liturgy and ritual, thereby holding an obvious place in relation to the divine. But for scholars in general throughout the entire period of cuneiform learning, knowledge was closely connected to texts, and the contents of scholarly texts were connected to the gods. This was not merely a theological claim, but had a historical dimension as well, in that the scribes were invested in what Alan Lenzi termed a "mythology of scribal succession."[24] By means of that "mythology," which forged an identity for scribes entrusted with the divine secrets, the *ummânū* were descended from the seven *apkallū*, the antediluvian sages of the *Abzu*, the cosmological domain of sweet waters and dwelling of the god of wisdom and patron of *ummânū*, Ea. The term *apkallu* served as an epithet, "wise" or "expert," for the gods Ea and Marduk, both associated with knowledge of healing magic and incantations.[25]

The uppermost echelon of scribes who wrote scholarly texts therefore identified themselves as the guardians, stewards, preservers, and users of a tradition of knowledge they claimed was handed down to them from the remotest antiquity before the Flood. A particularly vivid text, preserved in several copies from Assurbanipal's Nineveh, describes the sun-god's gift of knowledge directly to King Enmeduranki of Sippar, a fictitious figure known from the Sumerian King List as the seventh antediluvian ruler. The account stipulates that future savants will teach their sons "the secrets of the great gods" but only after an oath to the divinatory patron deities Shamash and Adad is sworn by the young scribe:

> Shamash in Ebabbarra [appointed] Enmeduranki [king of Sippar], the beloved of Anu, Enlil, and Ea. Shamash and Adad [brought him in] to their assembly . . . and showed him how to observe oil on water, a mystery of Anu, Enlil and Ea, they gave him the tablet of the gods, the liver, a secret of heaven and the underworld. . . . Then he in accordance with their word brought the men of Nippur, Sippar, and Babylon into his presence, . . . showed them how to observe oil on water, a mystery of Anu Enlil and Ea, he gave them the tablet of the gods, the liver, a secret of heaven and underworld, . . . The learned savant (*ummânu*, LÚ.UM.ME.A) who guards the secret of the great gods, will bind his son whom he loves with an oath before Shamash and Adad by tablet and stylus and will instruct him.[26]

This passage views knowledge as a matter of divine origin and divine transmission, thus its status as exclusive intellectual property, accessible only to the "learned savant," the *ummânu*.

The notion of the origination of knowledge from before the Flood is found in the *Epic of Gilgamesh* (XI 86), where Uta-napišti loaded the ark with *ummânū*, experts of all kinds ("I sent aboard . . . persons of every skill and craft"),[27] as well as later in the *Babyloniaka* of Berossus, in which the first of the antediluvian sages, Adapa-Oannes, the fish man and archetypal *ummânu*, brought knowledge to humankind.[28] The perpetuation of this topos into Seleucid times, both in the *Babyloniaka* and in the late text known as "The Uruk List of Kings and Sages," reflects its concrete rhetorical effects for the social, and possibly also political, station of scribes in Late Babylonia, much as it had such an effect for those of the Neo-Assyrian court.[29]

Rooted in this ancient Near Eastern topos, as transmitted during the Hellenistic period and further developed and transformed by such writers as Flavius Josephus in the first century CE, the idea that astronomy, specifically, was antediluvian in origins, a gift from God to the sages before the Flood, was an idea still at home during the European Renaissance. Concerning the belief in antediluvian astronomy during the sixteenth century, Daniel Špelda said:

> One of the basic forms of belief in antediluvian astronomy was that, at the beginning of history, humans received astronomy as a gift from God. This idea was easily comprehensible in the Renaissance context as it corresponded with the belief of Renaissance Platonists and Hermetics in the existence of *prisca sapientia*. This was the term used in the Renaissance period to mark the perfect knowledge given by God at the beginning of history to several chosen sages. Already in ancient times this knowledge had been lost, but Renaissance scholars hoped that it would be recovered through a study of ancient texts.[30]

Špelda documented a rich tradition of sixteenth- and seventeenth-century scholars, including Gabriele Pirovano (d. 1512), Joachim Rheticus (1514–1574), Petrus Ramus (1515–1572), and Hermann Witekind (1521–1603), who conceived of the history of astronomy as a decline from its pristine state in antiquity. As he pointed out, "In *Scholae mathematice*, Ramus states that the old Near East astronomy (before Eudoxus), passed on by the patriarchs, could dispense with hypotheses ('*sine hypothesibus fuisse*'), despite which the Chaldeans and Egyptians were still able to predict solar eclipses and other phenomena. That is why Ramus called for the creation of another system of astronomy: '*non ex fictis hypothesibus, sed ex ipsa astrorum veritate.*' This was the state of the ancient astronomy of the Chaldeans, Egyptians, and early Greeks, before it was spoilt by mathematicians and philosophers."[31] The attribution of great antiquity and traditional value to the divinatory parts of cuneiform scribal knowledge cannot be generalized to the whole of the scholarly corpus, or corpora. Some of the astronomical literature, for example the astronomical observation texts known as Diaries, for which colophons are extremely rare anyway, do not appear to belong to this category of antediluvian traditional knowledge. But other astronomical texts, such as the ephemerides, are sometimes designated as *nēmeq anūti* "the wisdom of Anu-ship." This term conveyed the idea that such a text represented the very highest level of knowledge, the substantive *anūtu* being constructed from the name of the sky-god, Anu. The following is from an eclipse ephemeris dated to 191 BCE. Note its inclusion of the injunction against the scribe who is not *mudû* "knowledgeable."[32]

> On eclipses of the moon.
> Tablet of Anu-bēl-šunu, lamentation priest of Anu, son of Nidintu-Anu, descendant of Sin-leqi-unnīnnī of Uruk. Hand of Anu-[aba-utêr, his son, scri]be of *Enūma Anu Enlil* of Uruk. Uruk, month I, year 12[1?] Antiochus [was king;]
> May whoever reveres the gods Anu and Antu [not remove (this tablet);] Computational table. The wisdom of Anu-ship, exclusive knowledge of the gods [. . .] Secret knowledge of the masters. The knowledgeable may show (it) to an[other one who is knowledgeable]. One who is not knowlegeable may not [see it. It belongs to the forbidden things] of Anu, Enlil [and Ea, the great gods].

As in this example, tablets containing colophons restricting access to the knowledgeable alone sometimes added that for the inexpert to see the tablet was a divine transgression. The term applied, that is, *ikkibu*/NÍG.GIG, is translated variously as "sin," "taboo," "transgression," or "restriction,"[33] generally giving the name/names of the god/gods against whom the offense would be committed, for example, Marduk, Ea, Anu, Enlil, the Igigu, the Anunnaki, and the DINGIR.GUB.BA "standing gods" of Ekur, Šullat, Ḫaniš, Shamash, and Adad, or simply "the great gods."

As represented particularly at Nineveh and elsewhere in the Assyrian Empire, the scribes referred to as knowledgeable held the professional titles *ṭupšarru*, *āšipu* (or *mašmaššu*), *bārû*, *kalû*, and *asû*. But rigid distinctions did not apply between these scribal specializations and the texts they wrote, copied, and interpreted. Especially in the Late Babylonian period, omens (including astral, abnormal birth, and human physiognomic omens) and astronomical texts were produced by *āšipu*s "conjurers," and *kalû*s "chanters," previously associated primarily with incantations and laments, with which to appease the gods.[34] The literature of the *āšipu*s and *kalû*s was at all times completely integral to the fields of knowledge belonging to *ṭupšarrūtu* as a whole, but their specialized knowledge, and authority to address the gods in ritual and incantation, places the textual contents of their practice most directly with what we classify as magic.

Even more readily than divination, magic has slowly been accepted within boundary lines defining early science. But this has come only after ahistorical adherence to certain ideas about science were gradually let go out of sheer pressure, so to speak, from the original sources. Under the entry "Magic and Science" in the *Encyclopedia of the History of Science, Technology, and Medicine in Non-Western Cultures,* the history of the separation of magic from science is explained in terms of a peculiarly Western intellectual history, in which the term *science* was gradually transformed to represent "a distinctive rationality valued above and against magic."[35] The opposition of the terms *science* and *magic* long encoded an opposition between the rational and the irrational. But with respect to the world of ancient Near Eastern antiquity the dualism implied by the terminology reflects less of a historical reality and more of a disjunct between the epistemic and cognitive values of the culture under study and those of the culture doing the studying.[36] A cultural history of science in the ancient Near East now acknowledges that magic, so-called today, with its special methods and textual reference corpora, was not only a legitimate but also a principal part of the ancient program of knowledge.

In the manner of all the cuneiform knowledge corpora, the medical disciplines of *asûtu* and *āšipūtu* produced systematic text compilations. The incantations of *āšipūtu* were collected and canonized in the same way as were divinatory texts.[37] The literature of *asûtu* focused on symptoms, diagnoses, and therapeutic techniques and included lists of medically efficacious plants (such as the pharmaceutical list titled *Uruanna*) and stones used in the fashioning of amulets to release the anger of the deities responsible for bodily affliction. Medical problems such as external skin infections, pain or numbness in various parts of the body, fever, gastrointestinal irritation, eye problems, nosebleeds, and gynecological problems could be treated by the *asû* "physician" with prescriptions. These are complex and varied. The following illustrates one such prescription from what is an extensive corpus: "2 shekels of roasted *murru*; 2 shekels of *nindan*; 2 shekels of *puruḫliban*; 2 shekels of juniper; 2 shekels of yellow sulphur; 2 shekels of wax; [2] shekels of (red) *kalgukku*-clay; 3 roasted . . . [. . .] . . . dates; 2 shekels of Egyptian roasted alum. You cook in 2 litres, 1 *seah* of olive oil. You rub it into the sore twice a day. It is effective (good) for a red sore or *rišûtu* that is suppurating."[38]

Because the etiology of disease was considered divine, demonic, from ghosts (*qāt eṭemmi* "hand of a ghost"), witches (*kišpu* "witchcraft"), curses (*mamītu*), or anything evil (*mimma lemnu* "whatever is evil"), the *āšipu* "exorcist" specialized in the literature used to appeal to deities who had the power to heal and protect the patient.[39] Incantations could be used in combination with other prophylactic and apotropaic acts such as fumigation, the topical application of salves, and the use of amulets as means to appease the divine sources of illness and pain. Equally as varied, complex, and technical as the therapeutic preparations were medical incantations, as in the following:

> Incantation: Šulak, who struck the young man and stole his life, angry *gallû*-demon, who spilled the blood of the young man, evil god or evil goddess wandering about the city square, who struck down that young man, consigning him to bed, who laid out for him his death bed . . . his corpse; accept his appeal! [. . .] behind him [. . .]. You touched him, and health he has no more. The one you struck with your right, may Išum cure! The one you touched with your left, may Kusu cure! May the healing spell of Ningirimma be cast over him! May the . . . of Ea and Asalluhi, which provides well-being, be cast over him! May Gula drive out his sickness with her pure spell! The incantation is not mine; the incantation is of Ea and Asalluḫi; the incantation is of Damu and Gula; the incantation is of Ningirimma, mistress of incantations.[40]

The medical corpus as such, including the texts of both *āšipūtu* and *asûtu*, has received much attention in recent scholarship.[41] Since 2003 under the editorial guidance of Annie Attia and Gilles Buisson, the *Journal des Médecines Cunéiformes* has provided a forum for specialized study of cuneiform medical texts. Whereas medical knowledge was a fully separate technical specialization within the scribal tradition, separate from divination and astrology, in the Late Babylonian period, astrology and medicine were brought together in innovative ways.[42]

KNOWLEDGE OF OMINOUS SIGNS

The use of divination in official contexts long antedates the Assyrian Empire. The Early Dynastic lexical list Lu, from the third millennium BCE, attests to professional titles such as *ugula-azu* "chief physician," *máš-šu-gíd-gíd* "diviner," and *ugula máš-šu-gíd-gíd*, "chief diviner."[43] On the basis of such titles, however, divination from the entrails of the sacrificial offering may have been present from earliest times, predating omen texts by many centuries. Urnanshe, an Early Dynastic king from the mid-third millennium BCE, for example, consulted an *ugula-azu* in connection with building a temple.[44] Sumerian terms for cultic functionaries associated with divination and dream incubation are also known in Ur III economic texts (ca. 2100–2000 BCE),[45] and late third millennium Sumerian literature attests to the association of divination and cult. In a hymn to Enlil is an enumeration of clergy, beginning with "the *en* priest of the house was a diviner."[46] Apart from the mention of various cultic offices for diviners in Sumerian contexts, Cylinder A of Gudea of Lagash confirms an acquaintance with dream omens, extispicy, and even celestial signs. This inscription also places divination in the context of a temple building ritual, and uses the word *giskim* "sign."[47] The reference to diviners in Sumerian lexical lists of professionals and other references to divination from animals,[48] and the entry in a lexical text for the Sumerian word *a-rá* (Akkadian *alaktu*) in the meaning "omen" or "divine oracular decision,"[49] are hints that divination by extispicy, at least, might have played a role in the earliest Mesopotamian society.

Indications that the practice of divination was a feature of the earliest literate cultures also come from Ebla, as indicated in administrative texts from that site.[50] There is still the lingering question, however, as to the nature and role of divination in the third millennium. Ulla Koch, for example, suggested that despite Sumerian evidence for inspection of livers for determining cultic matters, the lack of a Sumerian terminology built into the later Akkadian extispicy corpus, as compared with Babylonian mathematics, raises questions as to whether divination by extispicy in the third millennium was a systematic scholarly discipline.[51] Akkadian scholarly and/or didactic texts (including liver and extispicy models) of the Old Babylonian period offer the first evidence of a systematic and analytic approach to signs of all kinds.

Emerging in Old Babylonian are the compilations of omens from signs in the liver and exta of sacrificed sheep, physiognomy, malformed births (*Šumma izbu*), and astral and lunar phenomena.[52] These typically exhibit the characteristic relationship between sign and consequent forged by association or analogy, as in:

> If water secretes inside the gall bladder: The flood will come.[53]
>
> If the gall bladder is turned and has wrapped around the "finger": The king will seize the enemy country.[54]

In the case of extispicy, the reading of the lamb's entrails was performed in combination with a ritual offering of appeasement, as indicated by Old Babylonian and Kassite period reports on acts of extispicy. These reports have a general formulary beginning with statements such as "one lamb to Shamash for a favorable omen," and closing with statements such as "the omen (obtained from) the sheep for the angry (god) turned out favorable."[55]

As far as the history of the celestial omen series is concerned, a seventh-century exemplar contains a bilingual introduction, beginning (Sumerian) *ud an* [d]*en-líl* (= Akkadian *Enūma Anu Enlil*) "when (the gods) An, Enlil."[56] A version of this incipit was identified by S. N. Kramer in what is otherwise a catalog of Sumerian literary texts.[57] The opening paragraph of the astral divinatory corpus states in elevated poetic style that the great gods (by their Sumerian names) An, Enlil, and Enki (= Akkadian Anu, Enlil, and Ea) established by divine plan or counsel (*galga/milku*) the "great divine powers of above and below," translated into Akkadian as "the designs (*uṣurātu*) of heaven and earth."[58] It reads:[59]

> (Sumerian version) When Anu, Enlil, and Enki, the great gods, by their decisive counsel established the profound rites of heaven and earth, (and) the crescent-boat shape of the moon, the waxing of the crescent moon giving birth to the month, establishing an ominous sign of heaven and earth; the celestial boat (= gibbous moon) they made to appear, and it (too) came out to be seen in the sky.
>
> (Akkadian version) Alternatively: When Anu, Enlil, and Ea, the great gods, in their counsel, established the designs of heaven and earth, firmly established in the hands of the great gods (the power) to create the day and renew the month that humankind should see. (And) they (did) see the sun-god in his gate, in the midst of heaven and earth he shone forth regularly.

This passage depicts the cosmos as a coordinated system uniting above with below. Above, the heavenly bodies and their regularities were to be seen by humankind below. The moon, planets, and stars inhabited heaven as a benefit to human society. They were placed there by Marduk in the version of creation found in *Enūma Eliš*. The same idea of the interrelation and complementarity of cosmic realms is articulated in a unique text known as the *Diviner's Manual*: "The signs on earth just as those in the sky give us signals. Sky and earth both produce portents, though appearing separately, they are not separate (because) sky and earth are related. A sign that portends evil in the sky is evil on earth, one that portends evil on earth is evil in the sky."[60]

By the middle of the second millennium BCE, the importance of divination as knowledge was firmly entrenched in centers of scribal activity throughout the Near East. Omen texts have been found in sites to the west of Babylonia, such as at Emar, Harādum, Alalakh, and Qatna, as well as in the Hittite capital of Hattuša, and to the east at Susa.[61] Compared with later periods relatively few celestial omen texts from Mesopotamia proper are extant from Middle Babylonian and Middle Assyrian periods,[62] but the continuation of the tradition that would be *Enūma Anu Enlil* in its Neo-

Assyrian and Neo-Babylonian recensions is well attested from areas of the periphery, such as Emar, Qatna, Alalakh, Nuzi, Susa, Ugarit, and Boghazkoy.[63] Omen texts from these Middle period versions were further modified in the Neo-Assyrian period, where divination and its related activities composed the bulk and core of the highest level of attainment in ancient cuneiform scholarship, its practitioners and experts holding high positions in the Neo-Assyrian royal courts of the kings Sargon II, Sennacherib, Esarhaddon, and Assurbanipal at Nineveh (Kuyunjik in the eighth and seventh centuries BCE).

Assyrian scribes maintained close and regular communication with the reigning monarchs, especially Esarhaddon and Assurbanipal, as evidenced in their correspondence,[64] as well as in sometimes datable written reports of ominous phenomena, including citations from the appropriate omen handbooks.[65] Scholarly tablets featuring the major genres of omen, medical and ritual texts, hymns, prayers, literary and lexical texts, composed a collection, or several collections, in Assurbanipal's palace.[66] Short colophons contain a "catch-line" (first line of the next tablet in the series), title and number of the tablet in the series, a line count, and the identification, "Palace of Assurbanipal, king of the world, king of Assyria."[67] More extensive colophons attesting to the scholarly interests, if not literacy, of the king himself are known, as in "Assurbanipal, great king, mighty king, king of the world, king of Assyria, son of Esarhaddon, king of Assyria, son of Sennacherib, king of Assyria. According to writing boards, originals from Assyria, Sumer and Akkad, I wrote, checked and collated this tablet from the collection[68] of the scholars. I placed it in my palace for my royal viewing. Whosoever erases my written name and writes his name instead, may Nabû, scribe of the universe, erase his name."[69] The Nineveh palace was but one repository of texts for divinatory scholarship, among other scribal activities, but because in the nineteenth and twentieth centuries of our own era, the mound of Kuyunjik (Nineveh) yielded the mother lode of tablets for modern scholars, texts from Assurbanipal's palace are perhaps somewhat disproportionate in, but invariably central to, any study of ancient cuneiform scholarship. Other Neo-Assyrian cities were sites of scholarly activity, such as the Ezida temple of Nabû at Nimrud (ancient Kalhu, ninth to the seventh centuries),[70] as well as the more provincial site of Sultantepe (ancient Huzirina) in the western region of the empire.[71]

From the perspective of the social functioning of scholarship, divination played a role in the regulation of royal affairs. It was a strategy, or policy, that maximized future security in the realm. Today we might call this "foresight planning."[72] The overwhelming bulk of cuneiform knowledge, the divinatory and related texts, converges around the aim to determine the meaning of things established by the gods. High among the values of the cuneiform scholars, therefore, was divinatory knowledge. Within the practice of divination, the answers provided by Shamash and Adad, patron gods of divination, to questions posed by the diviner and read in the liver were held to be definitive, or certain, knowledge. Certainty, or confirmation (the *annu kēnu* "firm yes"[73]) in this framework derived, as did knowledge of the signs in general, directly from the gods. Special texts (queries to Shamash, *tāmītus*) contained the oracle queries for which divine confirmation was sought. In the case of the omen texts themselves, the ultimate aim was to obtain knowledge of the gods' "decisions" (*purussû*) that were manifest in the signs. Similarly, the aim of ritual was to correct the relation of divine to human signaled by ominous signs. But omen divination also provided a structure for thought about the phenomena of the external world. The way experience, perception, and imagination about the phenomenal world was structured so that order, norms, and anomaly were definable, will be taken up in chapter 4.

Following a gap after the fall of Assyria (612 BCE is the date of the fall of Nineveh), there is a return of textual sources for cuneiform scholarship in the Neo- and Late Babylonian period, from approximately the end of the seventh century BCE to the end of the cuneiform writing tradition in the Arsacid period (first century CE). During the second half of the first millennium, scholars were by and large not attached to the royal courts of Seleucid or Parthian rulers, as they had been in Assyria, but remained in the employ of major Babylonian temples such as the Rēš Temple of Uruk and the Esagil of Babylon. Neo-Babylonian and Late Babylonian scholarly texts have been found as well at Sippar, Borsippa, and Ur.[74] Traditional omen texts, such as the so-called terrestrial omens of *Šumma ālu* "If a city," *Šumma izbu* "If a malformed birth," the celestial omens of *Enūma Anu Enlil* "When Anu and Enlil," and magical texts such as the apotropaic rituals known as *namburbû* continued to have intellectual currency and, presumably, purpose in Late Babylonia.[75] But scholars also produced new forms of cuneiform knowledge, emerging from new methods and practices: mathematical astronomy, horoscopy, and astral medicine.

Natal astrology was introduced after circa 500 BCE. This development came after the invention of the signs of the zodiac, which provided a basis for extensive correlations and relationships to be made among the planets (including the sun and moon in eclipse), the twelve signs, and the twelve months. In contrast to the public omens of *Enūma Anu Enlil*, the object of the new astrology was the individual's character and life, as illustrated, for example, by LBAT 1593 in which a child born in the region (KI = *qaqqaru*) of a zodiacal sign is assigned various characteristics (a long chin, red hair, LBAT 1593: 3′) and life experiences (he will be widowed, LBAT 1593: 3′).[76] The correspondence between human life and the stars was not a new concept, nor was the idea of the efficacy of the stars in healing,[77] or the idea of medical treatment in accordance with calendar days (STT 300), but a new medical astrology, well attested at Uruk, now associated diseases with planets, constellations, or signs of the zodiac,[78] and treatments were determined by astrologically propitious times.[79]

Whether referred to as horoscopes[80] or protohoroscopes,[81] the natal astrological texts provide planetary positions in the zodiac and other astrological data required for forecasts about the life of an individual based on the situation of the heavens on the birth date. The cuneiform horoscopes range in date from the oldest at 410 BCE to the youngest at 69 BCE. The relationship between Babylonian nativities and their Greek counterparts is a matter of discussion with implications for the question of the ultimate origins of astrology.[82]

The importance of horoscopes for understanding the knowledge of the heavens in Late Babylonia is twofold. First, the celestial divinatory character of the texts is clear. Such forecasts as are preserved are formulated as omen apodoses, such as "he will see profit."[83] Second, the data themselves derive from several other kinds of cuneiform astronomical text types (GADEx, see below), principally Diaries and Almanacs. The several horoscopes in which planetary positions are given in degrees and fractions of degrees of zodiacal signs raise the possibility that the positions were calculated by means of methods known from the ephemerides, or from interpolations from them, though that has not been securely demonstrated and remains conjectural. These parallels, however, are enough to establish a thoroughgoing interdependence among many of the cuneiform astronomical text genres, thus a synthetic aspect to Babylonian astronomy and astrology in the Hellenistic period, much as the descriptive astronomy of *MUL.APIN* (see below) had a synthetic relationship to celestial omens.

While divinatory texts had discrete genre boundaries (celestial, physiognomic, birth, and so on) in previous periods, their interrelations begin to take shape in the last half of the first millennium, that is, in Persian (Achaemenid) Babylonia. The integration of astral with terrestrial divination seems to have been made possible by the development of astrology, that is, by the application of celestial signs for the human being (and the human body) in general, no longer focusing, as did *Enūma Anu Enlil*, on the king, the city, and its population. In one commented text, for example, omen series concerning human appearance, health, and births were brought into relation with celestial signs. Its opening lines: "(The omen series) 'If a Malformed fetus,' (the omen series) 'Symptoms,' (the omen series) 'Physical Characteristics.' Aries, Taurus, Orion are for predicting the appearance. When they (the planets?) reach (the various zodiacal signs) it refers to physical characteristics. Preserve the secret of heaven and earth!"[84] The laconic nature of the commentary leaves open the question of exactly what the connection between birth, medical, physiognomic and astrological phenomena were, yet a decidedly astrological, that is genethlialogical (birth astrology), turn has been taken.

The reference to the planets reaching the zodiacal signs, expressed with the phrase "when they reach" (*kî ikšudu*), relates no doubt to an important feature of astronomical Diaries and Almanacs, which was to track when the planets entered each zodiacal sign. The colophons of Almanacs identify these texts as "measurements of the 'reachings' of the (divine) planets" (*mešḫi ša kašādī ša ^{d}bibbī*). Although the statements of the "reachings" in Diaries and Almanacs do not in themselves denote astrological meaning, the arrival of a planet into a sign was certainly astrologically significant. Horoscopes confirm this by taking the zodiacal positions of the planets from Diaries and/or Almanacs, most likely from the sections in which the planetary "reachings" are enumerated.[85]

Astrology finds a connection to extispicy in the Late Babylonian period as well. A Late Uruk text associates traditionally ominous parts of the liver with a god, one of the twelve months, and a heliacally rising star, thus: "the path (of the liver) is Šamaš, Ajāru, Taurus; the gall bladder is Anu, Tašrītu, Libra," and so on.[86] The term "sign" or "ominous part," literally "flesh" UZU = *šīru*, used in liver omens is found again in a Seleucid astrological context where the term refers not to ominous parts of the liver but to ominous parts of zodiacal signs subdivided into microzodiacal parts, thus "the twelve-fold division of sign such-and-such" (12 UZU.MEŠ ḪA.LA *šá* . . .).[87] The practice of appeasing or appealing to the divine also establishes new connections to the zodiac, such as in a list of spells with their correlated "regions" in the zodiacal signs, for example, "changing one's mind (is the) region of Leo; overturning a judgment (is the) region of Aquarius; loosening the grasp (is the) region of Virgo."[88]

The list of spells and correspondences between signs of the zodiac and incantations to influence various people, or demons, or gods, represents innovative change in the epistemic landscape of Late Babylonia. Such innovation brought formerly separate fields of knowledge, such as omens, medical preparations, incantations, and mathematical methods for calculating appearances of planets, together in service to new forms of astro-medicine. Such changes are surely surface manifestations of other much deeper changes in the position of the human knower, the *mudû*, vis-à-vis the world as it was experienced and known in that period. And such changes surely have their explanation in the way the experienced and known world was understood in relation to human society and human life in the particularly complex social, political, and intellectual environment of Hellenistic Mesopotamia.

KNOWLEDGE OF THE HEAVENS

Knowledge of the heavens deserves a separate treatment because, despite its integral relation to other scholarly pursuits and other domains of knowledge, the scholar scribes' interest in celestial phenomena gave rise to textual forms that maintained a separate status and had a separate development. In fact, development of the various text types with astronomical content, culminating, perhaps, in Seleucid period planetary and lunar ephemerides and procedure texts,[89] reached a point where it becomes difficult to see how astronomy may have still been connected to its original context, that is, divination.[90] The category *astronomical* here serves to designate the texts, regardless of their divinatory use or motivation, that focus on observing and predicting the phenomena of the moon, sun, planets, fixed stars, and weather. This section will introduce the major sources in which knowledge of the heavens was codified. Further discussion of the contents and various aspects of these texts will emerge below in chapters 4, 7, and 8.

Knowledge of the rhythms of heavenly phenomena is important in a mixed agrarian and herding economy such as was early Mesopotamia; the names of stars and planets as well as awareness of their cyclical appearances undoubtedly antedate written evidence. As testimony to the antiquity of such knowledge, reference to the morning star and the evening star as appearances of Venus is already found in the third millennium BCE Sumerian divine names ^{d}Inanna.ḫud$_2$ (UD) "Inana of the morning" and ^{d}Inanna.sig "Inana of the evening." However, it was the sun and the moon that provided useful indicators of the passage of time. Indeed, ancient Near Eastern calendars reckoned the year by the sun and the month by the moon, and attention to the periodic nature of their appearances is implied by the existence of calendars in fourth millennium administrative documents.

The poetic text of a prayer from the Old Babylonian period, extant in two versions, made explicit the notion of the heavenly bodies as gods. In this text the "gods of night" (*ilū mušītim*) were called upon to stand by as a liver inspection was to be performed. Most of the stars enumerated in the prayer are circumpolar constellations, such as the Big Dipper (the Wagon GIŠMAR.GÍD.DA = *Eriqqu*), the bright winter stars of Canis Major (the Bow MUL.BAN = *Qaštu*), Sirius (MUL.KAK.SI.SÁ = *Šukūdu*), Orion (True shepherd of Anu SIPA.ZI.AN.NA = *Šitadallu*), and Lyra (the She-Goat ÙZ = *Enzu*), visible even when Shamash (sun), Sin (moon), Adad (weather-god), and Ishtar (Venus) have gone below the horizon to the underworld.[91] The role of the stars, constellations, and planets as celestial deities in divination, ritual, and prayer continued throughout the entirety of the cuneiform tradition.

The chronological distribution of Akkadian texts having to do with the observation and prediction of lunar, planetary, and stellar phenomena is the same as for other texts belonging to fields of scholarly knowledge, that is, from Old, Middle, Neo- and Late Babylonian, as well as from the Middle Assyrian and Neo-Assyrian periods. It is thereby clear that astronomical knowledge had an established place within the scholarly tradition of the *ummânū* from the beginning, in no way being a subject of use for practical calendar reckoning alone, or as background to mythological tales. The principal use of astronomy was for celestial divination, which rapidly took on central importance in Assyria as a tool for royal planning. After the fall of Assyria, when textual materials become numerous again in the Hellenistic period, temple archives of Babylon and Uruk show that astronomy had entered a new stage in its development. New forms of Babylonian astronomy are represented by observational records, ephemerides, and procedure texts, as well as new forms of astrology in horoscopes and natal omens. The last datable cuneiform tablets are Almanacs dated to the years 385 of the Seleucid era (SE), or 75 CE, and SE 390 or 79/80 CE, respectively.[92]

The omen series *Enūma Anu Enlil* is one source of early astronomical knowledge. The "Venus Tablet of Ammiṣaduqa," or Tablet 63 of *Enūma Anu Enlil*, is a case in point.[93] The association of this Tablet with King Ammiṣaduqa of the Hammurabi Dynasty is based on the appearance of part of the eighth year name of the king in place of one of the omen apodoses (omen 10).[94] Because of this, the Venus Tablet played a significant role in the establishment of a chronology for the ancient Near East because it was thought to preserve observations of the planet compiled in that period (the alleged Old Babylonian text recording such observations is not extant). Surviving exemplars of Tablet 63, all written during the Neo-Assyrian period or later, demonstrate an awareness that five synodic cycles of the appearances of Venus (as evening and morning star, that is, morning rise and set and evening rise and set) occur every eight years (that is, every ninety-nine Babylonian months minus four days). The following is the first line: "(Year 1) If on the 15th of Month XI (*Šabaṭu*) Venus disappeared in the West (Evening Last), remained invisible 3 days, and reappeared in the East on the 18th day of Month XI: Springs will open; Adad will bring rains, Ea will bring floods, one king will send greetings to another king."[95]

The omens of *Enūma Anu Enlil* Tablet 63 are constructed from a sequence of synodic phenomena of Venus over a period of twenty-one years (the length of the reign of Ammiṣaduqa) formulated as conditional statements "If Venus . . ." together with associated events "then . . .". In its extant form, as Erica Reiner and David Pingree showed, the tablet does not preserve a list of Venus observations from the Old Babylonian period but is a composite text and includes some computed values for the phenomena and the periods of invisibility that have been copied and corrupted in the manuscript transmission. Its value for chronology is thus compromised. Close analysis of the data, however, according to Teije de Jong, suggests that the first ten omens may indeed reflect genuine datable Venus observations.[96]

Although the Neo-Assyrian series *Enūma Anu Enlil* represents a comprehensive collection of celestial omens, including phenomena of the moon, sun, planets, fixed stars, and weather, nearly a third of the whole, Tablets 1–22, deal with lunar phenomena. The first thirteen tablets are the "visibilities of the moon" (IGI.DU$_8$.A.ME *ša* 30), concerned with appearances and disappearances, but mostly focus on the first appearance of the moon (*ina tāmartišu* "at its appearance"). These first tablets of *Enūma Anu Enlil* reflect the importance of the lunar syzygies, that is, around conjunction (between last and first visibility, or between the twenty-seventh and the first of the month) and the day(s) of opposition (usually days fourteen or fifteen).[97]

In particular, *Enūma Anu Enlil* Tablet 14 belongs to a group of early astronomical texts concerning lunar visibility.[98] It provides a tabulated arithmetical scheme for the length of visibility of the moon each night for the thirty days of an equinoctial month (when day and night are of equal length as the sun crosses the celestial equator). The interest in duration of lunar visibility is tied to the ominous nature of the moon when visible. This table, and a second table in the same tablet that gives a supplementary section of lunar visibility coefficients allowing calculation of lunar visibility in other months of the year, is underpinned by an arithmetical scheme for the variation in daylight throughout the year based on an ideal year of 360 days and a ratio of longest to shortest daylight of 2:1, consistent with the Astrolabes and *MUL.APIN* (see below).

Tablets 15–22 concern lunar eclipses,[99] followed by the solar omens, Tablets 23(24) to 29(30),[100] solar eclipses in Tablets 31–35(36), and 44–49 are the weather omens.[101] The stellar and planetary omens begin with Tablet 50(51),[102] and the remainder of the planetary omens are not well preserved.[103] Many tablets of *Enūma Anu Enlil* had commentaries (*mukallimtu*),[104] as well as a subseries of excerpts, called *rikis girri Enūma Anu Enlil* "guide to *Enūma Anu Enlil*," and a separate serialized commentary titled *Sin ina tāmartišu* "If Sin (the moon) at its appearance," of at least seven tablets.[105]

The Astrolabes, so-called, are an important source for early Babylonian interest in the fixed star heaven and in the phenomenon of the variation in the length of daylight. Astrolabes are attested in both circular and list form.[106] The earliest exemplar stems from the Middle Assyrian period (reign of Ninurta-apil-Ekur 1191–1179 BCE), while the latest are from the Seleucid period (third century BCE or later). As a designation for these texts the term *astrolabe* is misleading from the point of view that the cuneiform exemplars are not planispheric, but they do assign fixed stars, constellations, and even the planets to various parts of the sky for the twelve months of the ideal year, either listing them in groups, with the stars of the three "Paths" (see chapter 7, section "Reports on Ominous Celestial Phenomena" for this terminology) set alongside one another, or by arranging them in rings in a concentric circular form, three stars per month for a total of thirty-six stars. Each star represents a heliacal rising in its assigned month and in its assigned path. Part of so-called Astrolabe B contains a religious calendar. It assigns to each month activities dedicated to gods associated with that month, as in the following excerpt: "*Ajaru* (is the month) of the Pleiades, the seven great gods. (The month of) the opening of the earth, (month in which) the oxen go in procession, the water sluices are opened, the plows are flooded. The month of Ningirsu, the hero, the great *iššakku*-priest of Enlil."[107]

The month sections of the Astrolabe texts also contain numerical values (that can be read as time degrees [1 degree = 4 minutes]) for the variation in length of day through the year in accordance with a scheme that assigns the longest day, 4, to month III (= summer solstice), the shortest day, 2, to month IX (= winter solstice) and the mean values of 3 correspond to the equinoctial months VI and XII. This daylight scheme therefore assumes a ratio of longest to shortest day of 2:1, which is the established Babylonian value until the value 3:2 is introduced into late Babylonian astronomy. Even then, however, there are late texts that preserve the older standard.[108]

The most significant astronomical compendium of the early first millennium BCE is the two-tablet series titled *MUL.APIN*, "The Plow Star."[109] It catalogs and systematizes a wide variety of celestial phenomena, beginning with the names and relative positions of fixed stars in the Paths of Enlil, Anu, and Ea, the dates of their heliacal risings, simultaneous risings, and settings of certain stars and constellations, stars that cross the meridian (*ziqpu*), stars in the Path of the Moon, astronomical seasons, luni-solar intercalation rules with fixed stars, stellar calendar, appearances and disappearances of the five planets, periods of planetary appearance, length of day scheme, and a lunar visibility scheme. *MUL.APIN* might be considered propaedeutic to *Enūma Anu Enlil* in that it contains the astronomical knowledge necessary to use the omen series. Despite its primary interest in the phenomena themselves, hence our classification of the text as astronomical, the final section of *MUL.APIN* (II iii 16–iv 12) is devoted specifically to astral—not lunar, solar, or weather—omens.[110]

Typical of cuneiform scholarly texts is the long lacuna between the seventh-century copies of *MUL.APIN* and the next group of extant astronomical sources, from the fourth century and later. The state of astronomical knowledge before the Hellenistic period must therefore be inferred from the evidence of relatively few texts. As far as the Diaries are concerned, the interim period is represented by one Diary from the seventh century, one from the sixth century, and four from the fifth century. One key piece in this puzzle is the tablet known as "Strassmaier Cambyses 400," after its original publication from 1890.[111] The tablet contains astronomical data for the year 523 BCE, or year 7 of Cambyses, the Achaemenid ruler. Its contents show that already from the early sixth century BCE, Babylonian astronomers had been making and probably recording observations of the Lunar Six (see note 116)—the intervals between the rising and setting of the moon and sun around new and full moon—and of planetary synodic phenomena, and had also begun calculating the latter phenomena on the basis of Goal-Year periods. Confirmation of the early, seventh century, collection of Lunar Sixes is found in tablets devoted solely to these data.[112] Interestingly, among these sources is one from Nippur, possibly for the year 618 BCE.[113]

Other pre-Seleucid tables concern eclipse possibilities in accordance with the Saros cycle (223 lunar months = 18 years = 38 eclipse possibilities).[114] One such table, the Saros Canon, begins with an eclipse possibility in the reign of Cambyses (527 BCE) and ends in the early Seleucid period (257 BCE).

Seleucid tables represent a planetary theory aimed at the prediction of dates (months and *tithis*) and positions (degrees of the zodiac) of the visible planetary heliacal phenomena, that is, first and last visibility, first and second stations, and acronychal rising of the planets,[115] and a lunar theory aimed at predicting the moments of syzygies (new and full moons), eclipses, the variation in the length of daylight, and the Lunar Six phenomena.[116]

For the pre-Seleucid period, seventh-century tablets with observations of the synodic phenomena of the planets are sparsely attested, one from years 2 to 10 of Šamaš-šuma-ukīn and one from years 1 to 14 of Kandalānu, of first and last visibilities of Mars and of Saturn respectively.[117] Another Saturn observation text from sixth-century Uruk includes first and second stations to the first and last visibilities during the period from year 28 to 31 of Nebuchadnezzar II.[118] During this period, on the basis of the few extant planetary observation texts, the practice of citing planets with respect to a certain set of Normal Stars and the use of cubits, fingers, and degrees (UŠ) was still in the process of standardization.[119]

Finally, mention must be made of the observational texts (four in all, two in Babylonian script and two in Assyrian script, and all are fragmentary) concerning the synodic appearances of Mercury, giving in separate sections for the months and days the date of first visibility, duration of visibility, and date of last visibility.[120] Too fragmentary for dating the texts themselves, and having no colophons, Pingree and Reiner nonetheless judged the texts to antedate the seventh century. In addition to the dates of synodic appearances, the texts include the intervals between first visibility in the East and sunrise and between first visibility in the West and sunset in the following format: "[In MN on the nth day Mercury becomes visible in the West, the sun being n UŠ below, and remains in the West for n days, and then] disappears [in] the West [in MN on the nth day]; it remains invisible [for n days]."[121]

Late Babylonian (Seleucid) astronomical texts were classified by Abraham Sachs in what is still a classic paper in the field.[122] He divided the corpus into tabular and nontabular texts, and emphasized the fact that both text types contain predicted data. The distinguishing feature of the tabular texts was their use of linear functions (zigzag and step) in the form of difference sequences to generate positions in the zodiac (in degrees) or dates (in months and *tithis*). Later, Neugebauer would distinguish between mathematical and nonmathematical astronomical texts (frequently NMAT), nonmathematical astronomical texts being the same set of sources that Sachs classified as nontabular.[123] The nontabular, or nonmathematical astronomical texts were further designated GADEx texts, an acronym for Goal-Year texts, Almanacs, Diaries, and Excerpts, and were also collectively referred to as non-ACT texts (Astronomical Cuneiform Texts),[124] all of which terms are still standard.[125]

Tables for the planets and the moon are in late colophons designated *tērsītu* "computed tables."[126] This group was intimately related to a group of procedural texts stating the arithmetical rules (algorithms) used to calculate the various columns of the ephemerides.[127] Extant table texts and procedures date to the period from the mid-fifth century to the mid-first centuries BCE, with the bulk of preserved tablets dating to the second century BCE. Underlying the computational methods of the tables was the recognition of period relations such as the Saros, mentioned before, or the relations that govern the appearances of the planets and moon in their synodic cycle, *synodic* here pertaining to the relation between a celestial body and the sun. Two types of recursive mathematical steps (algorithms) were applied for calculation of the synodic arc: System A, classified by the application of the step-function, and System B by the zigzag function. The construction of both systems took place early in the Seleucid era, with chronological priority going to System A.[128] Each scheme entailed an understanding that heliacal appearances were the result of the distance (elongation) of the planet or moon from the sun, which is to say, the planet's direction in the sky in relation to the sun. Therefore, an intimate connection was found between synodic arc ($\Delta\lambda$), or progress in sidereal longitude made by the planet or moon per synodic phenomenon, and synodic time ($\Delta\tau$), or the time required for the body to complete a synodic cycle between successive phenomena of the same kind (for example, first visibility to next first visibility). This theoretical aspect of predictive astronomy seems to have little or no connection to a cosmology, or a spatial framework within which to conceive of the planetary or lunar positions. Both systems required the conception of the path of the sun, or ecliptic, albeit not precisely as it came to be defined in Greek astronomy as a great circle of the celestial sphere, and used the twelve signs of the zodiac as a reference tool for counting degrees around it.[129]

Predictive astronomy reached its fullest development and capacities in the period after circa 500 BCE, principally in the cities of Babylon and Uruk, where the astronomers came from a limited number of scribal families holding prebends in the Bīt Rēš sanctuary of the Anu Temple at Uruk, and most likely also in the Esagil Temple of Marduk.[130] Some held the professional title *ṭupšar Enūma Anu Enlil* "scribe of *Enūma Anu Enlil*," others the title *kalû* "lamentation (priest)," or *āšipu* "expert in ritual apotropaism."

The fact that the Late period astronomer scribes performed their duties within Babylonian temples should count for something in our attempts to understand the place of astronomical knowledge in the long history of cuneiform scribal scholarship. The purpose of Late Babylonian astronomy is made explicit by the rubrics or subscripts on the tablets themselves. To what extent the theoretical apparatus of Late Babylonian astronomy was motivated by the requirements of (what we call) astrology, is still difficult to judge. The most that can be said is that astronomy was the product of a culture that valued divination and the maintenance of the cults of the gods. These practices, however, far from limiting the development of astronomical knowledge, seem to have motivated and supported that development. Nevertheless, the contents of the late mathematical ephemerides and procedures exhibit an epistemic autonomy and integrity not explainable by any relation or connection to astrology or the cult.

Babylonian knowledge of the heavens had the most long-lasting influence of any other element of Mesopotamian intellectual culture, both to the east and west of Mesopotamia, where the bases for Arabic and European traditions of astronomy and astrology were formed. In Egypt during the Achaemenid period omens in the style of *Enūma Anu Enlil* appear in a Demotic papyrus.[131] Related to this text, although belonging more properly to the Greek inheritance, are the second century BCE celestial omens from Ptolemaic Egypt attributed pseudepigraphically to the Egyptian priest Petosiris in the reign of the Saite ruler Nechepso (= Necho II). Demotic horoscopes testify to the influence of Babylonian natal astrology in the Egyptian milieu.[132]

In Greco-Latin texts the term *Chaldean* began to emerge for writers wanting to claim an authoritative source on astrology. The term came to be synonymous with astrologer.[133] Greco-Roman attributions to the Chaldeans were sometimes spurious, but in general the widespread use of this term reflects the high reputation of the Babylonian scholars in *astrologia/astronomia* (for example, Strabo, *Geography* 16.1.6).[134] In the second century BCE Greek astronomy was influenced by Babylonian methods; the sexagesimal place value number system, astronomical parameters, and computational models (Systems A and B) are all eventually attested in Greek texts. A second-century treatise, the *Anaphorikos* of Hypsicles, on the rising times of the zodiacal signs, makes use of a Babylonian rising times scheme and introduces for the first time in Greek astronomy the 360° of arc and time.[135] In the first century BCE, chapter 18 of the treatise *Introduction to the Phenomena* (*Introductio astronomiae*) of Geminus reflects awareness of Babylonian lunar theory as preserved in cuneiform tabular texts and mentions the Chaldeans in the context of astrological aspect.[136]

The perpetuation of Babylonian linear methods is also evidenced in the Greek astronomical papyri from Oxyrhynchus, dating to the Roman imperial period, roughly from the first to the fifth centuries CE.[137] From these texts the practice of technical astronomy in the late Greco-Roman period can be seen to carry on the primary goals of Babylonian astronomy, namely, to determine zodiacal longitudes of the sun, moon, and five planets for a certain date (for the production of horoscopes), and to determine the dates of entry of the planets into the zodiacal signs (for the production of Almanacs). The kinematic spherical models of planetary motion exemplified in Ptolemy's *Almagest* and *Planetary Hypotheses*, with their geocentric cosmology and Aristotelian physics, represent only one of two methods of Hellenistic Greek astronomy, the other being the prediction of planetary and lunar phenomena by means of numerical algorithms originating with Babylonian mathematical astronomy. Another significant influence on Hellenistic astronomy was the Babylonian calculation of the rising times of the zodiacal signs (Greek *anaphorai*), for which values are constrained by a 3:2 ratio of longest to shortest daylight and the consequent canonical value for the *klima* of Babylon (latitude 32.5°). Finally, with the class of horoscope texts we see the most common practical application of technical astronomy throughout the period of the astronomical papyri. These too bear traces of cuneiform astrology in their use of the zodiacal signs, the attribution of masculine/feminine natures to the signs, benefic/malefic natures to the planets, the Lot of Fortune, the exaltations, and the Terms.[138]

Because classical Indian astronomy and astrology derived from Hellenistic Greek sources, Indian texts sometimes evince a mixture of Babylonian and Greek elements. The *Paitāmahasiddhānta* (early fifth century CE) is one such work, itself foundational for the Indian mathematical astronomical tradition.[139] Influx of Babylonian mathematical astronomy is also evident in the *Pañcasiddhāntikā* of Varāhamihira (sixth century CE). Varāhamihira's divinatory treatise, the *Bṛhatsaṃhitā*, was clearly influenced by Babylonian omens, and two works on genethlialogy, the *Bṛhajjātaka* and the *Laghujātaka*, also contain elements of Babylonian astrology.[140] Another important Sanskrit astrological text with parallels to Babylonian horoscopic astrology is the *Yavanajātaka* ("nativity of the Greeks") of Sphujidhvaja.[141] In a Sassanian context, Babylonian astrology and omens were also incorporated into second century CE Gnostic Mandaean astrology and preserved in the Mandaic work *Asfar Malwašia* "Book of the Signs of the Zodiac," where material clearly influenced by *Enūma Anu Enlil* and *Iqqur Ipuš* are found.[142] These examples of the impact of Babylonian astronomical knowledge and practices reflect the shared genetic material between the cuneiform world and its cultural-historical kin in the history of science, consisting largely of knowledge of the heavens for divinatory purposes.

The foregoing discussion of the knowledge of signs and of the heavens in cuneiform scholarship set out in a descriptive way the various text groups formed by Assyrian and Babylonian scholars independent of any program of natural knowledge or natural inquiry such as took shape in the West. Despite the disjunction in the context for inquiry, relationships can be made between the epistemic and predictive goals of this early knowledge of the heavens and the later history of astronomy, and history of science more generally. Other kinds of sources have been omitted that may have other kinds of relationships to later tradition, such as to later Western traditions of cosmology and the early roots of esotericism.[143]

In chapter 1 it was suggested that the conceptual scaffolding of the cuneiform scribes' world was not constructed with the idea of nature. The challenge to infer or describe what did provide such a scaffolding for the scribes' concept of the world is difficult because no unified framework served to structure the Assyro-Babylonian world order from a perspective of matter or the "physical," such as is provided by the conception of nature. Nevertheless, conceptions of order, norms, and schemata based upon such norms were central features of the scholarly corpus of texts dealing with the phenomena, and consequently of what was deemed in those texts to be knowable and significant. Before further considering just what the learned texts concerning divination, astronomy, astrology, magic, and medicine have to say about normative structure and aspects of order and anomaly in their conceptual world, a brief foray into the knowledge of various categories of terrestrial phenomena should be given in addition to what has been presented thus far. This is to take account of certain parts of cuneiform technical scholarship that, by its very subject matter, raises questions about the classification of some phenomena into what might seem on the surface to be natural kinds, or at least kinds, in particular, plants, animals, and stones.

KNOWLEDGE OF PLANTS, ANIMALS, AND STONES

Pursuing a little further the approach to cuneiform objects of knowledge, and having focused thus far on evidence for knowledge of signs and knowledge of the heavens, the textual compendia devoted to plants, animals, and stones warrant consideration as well. Three collections of this sort of knowledge are known by their incipits as *šammu šikinšu* "the plant, its appearance,"[144] *abnu šikinšu* "the stone, its appearance,"[145] and, although less extensive than its counterparts, a third "*šikinšu*" text is extant for snakes.[146] The *šikinšu* text for snakes is structured in the same way as those for the plants and stones, with lines beginning *ṣēru šikinšu* "the snake, its appearance," as follows:

> The snake whose appearance (is) like a mouse; that snake is called *mušgallu*; (it is the) snake of Anu.
>
> The snake whose appearance (is as follows): Its scutes are of pappardillu-stone, a snout of red stone, eyes of *mušarru*-stone, a face of lapis lazuli and gold, on a high . . . it stands, that snake is called *mušmaḫḫu*, (it is) the snake of the god [. . .].[147]

Erica Reiner saw these texts as precursors to medieval herbals, lapidaries, and bestiaries, speculating that the snake text is but a survival of a more comprehensive collection.[148] Some striking similarities are indeed to be found in those medieval handbooks. The late herbals compiled medicinal plants, described their appearances and uses, and made extensive use of illustration. Often found together with herbals, the lapidary similarly collected information on stones, describing appearances and properties. And because the classification system employed was not that of modern mineralogy, the lapidary, for example, included in the class "stone" substances that we would not classify as such, for example, pearl, amber, or coral.

The medieval bestiary included together with animals, both real and imaginary, birds and stones, all of which were subjects for descriptions and allegorical didactic lessons.[149] The *šikinšu* text for snakes, as in the above quoted example of the *mušmaḫḫu* and the *mušgallu*, similarly added to the descriptions of "real" snakes those for mythological snakes, or serpents. Benno Landsberger speculated as to whether in fact the names of the mythological serpents had also at one time been the names of actual snakes.[150]

Because it was aimed at the description of medicinal plants, *šammu šikinšu* relates to the cuneiform pharmacological list *Uruanna*. This text, as Barbara Böck has pointed out, shows that "issues of nomenclature, orthography, and synonymy played an unusually important role in cuneiform pharmacology."[151] *Uruanna* lists juxtaposed items (such as *inuš*-plant = *maštakal*-plant) as alternative or substitute plants, synonymous plants, and the names of plants in foreign languages.[152] How to prepare and use the drugs of *Uruanna* was the objective of the text *šammu šikinšu*, as in the following example:

> The plant whose appearance is (such) that it creeps along the ground like the *errû*-gourd,[153] whose tendrils are like (those of) the *qiššû*-gourd,[154] whose leaves stand far apart like (those of) the x[x], whose seed is like the seed of the *ḫurātu*-sumac, whose root is bitter and soft—that plant [is] called *imḫur-līmu*-plant; it is good against the Furious One, Deputy Po[wer] of Adad. You dry it, pound it and rub [him] with it in oil. It is (also) good against every kind of sores; you pound it, rub him with it in oil and he will be cur[ed].[155]

Many of the pharmacological substances (plants, stones, and animals) are to be found in lexical lists, and both *šammu šikinšu* and *abnu šikinšu* relate to parts of the lexical series *Ura*—namely, Tablet 17 on plants, and Tablet 16 on stones.[156] The snakes of *ṣēru šikinšu* similarly relate to the list of snake names in *Ura* Tablet 14, which also included undomesticated animals, reptiles, and insects. The content of these *šikinšu* texts on plants and stones, however, belongs properly to the study of Babylonian medicine. Their attention to plants and stones has to do with an interest in *materia medica*, and the selection criteria are therefore relative to medicinal usage, not plants and stones in general.[157] Some medical texts systematized both symptoms and therapeutic techniques for reference purposes, including lists of medically efficacious plants as well as lists of stones used in the fashioning of amuletic beads.[158]

Anais Schuster-Brandis discussed the principal terms in the use of stones for healing, especially to be worn by the sick or afflicted person as phylacteries, made effective with incantations and then placed around affected body parts, terms such as *kišādu* "necklace," *ṭurru* "string," and *takṣīru* "string of stones."[159] The word *abnu* "stone" has the meaning "bead" in medical contexts, and the text of *abnu šikinšu* provides descriptions of the colors of the substances from which, presumably, beads for healing were produced:

> The stone whose appearance is like that of unboiled ox blood is called *sābu*.
> The stone whose appearance is like that of the *ṭuḫītu* of an ox is called *sābu*.
> The stone that also looks like *sābu* but is flecked with iron is called *ittamir*.[160]

Additional passages of *abnu šikinšu* state the efficacy of the stone against specific conditions, illness, or demons, as in: "The stone whose appearance is like a green leek, its spots [. . .[161]], but has striations, this stone is called *ḫusīgu*. It is the stone with which panic[162] will not approach a person."[163]

The meaning of *abnu* in this context, as Schuster-Brandis emphasized, is specific to the use of substances for healing stones, whether that refers to "stones" made of stone, or indeed, of metals, or shell. As she pointed out with respect to metal beads, it is possible that the writing ^{NA_4}KÙ.GI, or other writings in which NA_4 is not a classifier for the substance "stone" but rather the object "bead," or the like, is to be interpreted as NA_4 KÙ.GI "a bead of gold."[164] Uncertainty about how to read such expressions relates to an uncertainty as to how the determinative NA_4 functioned as a classifier in those cases.

Much as the technical knowledge belonging to scribes of the professions *ṭupšar Enūma Anu Enlil*, *bārû*, and later, the *kalû*, had a divinatory purpose, the *šikinšu* texts constituted knowledge for the *āšipu* and the *asû*. The professions of the *asû* and *āšipu* were once understood to be differentiated according to types of treatment, the *asû* dispensing therapies consisting of drugs, lotions, and salves, bandages and poultices, the *āšipu* handling the magical methods of incantation and apotropaic ritual. This distinction, however, has already long ago been challenged, because it is not borne out by the texts themselves. Instructions for salves to apply or mixtures to ingest are often given in the same text detailing the incantations and rituals to be performed on behalf of the sick. In this context, what can be said about the nature and goal of the knowledge of plants, stones, and animals as contained in the *šikinšu* texts?

The noun *šiknu* in *šammu šikinšu*, *abnu šikinšu*, and *ṣēru šikinšu* has been variously translated "appearance/its appearance," "form/its form," and "nature/its nature," the last seeming to carry connotations of intrinsic features or essences gaining entry for a member in a particular kind, in other words, of kind essentialism.[165] Reiner argued for "nature" as a translation for *šiknu* on the basis of later European handbooks. She said: "If we take our cue from handbooks of a later period, we may choose the English term 'nature,' simply because medieval texts begin the description with this word (either with *natura*, if in Latin, or with the form equivalent to *natura* in the Romance language used). And so we may translate the opening words as 'the nature of the herb (stone, snake) is.'"[166] Nicholas Postgate also took *šiknu* to mean "the nature of *x*." But Postgate came to his understanding of the meaning of *šiknu* by the interrelation of the archaeological record with the textual record of stones and stone artifacts, finding conceptual "awareness of natural properties as distinct from functions" as well as a "concept of class."[167] He added that "for the Mesopotamian it seems likely that there was no hard and fast line between what we would consider the 'natural' and the 'supernatural' properties of the stones they were classifying."[168] Certainly the difference between natural and supernatural properties of things would not carry meaning, if only because nature, both in the sense of the inherent features or properties of things, and of the extended sense of the sum total of the entities we consider "physical" was not a relevant concept for the cuneiform world. The translation "nature" for *šiknu*, for this reason, is difficult to justify by semantic criteria in Akkadian. To sustain it, with Reiner and Postgate, one must justify its conceptual implications, specifically that the idea of inherent properties expressed by the Greek word *phusis*, which is of a piece with the rest of the framework constructed by the idea of nature, has a counterpart in Akkadian.

Akkadian *šiknu* belongs to the particular noun formation *pirs*, which derives, as do most Akkadian nouns, from a verb, and, as a *pirs* formation, takes on a passive nuance.[169] But pinpointing the meaning of *šiknu* from its related verb *šakānu* is made difficult by the extreme semantic complexity of that verb, whose lexical entry occupies over forty pages in *The Assyrian Dictionary* and is there given nearly one hundred separate usages. The first meaning of *šakānu* is given as "to place something for a particular purpose, with a particular intention." Indeed, if a basic meaning can be deduced from all its various usages, it is in the semantic range of putting or establishing something somewhere for some reason. On this basis, one of *šakānu*'s meanings is "(mostly in the stative) to be present, exist . . . to be provided with, have a feature, a characteristic."[170] Thus *šiknu* could have the nuance of something that was put somewhere for some reason. Under *šiknu* A, *The Assyrian Dictionary* gives as its principal meaning "outward appearance, shape, structure," and its usage is found in contexts pertaining to the facade of a house, the appearance and surfaces of divine statues that have gone dull, of stones, plants, animals, human facial features, parts of the exta (in the description of ominous features), and even the extent of an eclipse (again, a feature of its appearance). Landsberger's German translation for *šiknu* was *Gestalt*,[171] connoting appearance, form, shape, or guise, thereby emphasizing the external attributes with which a thing has been in some sense "provided."

In the face of parallels in phraseology from the *šikinšu* texts to later European herbals, and the translation of *šiknu* by "nature" in the literature, it seems worthwhile to investigate a little more closely the semantics of *šiknu* as compared with *phusis/natura*. It is equally important to introduce the related term *šīmtu*, which sometimes refers to the qualities or character of something. Indeed, where properties of medicinal plants are concerned, perhaps *šiknu* is not the relevant word, or concept, as much as is *šīmtu*.

For the conceptual counterpart to "nature" in the sense of the inherent rather than the outward features of something, Oppenheim pointed out that the noun *šīmtu*, from the verb *šâmu* "to allot (qualities, character)" or "determine (fate, one's lot in life)," had a basic meaning of "share," as in one's share or portion, and appealed to the Greek terms *moira* "portion," or "lot," hence "fate," and *phusis* "nature" for clarification.[172] He adduced the single attestation of the word *phusis* in all of Homer, which appears in the Odyssey Book 10.[173]

While on the mythological island of Aeaea, Odysseus's crew falls prey to the witch goddess Circe, who offers the men "baneful drugs" (*pharmaka lugra*) that turn them into swine. On his way to the rescue, Oddyseus meets Hermes (Argeiphontes), who gives him an antidote (*pharmakon esthlon*), called moly. "But for the plant to work," as Gerard Naddaf explained, "Odysseus must in some sense understand its *phusis*. Thus, after drawing the *pharmakon* from the ground and giving it to Odysseus, Hermes proceeds to show/explain/reveal its *phusis* to him: *kai moi phusin autou edeixe* (10.303) ['and showed me its nature']."[174] What is of interest for our understanding of the *šiknu* of plants, stones, or snakes is Naddaf's further exposition of the relation of *phusis* to the terms *eidos* "form," or " appearance," *morphē* "form" or "shape," and *phuē* "form" or "physical constitution of a thing." He says,

> At first glance, the term *phusis* seems to be employed synonymously with *eidos*, *morphē*, or *phuē* (all of which are found in Homer), insofar as the moly plant is identified by its form. . . . However, that Homer does not employ the terms *eidos*, *morphē*, or *phuē* suggests the possibility that the term *phusis* means something quite different from "form" or "exterior aspect" . . . in Homer *phusis* can be defined as "the (completed) realization of a becoming" and thus as "the nature [of the thing] as it is realized, with all its properties." In other words, while *eidos*, *morphē*, and *phuē* designate the form or the physical constitution of a thing, *phusis* designates the process by which the object becomes what it is.[175]

Whereas the semantics of *phusis*, by its etymology from *phuein* "to grow," or "bring forth," had from the beginning, as in the Homeric usage, the force of accounting for the physical origin of a plant from a process of growth, the meaning of *šīmtu*, derives from a verbal root meaning "to decree" or "determine," thus "that which is decreed to occur." Something's *šīmtu* is the result of divine decree, whether its share, quality, or character.[176] In this it bears relation to *šiknu* in being the (largely) permanent result of action from an external agency. In Oppenheim's words, *šīmtu* "denotes a disposition originating from an agency endowed with power to act and to dispose, such as the deity, the king, or any individual may do, acting under specific conditions and for specific purposes."[177] Both are *pirs* formations, passive in meaning, but where *šiknu* points to the character, appearance, or composition of something by virtue of its having been put there, *šīmtu* points to the acquisition of characteristics or lot in life by virtue of decree.

Oppenheim elucidated his understanding of *šīmtu* by reference to the Sumerian literary text *Lugale*, in which the god Ninurta endows (nam . . . tar) the stones with their nam (the Akkadian lexical equivalent of which is *šīmtu*),[178] thus concluding that *šīmtu* is the "nature" of the stones.[179] In this case the divine act is one of apportioning, or deciding (tar = *parāsu* "to apportion, decide") such qualities. In terms of the etymology of Sumerian nam in nam . . . tar, Piotr Steinkeller pointed out that "Nam is identical with the nominal element nam, which serves to form abstract concepts in Sumerian . . . the latter element should perhaps be explained as a frozen verbal form, consisting of the affirmative preformative na-, which means 'truly, indeed,' and the verb-am_3 'to be.' If this is correct, *na-am_3 > nam would mean 'that which indeed is,' therefore 'essence' (*essentia*) or 'essential nature.'"[180] Notably, however, the idea of "essence," at least in the Sumerian idiom (not, I would suggest, necessarily transferable to the Akkadian) is constructed with a verb to convey the action of a divine force, apportioning, or deciding/determining the nam of things. Another Sumerian literary composition, *Enki and Ninhursag*, develops a motif of the generation of medicinal plants by the semen of the god Enki implanted in and then generated from the womb (šà, literally, "heart" or "interior") of the goddess Uttu. As Cale Johnson explained, "'Enki determines the fate' of each plant ([^{d}en.ki].«ke_4» ú nam.bi bí.in.tar) and then embeds these properties in its interior (šà): 'he made it (= the plant's fate) known in its (= the plant's) interior/heart (šà.ba ba.ni.in.zu).'"[181] Interestingly, the semantics of nam . . . tar in this context involve growth and generation, as in Greek *phuein*.

When Aristotle defined nature in the *Metaphysics* (Bk 5, 1014b), he gave as its first sense "the genesis of growing things" and the second sense as "that immanent thing from which a growing thing first begins to grow." This passage continues with the definition of the word in another several senses, the most important for the present context being the following: "Again, 'nature' means the primary stuff, shapeless and unchangeable for its own potency, of which any natural object consists or from which it is produced; e.g., bronze is called the 'nature' of a statue and of bronze articles, and wood that of wooden ones, and similarly in all other cases. For each article consists of these 'natures,' the primary material persisting." But the Sumerian narrative also formulated a relation between characteristic properties of plants and the internal states of those who ingested them by connecting the "interior" of the plant with the "interior" of the human body. Johnson drew attention not only to this "container metaphor" but also to the significant component of exegesis by orthographic elements, by means of which this relation was constructed.

This integrated orthographic-semantic method, or etymography, has become of increasingly recent interest among cuneiformists.[182] Johnson said,

> In simple referential terms "šà" in Sumerian can, for instance, refer to a part of the body ("womb, heart, innards, guts") or to the "content" or "significance" of an event of semiosis such as a dream. At the same time, however, "šà" also acts as the nominal component of [*sic*] in a number of compound verbs in Sumerian that express internal states (ranging from pain and illness to elation). And, last but not least, when Enki places each plant's "fate" ("nam.tar"), or what we might call its "nature" in its šà, it strongly suggests that the pharmacological properties of these plants also reside in their šà. Given the Mesopotamian predilection for using orthographic form as a guide in formulating theoretical models, we have good reason to suppose that the authors and redactors of *Enki* and *Ninhursag* are postulating a semiotic model that links the "properties" ("šà") of individual pharmacological plants to the "internal states" ("šà") that they produce, when they are consumed.[183]

The Akkadian *šikinšu* texts selected and described stones and plants that had medicinal use. The focus there seems, however, to be less in terms of how properties come to inhere in a body and more to do with how to recognize them by their appearances, also a useful kind of information for the *āšipu* and *asû*. As Johnson implied, Sumerian literary passages about stones (*Lugale*) and plants (*Enki and Ninhursag*) focus on properties, but, as he showed, the manner by which these properties took on meaning had as much or more to do with orthographic linkages as with the investigation of those properties from an empirical viewpoint. Ultimately, in reading nam.tar as *šīmtu*, the Babylonian scribes made the connection to the Sumerian idea of properties of things, but as being endowed by (divine verbal) decree from without. This seems to me quite different from the Greek formulation about the generation of things in a material way from within. The *šikinšu* texts seem to have a particular emphasis on description, no doubt for the purpose of recognizing such plants, stones, or snakes.

Three distinct idioms concerning the appearance (Akkadian), property (Sumerian), or nature (Greek, followed by Latin, and so on) can therefore be delineated in the textual record of each linguistic culture. Despite the relationships that can be made from one to another, nuances of difference are equally important as they clue us into fundamental distinctions among cultural ways of viewing and linguistic ways of describing and classifying the world.

BEFORE NATURAL KINDS

Another aspect of the *šikinšu* texts germane to the subject of the nature of the knowledge of plants, stones, and animals is what they reflect of a taxonomy of things. The *šikinšu* texts share with each other and with lexical lists the use of a feature of the cuneiform script—namely, silent graphemes dubbed determinatives, to classify members of a group of words and also to structure the texts themselves. Such determinatives are unpronounced classifiers either before or after a noun (or a proper name) that indicate the category to which a word (or name) belongs, such as LÚ before names or professions of men, DUG before vessels, TÚG before garments, URUDU metals, GIŠ trees and wooden objects, GU_4 cattle, KU_6 fishes (as a postposed determinative), URU before names of cities, MUŠEN birds (postposed), Ú plants, and NINDA foodstuffs.[184] A determinative for divine names (of gods, planets, and stars) and divine things (divine emblems or standards, such as ^{d}*šurinnu*), DINGIR, points to the existence of a superordinate category of "the divine," which, at least for us, seems to refer to a different sort of category than the determinatives named before. The use of the divine determinative might suggest that a divinity as a "divine thing" was as much an object, a thing *in* the world, as the referents of other determinatives. Alternatively, the use of the determinative to mean "divine," especially before divine names such as ^{d}*Aššur*, as opposed to emblems (^{d}*šurinnu*) or stars (^{d}*Gamlu*, the astralized divine crook of Marduk) and planets (^{d}*Ištar* "Venus," ^{d}*Šiḫṭu* "Mercury"), might suggest some further abstraction from the concrete entities viewed as divine.

Determinatives offer up a kind of nomenclature devised to indicate or "determine" the taxonomic category to which a written word, and the thing (or idea) it represents, belongs. As such, determinatives afford a look into some of the fundamental organizational categories of a conceptual world as it existed for the ancient scribes, as well as the criteria they may have used to create such categories. Cuneiform determinatives typically apply to substances as well as to objects made of such substances where the "substance" referred to is sometimes a function of the object itself, as in "stones" that refer to beads made from a variety of hard substances (stone, shell, metal). The determinative as a classification is, therefore, thoroughly cultural.

Functionally similar but more numerous and complex are the uses of determinatives in Egyptian hieroglyphs. Orly Goldwasser argued that the Egyptian system reflects the conceptual categories of the Egyptian scribes' understanding of the world.[185] She said: "The determinative phenomenon of the Egyptian script consists of and reflects a knowledge organization. This organization is not at all arbitrary or exclusively context-bound, but represents the deep structure of world classification of the Egyptian culture or what may carefully be termed the ancient Egyptian 'collective mind.'"[186] Thus Goldwasser took a realist position on determinatives in Egyptian, that is, that they are not arbitrary inventions, but rather direct reflections of the Egyptians' perceived world.

The cuneiform use of such classifiers may equally be the product of the cuneiform scribal categories that carved their world at its own conceptual joints. But the "world" being thus carved, as Veldhuis points out with respect to the Archaic lexical corpus, is relative to administrative practice, not to an empirical description of the perceived universe as a whole. As he put it, "Roughly, the lexical corpus covers these same topics [as in early administrative account texts]: titles, domestic animals, fish, birds, food, grain, numbers, etc. Subjects that are not relevant for the administrative system, such as stars or wild animals, tend not to be treated in lexical texts."[187] The thematic matter of the earliest word lists, as tallied by Veldhuis, corresponds to a large extent to the categories indicated by determinatives, as just enumerated. This is only one dimension of the Archaic lists' complex structure, not to mention their purpose at the time of composition. The very subject of classification, and how to analyze it through the use of determinatives in word lists, however, not to speak of whose "collective mind" is engaged, is highly problematic.

The idea that the lexical lists of the Archaic period were motivated by "the Sumerians'" desire to categorize the things of the world was promoted in the so-called *Ordnungswille* thesis put forward in Wolfram von Soden's paper "Leistung und Grenze sumerischer und babylonischer Wissenschaft" of 1936. Subsequently, the "will to order" became a loaded concept in Assyriology, as has the idea that one can identify through the texts a racially identified population group. The claim that "the Sumerian mind" characteristically ordered the world in the form of word lists and that those lists evoke theological and cosmological ordering principles was rejected first by Leo Oppenheim,[188] more recently by Markus Hilgert and Niek Veldhuis. Veldhuis exposed both the essentialist and racialist, even racist, underpinnings of von Soden's notion of "Ordnungswille," as well as the untenable nature of his "order of the world" theory:

> The publication of the entire archaic lexical corpus in Englund and Nissen 1993 and the resulting progress in our understanding of the contents of these lists has made clear that archaic lexical texts are not one homogeneous group and that not all of them may be described as thematic—or only in a very weak

> sense. The individual lists may come from different backgrounds, do different things, and use different styles of organization and standardization. Modern research may have been misled by (modern) titles and labels. Lists such as Plants or Vessels are hardly inventories (or classifications) of all plants or all vessels—they are both less and more. If, however, we deny the archaic lexical compositions their classificatory nature, the Order of the World theory loses whatever plausibility it had still left. However, even if we allowed Plants and Vessels to stand as classificatory, thematic lists in a weak sense, we end up with an odd and strangely skewed classificatory system that includes lots of professions, vessels, place names, and metal objects, but no stars, no wild animals, and no gods. The kinds of things that are treated in the lexical corpus do not describe a universe in any accepted sense of the word.[189]

There is yet a further problem with von Soden's view of the archaic lists as a reflection of Sumerian world organization. A claim to world classification from word classification raises questions about the intersection of categories (or sets) and kinds, and whether the categories represented by determinatives provide a basis for inferences not only about taxonomy but also ontology.[190] How we read and critique von Soden has also to do with what we expect of cuneiform classification systems in general. Even if we do not accept that the purpose of the lists was to "order the world," we are left with categories of words (the product of Veldhuis's "weak sense" of the word lists as classificatory), if not categories of things. The lexical lists themselves, as already indicated, are beyond the scope of this study, but some of the principles involved in their classification scheme(s), it seems to me, are propagated into many other texts through the spellings of words with determinatives. This, then, bears on the understanding of the texts concerning plants, stones, animals, and woods as *materia medica*, as well as, perhaps, omen texts from which we might also extract taxonomic structure.

To speak of the classification of words and things in the cuneiform world and to consider the relationship of taxonomy to ontology is to immediately raise the very difficult and complex matter of "kinds." As W. V. O. Quine pointed out, "surely there is nothing more basic to thought and language than our sense of similarity; our sorting of things into kinds. . . . Kinds can be seen as sets, determined by their members. It is just that not all sets are kinds."[191] It is an interesting question to consider the classificatory function of determinatives and in what way they did their work on the basis of sorting by similarity. Nelson Goodman was not sanguine about the utility of the relation of similarity: "When, in general, are two things similar? The first response is likely to be: 'When they have at least one property in common.'

But since every two things have some property in common, this will make similarity a universal and hence useless relation."[192] How are we to understand the way the cuneiform script developed and used the determinative? And did they construct kinds, or sets, both, or neither?

There is no classification of things without some conception of the world, and thus classification systems are as laden with theory as is observation. A classification system that uses the notion of a natural kind bespeaks a view of the world that has nature as an available category. A clear division between classification systems of an entirely arbitrary or invented nature and those grounded in structures discoverable in the natural world seems to be important to those of an essentialist orientation. The decision as to whether categories are "natural" or "cultural" entails commitments both semantic (to what do *kinds* refer?) and metaphysical (how are kinds constituted ontologically?). Each is relevant to understanding the classifications cuneiform determinatives represent, and how we think they intersected with the perceptual and conceptual world of the scribes.

At the very least, cuneiform determinatives, which principally classify words and not things, do not have a straightforward relationship to natural kind terms. For example, the determinative ÍD "river" was used to indicate the names not only of rivers (a natural kind) but also of canals (an artifact). And the determinative NA_4 "stone" was used to indicate not only stones (a natural kind) but also beads (an artifact), and also not exclusively beads made of stone, but of metals (gold, silver), and of shell. This underscores the fact that it is only in terms of *taxa*, not of "mind," that the cuneiform scribes differ from "us," or from many others, in the sense that they had established a different "world."

Similarly, with respect to the ancient Romans, Daryn Lehoux noted that "because of the depth and invisibility of classifications, and because of their long reach into ontology and possibility, people with significantly different classifications can best be said to inhabit different worlds."[193] Lehoux reminds us that Plutarch lived in a world different from ours, "one with different entities, different rules, and different possibilities."[194] The same is true of the ancient Assyro-Babylonian scribes in their world where the classifications of words and things, as indicated by determinatives at least, did not (could not) reflect a concern for natural kinds as such, and so did not "carve nature at its joints." Even more fundamentally, however, is the question of whether kinds can ever be said to be natural, to inhere in nature, and thus to be discoverable, or whether all categorization of things are products of culture.

Whether kinds can be said to be natural or cultural has been and still is a subject of ongoing and recent debate, from Quine's 1970 "Natural Kinds,"[195] to the articles of Paul Churchland in 1985 and Ian Hacking in 1991, to the anthologies of Helen Beebee and Nigel Sabbarton-Leary, *The Semantics and Metaphysics of Natural Kinds* of 2010, Joseph Keim Campbell, Michael O'Rourke, and Matthew H. Slater, *Carving Nature at Its Joints: Natural Kinds in Metaphysics and Science* of 2011, P. D. Magnus, *Scientific Enquiry and Natural Kinds: From Planets to Mallards* of 2012, and Muhammad Ali Khalidi, *Natural Categories and Human Kinds: Classification in the Natural and Social Sciences* of 2013. Tracing the history of the term *natural kind*, and establishing the discursive ground for what followed, Hacking said, "the nominalist 'left' says that all kinds are human, or at any rate, there are no kinds in nature. The realist 'right' says that there are indeed natural kinds, and that human kinds—at any rate those susceptible of systematic study—are among them."[196] Magnus, on the other hand, opens his 2012 book with the following: "In many scientific domains, the world constrains our taxonomy. We cannot approach phenomena using just any categories and expect to achieve predictive and practical success. Where our choice is strongly constrained, it is tempting to say that our categories correspond to structures in the world. I think it makes sense to use the label 'natural kinds' to pick out such structures."[197]

Of the many issues at stake in the discussion of natural kinds, as enumerated by Churchland, are "the aim of science and the problem of scientific realism in general . . . of essentialism, meaning, and reference." Magnus's statement, above, flags the aspect of the success of science, situating him to the right of center in accordance with Hacking's "political spectrum" on natural kinds. Churchland, on the other hand, is a leftist, who rejects "the idea that there is a theory-neutral or intension-independent relation that connects words to unique natural sections of the world."[198]

To use the success of science, in particular its predictive and explanatory success, as an argument for the reality of natural kinds is a position unfavorable to historical evidence falling outside the boundaries of science conceived as the study of nature, and therefore outside the boundaries of the discussion of cuneiform taxonomy. Naturalism on an ontological level is utterly excluded in this context, even implicitly, as the purpose of cuneiform classifications for words and their underlying categories of things was not about what exists, what is real, or what is discoverable by science in nature. Rather, the taxonomy and nomenclature of the cuneiform texts devoted to lists of plants, trees, and stones aimed to describe observable, and imaginable (as seen in the list of snakes), phenomena for medical and magical purposes, not to inquire into how the world was physically constituted.

It is not only in the context of cuneiform culture that kind essentialism proves problematic, as shown in the recent philosophy of biology.[199] Geoffrey Lloyd's comments are pertinent here:

> Where humans find *order* in a domain, that is often expressed in terms of the taxonomic structure of genera and species exemplified paradigmatically in the world of living things. For those purposes it does not matter unduly how much detailed knowledge of different natural kinds is available to underpin the classification. Plants and more especially animals may be recognized, as they often are, by what are represented as their stereotypical characteristics, as the tiger by its possessing stripes, being furry, having four legs and so on. An account may be built up by accumulating such features without any commitment to which may be essential, which merely accidental.[200]

Such classifications by similarities, characteristic features, or functions (what is edible, for example), are found in cuneiform texts, especially in the word lists. And in those lists the sense of how similar properties or characteristics were understood seems to stem from the idea to classify certain words together and not to investigate whether their properties are a manifestation of a material world order. As Lloyd said, "those who claim that essentialism is a universal human trait have usually, though not always, been careful to distinguish between, on the one hand, the ways in which essences figure in humans' representations of the world ('psychological essentialism' in the vocabulary of Medin and Ortony 1989), and, on the other, the question of whether essences actually exist in any given domain."[201]

Claims to essences, or internal properties, that define membership of things (and the words that represent them), as discussed above in relation to the translation "the nature of" certain animals, stones or plants in cuneiform texts, are tantamount to a claim to the existence of natural kinds within the cuneiform classification system. On the basis of the *šikinšu* texts this position seems to me untenable. It is of interest in this context to see that Landsberger already emphasized the cultural force of kinds (and species) in cuneiform word lists. He said: "An all embracing order of classifying concepts of kinds and species is unknown. Avowedly, concepts of species *are* formed, but only where practical needs require it, as where a great number of species difficult to distinguish occur, like bird, snake; but no concepts of species are formed when there are only few distinctive species like grain or metal. In this case the designation *a potiori* of the one that occurs most frequently serves, e.g. barley for grain."[202]

Another aspect of classes and kinds concerns the manner in which classification is made and how membership is defined. The definition of membership will of course reflect both the idea and the purpose of classification itself. In her study of the taxonomy of animals in the ancient Near East, Paula Wapnish noted the polythetic nature of Sumero-Akkadian classifications.[203] In contrast to monothetic—an Aristotelian—classification, where features of the members of a class are both necessary and sufficient, criteria for polythetic classification are neither necessary nor sufficient. Where members of monothetic classifications are all identical as a result of their possessing both necessary and sufficient properties, those of polythetic classes belong to the group by other criteria of likeness.[204]

Polythetic classification is reflected in the attachment of determinatives to the writings of words. Accordingly, the determinatives that acted as organizing principles for the lexical lists and for the *šikinšu* texts stand for classes of things with multiple properties, not all of which are held in common by every member of the class. The category marked by the determinative GIŠ "wood," therefore, included trees—as in the writing of *erēnu* "cedar" as GIŠ.ERIN—and artifacts, or things made of wood—as in the writing of *narkabtu* "chariot" as GIŠ.GIGIR, or the writing of *eriqqu* "wagon" as GIŠ.MAR.GÍD.DA. The determinative indicating things having wood as their primary "substance" covered trees, wagons and chariots, together with doors (*daltu* written GIŠ.IG) and other objects. Indeed, the members of a class of words written with a certain determinative do not necessarily resemble one another; they are more unlike than like. It is worth emphasizing that the classification of things into categories of words marked by a given determinative effected the writing of words for things; it was not aimed at systematizing morphologies or even behaviors of living things, nor of establishing specific families of inanimate things based on physical characteristics.

Veldhuis provided a vivid illustration of the nature of lexical classification and organization in the lexical lists with respect to the category pigs (šáḫ).[205] He noted that while the separation of the categories domestic and wild animals was made in the earliest version of *Ura* from Nippur, a problem arose for the pig because the domestic and the wild pig were not differentiated in the language. He said that "the list does not represent (biological) taxonomy, but follows an ordering system that is based on cultural and linguistic principles. Thus am (wild ox) is not found in the section gud (ox), but rather with am-si (elephant) and am-si kur-ra (camel). For (domestic) pig and (wild) boar, however, there are no separate words in Sumerian, so that all of the pig terminology (wild and domestic) is found in one place among wild animals."[206] The Nippur text of *Ura* put pigs together with rodents, turtles, and lizards. Veldhuis noted that both rodents and pigs were kept, or raised, in the ancient Near East as food. Turtles and lizards, while not "farmed," were no doubt also regarded as comestibles, as they still are in many cultures. The content of the pig section equally reflects considerations not biological in nature. The Nippur list, for example, includes pig, wild pig, daily pig, fattened pig, Magan pig, pig of the Oath(?), pig owned by a lord, piglet, and sow.[207] The criteria for membership in the group of pigs, therefore, are what people did with pigs, and what pigs were for. These criteria are specific to the context within which the list was made, that is, an administrative one, and as such did not employ principles of organization aimed at a universal description of things in some universal order of things. As a result, the choice of words to be used as determinatives reflects similarly the context of their first use, decidedly not that of a universal taxonomy, but a convenient marker for the words organized within lists useful to the scribes producing administrative texts.

Cuneiform determinatives as classificatory nomenclature do not point to a world taxonomy in any other sense than a cultural one. The cuneiform determinatives may perhaps be marshaled as evidence for Hacking's claim that "some classifications are more natural than others, but *there is no such thing as a natural kind*."[208] This, as he says, goes back to Nelson Goodman, whose concept of relevant kinds as an alternative to natural kinds arose from his view that "the uniformity of nature we marvel at or the unreliability we protest belongs to a world of our own making."[209] Recent work on the question of natural kinds has explored ways to further define the kinds that seem to depend upon a world that is independent of human interaction with it from those that are clearly a function of cultural categories. The only consensus seems to be that the idea of a universal essence of things as a defining property for kinds, which science is equipped to discover, has become untenable even for those who maintain stakes in natural kinds. The following comment of Thomas Reydon sums it up: "Natural kinds no longer are what they traditionally were thought to be, i.e., groups based on essences shared by all and only the members of a kind, or kinds that have an objective existence in the world independently of human concerns. But then again, natural kinds never actually *were* such kinds in the first place, as the essentialist and objectivist conceptions of natural kinds have always been just some among a number of available views."[210]

When we consider the categories of things referred to by cuneiform determinatives, Goodman's original argument for relevant rather than natural kinds has particular resonance: "Worlds differ in the relevant kinds they comprise. I say 'relevant' rather than 'natural' for two reasons: first, 'natural' is an inapt term to cover not only biological species but such artificial kinds as musical works, psychological experiments, and types of machinery; and second, 'natural' suggests some absolute categorical or psychological priority, while the kinds in question are rather habitual or traditional or devised for a new purpose."[211] Although Hacking's response is that "Goodman's expression, 'habitual or traditional or devised for a new purpose' does not do justice to the variety of kinds of relevance to students of nature," it opens possibilities for understanding the classificatory nature of cuneiform determinatives. Determinatives as classifications were devised for the purpose of cuneiform writing, emerging within the confines of the scribes who listed words in Sumerian in an administrative context. Later, of course, the continued use of determinatives, as seen in the *šikinšu* texts and throughout scholastic Akkadian with its use of logograms, took on a habitual or traditional status with the script.

To sum up, the nature and goal of the knowledge of plants, stones, and animals as contained in *šikinšu* texts is a function of the purpose of those texts as compilations by and for scribes (presumably the specialists *āšipu* and *asû*) who dealt with medical and magical substances. The texts provide descriptions of the appearances of substances belonging to a class of materials defined by their use, in this case as *materia medica*. Presumably the *šikinšu* texts were guides to the recognition of the substances classed as useful for medical and magical procedures. The stone list describes those substances appropriate for use as healing beads, made from a variety of substances, not all of which we would classify as stone. As it was primarily the function and not the substance itself that mattered, at least as far as the classification NA_4 in the *šikinšu* texts, the taxonomy reflected there is not amenable to the idea of natural kinds, but rather evinces theoretical, or perhaps practical, certainly relevant, kinds and is characterized by polythetic (or polytypic) classification.

In the preface to *The Order of Things: An Archaeology of the Human Sciences*, Michel Foucault referred to the classification system of the fictive Chinese encyclopedia that Jorge Luis Borges titled the "Celestial Empire of Benevolent Knowledge," where animals are classified as "(a) belonging to the emperor, (b) embalmed, (c) tame, (d) sucking pigs, (e) sirens, (f) fabulous, (g) stray dogs, (h) included in the present classification, (i) frenzied, (j) innumerable, (k) drawn with a very fine camelhair brush, (1) et cetera, (m) having just broken the water pitcher, (n) that from a long way off look like flies."[212] Foucault said, "In the wonderment of this taxonomy, the thing we apprehend in one great leap, the thing that, by means of the fable, is demonstrated as the exotic charm of another system of thought, is the limitation of our own, the stark impossibility of thinking *that*. But what is it impossible to think, and what kind of impossibility are we faced with here?"[213] Zhang Longxi critiqued Foucault's analysis of Borges's fabulous taxonomy on the grounds that Foucault did not realize

> that the hilarious passage from that "Chinese encyclopedia" may have been made up to represent a Western fantasy of the Other, and that the illogical way of sorting out animals in that passage can be as alien to the Chinese mind as it is to the Western . . . in fact, the monstrous unreason and its alarming subversion of Western thinking, the unfamiliar and alien space of China as the image of the Other threatening to break up ordered surfaces and logical categories, all turn out to be, in the most literal sense, a Western fiction.[214]

Certainly one of the striking features of Borges's fictive list of categories is the way in which it seems to focus on something that will not submit to the order of nature, but rather has been given its own orientation to "something else." Therein lies its strangeness. On the other hand, the classifications inherent in Sumerian, Sumero-Akkadian, and Assyro-Babylonian lexical and divinatory texts are not fictive, but historically real. The criteria for classification and making connections between elements of various categories found in those texts, however, because they do not reduce to a desire to know and classify nature, can have a similar effect.

Unlike China, ancient Babylonia and Assyria have not always played the role of the Other in the Western imagination so much as they have been conscripted into the role of precursors of Ourselves, of Western civilization. And yet when it comes to the analysis of cuneiform corpora of knowledge, where the intellectual history of the ancient Near East merges with the beginnings of Western science, we find ourselves confronted with classifications and categories, even phenomena, that sometimes confound our own sense of the order of nature. Still, as conceived in the cuneiform world, the overriding goal of the observation and interpretation of phenomena was to determine norms and anomalies within meaningful categories, and using those categories as vehicles, to find an order of things. Modes of establishing norms, in relation to which anomalies may be defined, as well as the models devised to describe them in cuneiform texts, are the concerns of the next chapter.

CHAPTER FOUR

A Cuneiform Modality of Order

uṣurāt šamê u erṣeti "The designs of heaven and earth"

> The fundamental codes of a culture—those governing its language, it schemas of perception, its exchanges, its techniques, its values, the hierarchy of its practices—establish for every man, from the very first, the empirical orders with which he will be dealing and within which he will be at home. . . . Thus, in every culture, between the use of what one might call the ordering codes and reflections upon order itself, there is the pure experience of order and of its modes of being.
>
> MICHEL FOUCAULT, preface to *The Order of Things*

The late period (after 500 BCE) was particularly productive of innovation and intellectual change in Babylonian astronomy and astrology. In the context of seemingly radical shifts in interest among the cuneiform *literati*, such as the focus on new methods of mathematical prediction of astronomical phenomena and new ways of construing the relationship between the individual and the heavens through horoscopy, certain continuous elements can still be identified. This section returns to the subject of the knowledge of signs, specifically to the question of whether the way in which signs had meaning might point to a particular understanding of the world and thereby account for an element of continuity in that tradition, indeed, giving it its particular epistemic slant.

If a certain way of understanding the world is embodied in the Assyro-Babylonian idea of the sign, we might regard this as a function of the cuneiform culture that Karen Radner and Eleanor Robson defined as having "bound the ancient inhabitants [of the ancient Near East] into a shared set of ways of understanding and managing their world."[1] While there may have been many cuneiform cultures within the various contexts of use of the cuneiform script, the culture of the elite *ummânū* perhaps retained a certain continuity based on the interpretive structures of divination. How the understanding of norms, ideals, malformations, and anomalies was framed within that tradition and way of understanding the world is the subject of this chapter.

James Allen pointed to the essential fact that "our term 'sign' comes, of course, straight from the Latin *signum,* which in turn renders the Greek σημεῖον, whose range of uses it tracks pretty closely. Not only the term, but the idea or complex of ideas for which it stands are an inheritance from Greco-Roman antiquity."[2] John Deely, however, saw in the transfer from Greek σημεῖον to Latin *signum* a discontinuity, suggesting that "the move from Greek σημεῖον to Latin *signum* . . . is not primarily a linguistic one," but something "of a conceptual saltation rather than of a linguistic translation."[3] He saw a conceptual jump already between the ancient Near East and Greece, saying that "in ancient Mesopotamia this attitude toward a σημεῖον was rooted in divinatory practices, but not in ancient Greece, where it was rather rooted in a rational attitude toward natural phenomena and made the basis of Greek medical practice,"[4] implying that cuneiform divination is an exercise in the irrational. The difference in question is not, as I will argue in chapter 5, rationality, but rather the conception of nature and the development to make the conception of nature a requisite part of rational science. For the present, it suffices to note that Deely went on to characterize Latin *signum* as having come into its own through a late antique discourse originating with Augustine, where signs revealed the will of God.[5]

Underlying cultural meanings aside, the persistent preoccupation with phenomena as signs is exhibited from cuneiform to Greek and Latin scientific texts, to those of later antique and medieval science and philosophy, with descendants in western European, eastern Byzantine, and Islamic traditions, not to mention Indian science. Peter Harrison, in a striking statement, said, that "for virtually the first fifteen hundred years of the common era the study of natural objects took place within the humanities, as part of an all-encompassing science of interpretation which sought to expound the meanings of words and things."[6] Divination and astrology found a central place among various ancient and medieval cultures of knowledge, from the point of view of prognostication as well as of the philosophy of inference-making from signs.

Ancient cuneiform knowledge of what Harrison referred to as "natural objects," roughly for the fifteen hundred years before the Common Era, constitutes very much the same thing that he identified for the first fifteen hundred years of the Common Era, that is, it was "part of an all-encompassing science of interpretation which sought to expound the meanings of words and things." This is represented in the cuneiform tradition of knowledge in the overwhelming focus by the scribes on the systematic and interpretive science of divination from signs. The principal qualification must be in designating the objects of this knowledge not as "natural objects" but as observed, imagined, and conceived objects in relation to physical as well as imagined things, and for the focus not on observation of the signs alone, but on their interpretation according to systematic codes embodied in textual compendia (some might argue it was really one code with particular variants for different domains of phenomena, say, exta, or births, or the stars and planets). Both the idea of a sign and the hermeneutics of the texts constituted the science of signs in the culture of cuneiform knowledge.

In the West, the Hellenistic period saw a new focus on signs from astronomy to philosophy. Already in the third century BCE Aratus began his poetic star catalog the *Phaenomena* (lines 5–6), as Katharina Volk noted, by reference to Zeus as giver of "propitious signs to humans,"[7] thus framing the composition in terms of signs. Volk further explained:

> In addition to announcing the poem's topic, this proem neatly states the *Phaenomena*'s conception of the world as a cosmos full of benevolent signs from an omnipresent god who has the welfare of human beings at heart. . . . The idea of the sign is central to the *Phaenomena*, as is apparent from the fact that forms of the noun σῆμα (*sēma*, pl. *sēmata*) "sign" appear 47 times in the course of the poem, those of the verb (ἐπι)σημαίνω "to signal" 11 times. The repetition of these and similar keywords . . . drives home the message that Aratus is *not interested in natural phenomena* (e.g., the constellations) as such, but only in as much as they are part of the cosmic system of signs that has its origin in the benevolence of Zeus.[8]

Despite the cultural track running from Rome to Greece and farther on to the ancient Near East, many particulars of ancient Greek (or Roman, or medieval) ideas about signs differ from those of cuneiform texts. The ubiquity of the importance of signs throughout the cultural worlds of the ancient Near East and Mediterranean should not be mistaken for a thoroughgoing similarity or unity in how signs were understood from one cultural milieu to another.[9] What a sign was in the cuneiform context, even as it may well have changed over the millennia-long span of the tradition, reflects nonetheless within bounds an Assyro-Babylonian way of seeing the world where the portentous, the anomalous, and the prodigious differed, however subtly, from the preternatural, the monstrous, or miraculous in the worlds of later Hellenistic (Greek, Greco-Roman, Greco-Egyptian, Indian) diviners, (Platonist or Christian) theologians, or Late Antique and medieval natural philosophers who were interested in these things. In particular, once God and nature entered into the matrix of ideas that defined the world, signs would begin to signify specifically in terms of that matrix, in which sometimes there was an equivalence of God and nature, sometimes a tension between Divine will and the laws of nature. Consider the statement of Augustine, where God's will works within nature for His purposes: "So, just as it was not impossible for God to set in being natures according to his will, so it is afterwards not impossible for him to change those natures which he has set in being, in whatever way he chooses. Hence the enormous crop of marvels, which we call 'monsters,' 'signs,' 'portents,' or 'prodigies.'"[10] The explanatory rhetoric of God and nature, or natures, is evidence of a new conceptual foundation for prognostication through signs, and for science, differentiating it from anything that developed in the Hellenic cultural realm, and certainly from that of the ancient Near East.

In the most general of terms, signs are communicative. They point to things beyond themselves, conveying information in a multiplicity of ways, as is readily seen in the fourteen meanings of "sign" in *The Oxford English Dictionary*. Signs can be read and understood, or variously interpreted. Signs can be linguistic and orthographic, and thus can constitute a form of writing, literally (cuneiform or any other script) or figuratively (the liver, the stars and planets, or the Book of Nature). Like its English counterpart, Akkadian *ittu* had a range of meaning from "mark, feature, characteristic," or even "diagram," to "omen," "password, signal, notice, acknowledgment," and "written proof."[11] Signs entered into the Western cultural-historical discourse on various levels, including the linguistic, the theological, the philosophical, the divinatory, and the medical diagnostic. The last two in this enumeration played a central part in the discourse of the Assyro-Babylonian scholars.

From these general statements, many distinctions are to be made among the forms, the functions, and the responses to ominous signs in cuneiform texts. In addition to the various kinds of signs, some discussed above, another important distinction can be made between signs that were seen and/or reported and those that are found as entries in written compendia, as in the series *Enūma Anu Enlil*,[12] *bārûtu*,[13] *Šumma izbu* (signs from anomalous births), and others. The compendia served as vehicles for organizing the signs together with their portents in complex lists of antecedent–consequent statements, the conditional statements "If *P*, then *Q*." The phenomena are presented in a way that follows a fundamental method of interpretation, more or less employed in each series. This method has been variously referred to as a code,[14] or a hermeneutic strategy.[15]

The relationship between the antecedent and consequent clauses allowed the development of thinking about signs to encompass the observable, the possible, and the conceivable, including, within the category of conceivable signs, those that cannot occur in actuality. Actuality, as we might define it by what is permissible by nature, was not the sole focus of the scholarly imagination working within the sources in question here. The so-called impossible phenomena have been a puzzle to Assyriologists for a long time. The puzzle has been based on the fact that these "impossible phenomena" are impossible from the perspective of the violation of the normal order of nature, the normal functioning of phenomena in accordance with nature's laws, such as the lunar eclipse on the twenty-first day of the lunar month, or any other variety of such nonoccurring phenomena. Assyriological consensus on the invention of these impossibilities was that they completed interpretive schemata. However, this leaves open the question as to the nature of the framework in which the interpretive schemata had validity. Perhaps the reason for our puzzlement is that for too long we have failed to see how a notion of the order of nature was fundamentally absent from and irrelevant to cuneiform divination. This is not to say that certain realities did not constrain what the Assyro-Babylonian scribes actually observed. However, the omen series explored the world in a different way, a way that valued the schematic possibilities of ominous phenomena.

What were the characteristics that rendered phenomena ominous? Although many details of the appearances of stars, moon, sun, planets, animals, birds, and insects; human appearance and behavior; as well as sounds and light phenomena, and things seen in dreams, were ominous, not every single phenomenon was an omen. It is with respect to some conventionally established system of reference (the code or hermeneutic method) that something was interpretable as ominous, and even though many unreal and purely imagined phenomena were included in the schematic compilations of ominous phenomena, the system supported a notion of norms and a sense of normal and abnormal. Georges Canguilhem noted the ambiguity of the term *normal*: "Sometimes it designates a fact that can be described through statistical sampling; it refers to the mean of measurements made of a trait displayed by a species and to the plurality of individuals displaying this trait—either in accordance with the mean or with certain divergences considered insignificant. And yet it also sometimes designates an ideal, a positive principle of evaluation, in the sense of a prototype or a perfect form.[16] Canguilhem saw these two meanings as linked, therein finding the ambiguity of the term normal at the root of medical thinking about the pathological. In the realm of cyclical physical phenomena, such as those of the sun, moon, planets, and ecliptical star phases, the idea of a mean stemming from a measured standard is a related concept. I submit that the conceptual link Canguilhem drew attention to for the life sciences is also manifested in cuneiform scholarship, from divination—which employed the sense of an ideal[17]—to astronomy, which began with the use of the mean as an ideal, to a later approach that focused on anomaly as defined in relation to a numerical mean.

Referring to the turn of the nineteenth century anatomist and physiologist (and father of histology) Marie François Xavier Bichat, Canguilhem noted that "in his *Recherches sur la vie et la mort* (1800), Bichat locates the distinctive characteristic of organisms in the instability of vital forces, in the irregularity of vital phenomena—in contrast to the uniformity of physical phenomena."[18] Further, he defined Bichat's vitalism in his idea that "there is no pathological astronomy, dynamics, or hydraulics, because physical properties never diverge from their 'natural type.'"[19] The integrity of inanimate physical forms, therefore, did not permit the appearance of "monstrosities" among such phenomena as, say, the moon and planets. Canguilhem stressed the distinction between living organisms' capacity for monstrosity and the fact that "there is no machine monster,"[20] saying, "the distinction between the normal and the pathological holds for living beings alone."[21]

It seems relevant in this context to observe that across the various omen text categories a qualitative distinction between anomalous features of physical phenomena and monstrous features of births does not seem to apply. On the other hand, if we search for conceptions of the normal over a range of Akkadian divinatory texts, the same ambiguities as Canguilhem described for the concept may be found. That is, normal can be gauged in terms either of a "mean of measurements" or an ideal, a "positive principle of evaluation," where that ideal is determined by the divine scheme of things.

The adjective *kajamānu* (SAG.UŠ) "normal" is found in omens of the *izbu* and *ālu* series, as well as in extispicy, as a description of, or a feature of a phenomenon. In addition to the passages cited in CAD s.v. *kajamānu* usage a 1′ and 2′, a number of additional passages from liver omens can be adduced, referring to the "Presence" (*manzāzu*) of the liver, meaning the feature of the liver associated with the presence of a deity. For example, from early exemplars (second millennium Middle Babylonian and Middle Assyrian)[22]: "If the normal Presence is there and a second one is placed on the left: The king will resettle his abandoned territory." And from another Middle Babylonian source:[23] "If the normal Presence is there and a second one descends to the River of the Pouch: The gods of your army will forsake it at its destination." We might cite, additionally, the statement from an extispicy ritual, "for his well-being let there be a normal *naplastu,* let there be a normal *manzāz ilim,*" referring to the Presence (of the god) on the liver.[24]

In light of Canguilhem's reference to the "mean of measurements" being a defining basis for a conception of the "normal," a passage from a Late Uruk commentary may be noted that explains "the measurement of a normal Presence" as of three fingers length.[25] Ulla Koch cites another, Neo-Assyrian period, text that also describes norms in the features of the liver in terms of sizes:[26]

> The Presence, the Path, the Pleasing Word, the Strength, the Palace Gate, the Well-being, the Gall Bladder, the Defeat of the Enemy Army, the Throne Base, the Finger, the Yoke and the Increment, the designs (subsections) of the Front of the Pouch are three fingerbreadths each measured in the "large finger," the finger of the diviner or the *asli*-measure. Seven Weapons, five holes, three Fissures, you count as *niphu*s. The Foot is one fingerbreadth long, the Fissure is half a fingerbreadth long, the cleft is two fingerbreadths long, the *šitḫu* is three fingerbreadths long, they affect the consecrated place. The circumference of the liver is one cubit six fingers, 14 fingers its diameter(?).

As Koch noted, "The liver may undergo morphological changes or changes due to diseases or parasites. Also external influences can cause changes in the appearance of the liver in the form of lesions and contusions, and different causes may have the same symptoms on the liver. All this was obviously irrelevant to the Babylonians; only the visible symptoms were of interest. They did note the healthy and normal appearance as a favorable sign."[27] Despite the fact that the health and wholeness of the liver were regarded as of positive divinatory value, the emphasis on deriving positive and negative values for features of the exta overrides the value of the norm in a biological or anatomical sense. The evidence shows that from the seventh century to the later Babylonian period the system was relatively unchanged, and did not reduce solely to a binary of normal and abnormal, but employed many schemes for determining positive and/or negative outcomes of a liver inspection. As Koch implied, the designation of what was normal did not relate to an investigation of the physical causality of malformation. The interpretive scheme did not function around the understanding of what makes for biological normality, but rather what could be observed of regularity and irregularity from a visual standpoint. Nor did it work in this way in the omens from the twenty-four-tablet *Izbu* series, which itself seems to be based, by definition, in the *ab*normal.[28]

MALFORMED BIRTHS AND MONSTERS

The omen series *Šumma izbu* "if a malformed birth" seems to be the right place to raise the question of whether the scribes thought in terms of "monsters." To put the notion of an *izbu* in the context of monsters requires reference to later history.[29] In later antiquity the understanding was that it was in the power of God to act within and against nature to produce any conceivable, or inconceivable, phenomenon so as to communicate with humankind. Indeed, by its etymology, a monster is something that "warns" (Lat. *monere*) and is therefore a portent (Lat. *monstrum*).[30] Isidore of Seville, for example, said:

> Portents, according to Varro, are those things that appear to be produced against nature. But they are not against nature, since they happen by the will of God, since nature is the will of the Creator of every created thing. For this reason, pagans sometimes call God nature and sometimes, God. Therefore the portent does not happen against nature, but against that which is known as nature [*contra quam est nota natura*]. Portents and omens [*ostenta*], monsters and prodigies are so named because they appear to portend, foretell [*ostendere*], show [*monstrare*] and predict future things . . . for God wishes to signify the future through faults in things that are born, as through dreams and oracles, by which he forewarns and signifies to peoples or individuals a misfortune to come.[31]

While Isidore's reasoning may not be totally incompatible with what can be reconstructed for Assyro-Babylonian thinking on the matter, his explanatory rhetoric is. On the grounds of the attributions in prayers to the unlimited power of the gods as well as from the omens themselves, the Assyro-Babylonian gods were viewed as producing any conceivable phenomenon to signal yet another event, but the key element of explanation, as Isidore related it, either that God acts against nature or that God's will is tantamount to nature, departs from the framework within which omens would be understood by the cuneiform scholars.

An *izbu* is clearly a birth, and in the omen series, *izbu*s can be of animal (dog, pig, bull, cow, sheep, goat, donkey, or horse) and human births.[32] *Izbu* is defined in a bilingual lexical commentary as "a prematurely born fetus that has not completed its months."[33] The description of human *izbu*s may be found in the first tablet of the series, where a woman gives birth to newborns with various sorts of impairments (blindness, *Izbu* I 60, and deafness, *Izbu* I 63)[34] or deformities (mental, as in a *lillu* "fool,"[35] physical, as in *akû* "deformed,"[36] and various kinds of conjoined twins). Included among the omens from human *izbu*s are descriptions such as "if a woman gives birth to a lion/wolf/dog/pig/bull/elephant/ass/ram/cat/snake/tortoise/bird,"[37] as well as "to membrane,"[38] or "spittle."[39] There are also omens for multiple births,[40] up to "eight or nine," in the last omen of Tablet I: "If a woman gives birth to eight or nine (children): A usurper will attack; [. . .] the land will become waste."[41]

Overall, including the *izbu*s from animal births, Nicla de Zorzi has shown how the conception of deformity manifested itself in the categories of (1) malformations resembling animal features, (2) absence of body parts, (3) deformed or incomplete body parts, (4) misplacement of body parts, and (5) presence of excess body parts.[42] Some of the *izbu*s are vividly imagined, as for example, "if a woman gives birth, and (the child) is half a cubit tall, is bearded, can talk, walks around, and his teeth have already come in, he is called '*tigrilu*': Reign of Nergal; a fierce attack; there will be a mighty person in the land; pestilence; one street will be hostile to the other; one house will plunder the other."[43] On the other hand, further evidence that *izbu*s were not conceived of as monsters is that breach birth,[44] and twins, both identical and fraternal,[45] are also found in the series, neither of which would classify as "monsters" today for their irregularity.

Erle Leichty noted in his introduction to the series' *editio princeps* that "the ancient Mesopotamians had no interest in the scientific study of anomalies to seek out their cause or cure. Their interest was centered on the apodosis, or prediction, and not the anomaly itself. . . . His major concern with the anomaly itself lay in description, and he classified anomalies only to enable himself to find them within the series in order to ascertain their significance."[46] At the time of writing (1970), of course in order to qualify as science, the study of birth anomaly had to have explanatory and causal components as to why such malformations occurred biologically, or from the point of view of the genetics of the developing embryo. As a consequence of this approach, Leichty emphasized the systematic nature of the omen series' compilation of so many malformed births, even though they lacked the notion of physiological deformity having determinant causes. Despite his underlying sense of the nonscientific character of the anomalous birth omens, Leichty rightly observed the importance of the *izbu*'s description and in what sort of interpretation the entity was given, rather in line with Harrison's "science of interpretation which sought to expound the meanings of words and things." The cuneiform study of anomalies at the births of animals or humans was based on the same kinds of relationships between features of other phenomena construed as positive or negative in accordance with an idea of the norm. A binary interpretive system in which right has positive value and left negative, enabled an anomaly on the left side of an "anomaly" to be positive. Thus, "if a woman gives birth, and the right foot is twisted: That house will not prosper." But, "if a woman gives birth, and the left foot is twisted: That house will prosper."[47] De Zorzi discussed the binary oppositions of above and below, front and back, inside and outside, large and small, right and left, male and female, dead and alive, as well as normal and abnormal in the context of the *Izbu* series. She said:

> The most common form of binary opposition in the protases is the opposition right/left. The corresponding apodoses fall into the opposing categories of favorable/unfavorable predictions, thus combining themselves with the protases to form pairs of omens based on a structure of symmetric oppositions. While this organizing principle is in evidence in all divinatory disciplines, in *Šumma Izbu* a malformation on the right side (normally the *pars familiaris*) is considered negative, a malformation on the left (normally the *pars hostilis*), positive. This is owed to the context of the observation: a malformation being *eo ipso* a negative sign, the normal meaning of the opposition right ("favorable") / left ("unfavorable") is inverted.[48]

The same interpretive reasoning is also found with respect to planetary phenomena in which the binary pair bright/dim is applied to planets taken to represent benefic (Venus and Jupiter) or malefic (Mars and Saturn) qualities. Brightness is usually a positive indication, and dimness a negative. The brightness of a malefic planet, either Saturn or Mars, is therefore judged to be negative, while its dimness is positive, and vice versa for the benefic planets Jupiter and Venus.[49] It was no doubt in relation to the degree of brightness that the planets came to represent benefic or malefic qualities in the first place. In relation to the system of analyzing *izbu*s, which were abnormal and unpropitious in and of themselves, in much the same way as malefic planets were "bad" and unpropitious in and of themselves, the parallel in Late Babylonian texts concerning planets shows that such associations had nothing to do with physical essences, but rather with the value of the phenomenon as a portent, propitious or unpropitious. In the context of the planets, nothing can be inferred as to the planets' nature as physical phenomena from the Babylonian standpoint. Far from representing Canguilhem's "machine monster," or Bichat's "mechanical pathology,"[50] malefic planets had "by definition" negative interpretive value within a divinatory schema.

Thus the norm for an *izbu*, as for a malefic planet, was simply that untoward events were signaled in each case. Their appearances could, however, signal propitious events if an inversion of the binary values right/bright = good or left/dim = bad, or the like, occurred. Consistent with Koch's observation of what was of chief interest to the diviner's inspection of the liver, that is, in visual description rather than underlying causes of variation or deformation, the *izbu*s were a focus of interest because they represented a class of negatively evaluated forms.

As in extispicy, implicit in the *izbu* omens was the notion of a norm against which *izbus*, as a class of phenomena, were judged abnormal, and in relation to which the scholarly imagination spun its variations on normal. Indeed, *izbu* omens occasionally use the term "normal" to refer to a part of the newborn not construed as anomalous, however attached it was to the anomaly. Thus:

> If there are 2 *izbu*s and they are normal (*kajamānu*) except the second one protrudes from his (the first one's) mouth: The king will be defeated, and his army . . . his troops and his suburbs will be devastated.[51]
>
> If an *izbu* has 2 heads, and the second one rides (above) the normal (SAG.UŠ) one: Rebels will revolt against that prince.[52]
>
> If an *izbu*'s eyes are normal (SAG.UŠ.MEŠ), but it has a third one on its forehead: The prince . . .[53]

The malformed birth omens attest to the keen study of the morphological variation of animal and human births alike, in which was embedded the idea of a norm. As Canguilhem also pointed out, morphological anomaly is not, by definition, pathology, which seems to have first been conceived by Aristotle in the *Physics*,[54] where a monster is an error of nature. The separation of monsters from prodigies, according to A. W. Bates, did not occur until the sixteenth century: "Neither classical embryology nor its medieval interpretation required it [the separation between monsters and prodigies] to be made. In medieval times monsters were *peccata naturae* (slips of nature) and in common with other rare or unusual happenings they were 'unnatural': to the medieval mind expressions such as *praeter ut in pluribus* (outside that which occurs frequently) and *praeter naturam* (beyond the range of nature) were interchangeable.[55] Consistent with this remark, Lorraine Daston and Katharine Park pointed out that in the Middle Ages "the explanation of monsters by natural causes" could be found side by side with the idea that monsters were divine portents, sent by God as a warning for sinners.[56] In their words, "[monsters] were suspensions of that [natural] order, signs of God's wrath and warnings of further punishment."[57]

Daston and Park's *Wonders and the Order of Nature* devoted a chapter to the phenomenon of monstrous birth in the early modern period, presenting a case study of a monster born in Ravenna of the early sixteenth century. This birth reads like an exaggerated distant cousin of an *Izbu*, as it "had a horn on its head . . . and instead of arms it had two wings like a bat's, and at the height of the breasts it had a *fio* [Y-shaped mark] on one side and a cross on the other, and lower down at the waist, two serpents, and was a hermaphrodite,[58] and on the right knee it had an eye."[59] Shortly after the creature's birth, enemy troops came and sacked the city of its birth. As the contemporary source remarked further, "It seems as if some great misfortune always befalls the city where such things are born; the same thing happened at Volterra, which was sacked a short time after a similar monster had been born there."[60] The similarity between the ancient Near East and western Europe in prognosticating from monstrous births could be due ultimately to the Greco-Roman cultural bridgehead that enabled material of Near Eastern origins to penetrate western Europe.[61] The idea that it was God's work within, or against, nature that provided an explanation for monsters, however, is altogether different from what was conceptually available in cuneiform texts.[62]

During the Neo-Assyrian period, the untoward consequences of *izbu*s, as well as those of many other signs, for those given by the gods in the observable world as well as signs from extispicy,[63] were dealt with by means of rituals called *namburbi*, performed by an *āšipu* or *mašmašu*.[64] As is the case in many technical terms in the Akkadian scholarly corpus, *namburbû* is a loan from Sumerian NAM.BÚR.BI, meaning "its BÚR," with NAM acting to nominalize the verb BÚR. The Akkadian equivalent for Sumerian BÚR is *pašāru* "to loosen," or "undo," "release," even "exorcise."[65] These rituals were used against the evil (HUL/*lumnu*) portended by ominous signs, as well as other potential dangers (e.g., temple offices not carried out properly, headache or disease among the army and horses going on campaign, the effects of sorcery and witchcraft, the evil of fungus).

In his full-length treatment of the *namburbi* ritual,[66] Richard Caplice discussed the semantics of *pašāru* in order to specify the purpose of the ritual. He pointed out that among the fundamental senses of this verb is that of a restoration to order, and is used to mean "untangle" or "unravel," that is, to a state of right order. He cited a passage from the incantation series *Šurpu*, which states that the evil of sorcery may be unraveled by "the symbolic and magically efficacious act of unraveling a tangle of matted material."[67] He concluded that it is this sense that applies in the *namburbi* ritual against ominous signs. What is being untangled, or set to rights again, is the evil portended by signs, not the sign itself.

A separate compilation of ritual actions required on the dates of lunar eclipses throughout the calendar year supports this picture of the ritual ridding of evil due to the occurrence of an eclipse.[68] It begins: "If in the month of Nisannu on the 12th, 13th, or 14th day there is an eclipse of the moon, that evil will not touch the king." However, should an eclipse occur at this time, further action is prescribed: "After the eclipse has become light he prostrates himself to the south. You mix swallow's blood with cypress oil and smear it on the bed. The evil will be released (HUL DU_8-*ár*)."[69] The clearing of the moon's eclipse is a turning point in the procedure. Thus the dispelling of evil and the dispelling of the sign are here interrelated.

In contrast to the hemerological instructions for calendar days, some favorable and some unfavorable, the *namburbi*s targeted evil portended by a wide range of signs. There are, for example, *namburbi* texts against the evil portended by certain birds.[70] Expressions used in the sources are unequivocal in saying their purpose is to "make the evil pass by"—a phrase used as well in the context of lunar eclipses portending untoward events—or "so that the evil not approach (the man)." Indeed, the undoing of HUL/*lumnu*, "evil," is the goal. However, as in the passage quoted at length below, the *izbu* itself will also be destroyed in the process of undoing its evil. Similarly, in reference to a lunar eclipse, a *namburbi* is performed against its evil portent, but the eclipse itself is also in effect undone, as the lunar disk becomes bright again. In each case, the purpose of ritual action is to restore the order of things threatened by the appearance of a bad sign.[71]

The following is a series of *namburbi* rituals for dispelling the evil of an *izbu*. To dispel the portended evil, the supplicant went symbolically before the divine judge, the sun-god Shamash, and by means of plants, the river, or strings of beads, cast it out. In the following translation, the lines of the tablet have been made continuous for space saving, and words or letters inside square brackets means that the broken tablet has been restored:

> If in a man's house there was an *izbu*, whether of cattle, or of sheep, or an ox, o[r a goat], or a horse, or a dog, or a p[ig], or a human being, in order to avert that evil, [that it may not approach] the man and his house:
>
> You go to the river and construct a reed hut. [You scatter] garden plants. You set up a reed altar. Upon the reed altar you pour out seven food-offerings, beer, dates, and *sasqu*-flour. [You set out a censer] of juniper. You fill three *lahḫannu*-vessels with fine beer, and [you set out] . . .-bread, DÌM-bread, "ear-shaped" bread, one grain of silver, (and) one grain of gold. You place a gold ZU on the head of that *izbu*. You string a gold breast-plate on red thread. You bind it on his breast. You cast that *izbu* upon the garden plants. You have that man kneel, and recite thus:

Incantation: Shamash, judge of heaven and earth, lord of justice and equity, who rules over the upper and lower regions, Shamash, it is in your hands to bring the dead to life, to release the captive. Shamash, I have approached you; Shamash, I have sought you; Šamaš, I have turned to you. Avert from me the evil of this *izbu!* May it not affect me! May its evil be far from my person, that I may daily bless you, and that those who look on me may forever [sing] your praise!

You have him recite [this] incantation three times. The man's house [will (then) be at peace] . . . and before the river [you recite] as follows:

[Incantation: y]ou, River, are the creator of ev[erything].

. . . the son of Zerūti, whose [personal god is Nabû, whose personal goddess] is Tašmētu, who [is beset by] an evil *izbu*, is therefore frightened and terrified. Avert [from him] the evil of this *izbu*! May the evil not approach (him), may it not draw near, [may it not press upon him!] May that evil go out from his person, that he may daily bless you (and) those who look on [him] may forever sing your praise! By the command of Ea an Asalluḫi, remove that evil! May your banks not release it! Take it down to your depths!

Extract that evil! Give (him) happiness (and) health! You recite this three times, and cleanse the man with water. You throw tamarisk, Dilbat-plant, *qān šalali*, a date-palm shoot, (and) the *izbu*, together with its provisions and its gifts, into the river, and you undo the offering-arrangement and prostrate yourself. That man goes to his house.

[You string] carnelian, lapis-lazuli, serpentine, *pappardillu*-stone, *pappardildillu*-stone, bright obsidian, *ḫilibû*-stone, [. . .] (and) breccia on a necklace. You place it around his neck for seven days . . . the evil of that *izbu* will be dissipated.[72]

The dissipation of the *izbu*'s evil is expressed here directly as the return to the norm, or the normal, taken to mean happiness and health.

The final point to consider is whether *izbu*s were understood as an expression of divine wrath, in the manner argued in the context of later European monstrous births. This is an interpretation deeply rooted in Assyriological literature, going back, according to Caplice, to Julian Morgenstern in the early twentieth century.[73] We find it again in Stefan Maul,[74] and in Amar Annus's introduction to the volume on divination, where he said, "according to Namburbis, the person to whom the evil omen was announced had to placate the anger of the gods that had sent it to him and effect the gods' revision of their decision," thereby achieving "a correction of his fate which the gods had decreed."[75]

Divine anger is attested as a source of evil, of evil portents, of disease, often as the otherwise inexplicable nature of these things. Omens and ritual texts against evil portents do not always convey this attribution of the ills of humanity directly to the anger of a god, or of gods, although the theme is developed in literary texts, notably in the Akkadian "Poem of the Righteous Sufferer," or *Ludlul Bēl Nēmeqi* "Let me Praise the Lord of Wisdom."

There the first person narrator explains that,

> From the day Bēl punished me, and the hero Marduk was angry [wi]th me,
> . . . Portents of terror were established for me.[76]

In relating further the course of his suffering, the narrator tells of his imploring his personal god and goddess to no avail, and the fact that:

> The diviner could not determine the situation with divination. . . .
> The exorcist with his rituals did not release the divine anger against me.[77]

As the horrible decline progressed, the sufferer nears death and describes his condition:

> My limbs were splayed, just hanging apart.
> I spent the night in my filth like an ox.
> I wallowed in my own excrement like a sheep.
> The exorcist (*āšipu*) was scared (?) by my symptoms.
> The diviner (*bārû*) confused by my omens.[78]

The connection between divine anger and bad portents is thus vividly conveyed in literary texts. However, the *namburbi* rituals for the *izbu* quoted above do not address the anger of the gods, but have the person in whose house an *izbu* appeared present himself before the sun-god and say the incantations that ask Shamash, the divine judge of heaven and earth, to rid him of the evil omen and prevent the evil of that *izbu* from approaching. The ritual does not specifically involve appeasement of the gods either on the part of the supplicant or the *āšipu* in charge of the ritual performance, but consists of various symbolic acts of casting off (onto the plants, into the river) and cleansing, as well as the request through incantation for restitution by Shamash. Shamash is not to be placated in the ritual, but to receive the plea and make a decision.

Just as the *izbu* omens' interpretive structures had to do with norms and abnormality, the ritual against an *izbu*'s portended evil acted to remove or keep evil away and reestablish the norm. That the *izbu* was a sign of evil to be dispelled by the judicial decision of the sun-god is clear in the Sumerian incantation included in another *namburbi*: Incantation: The sign that is evil shall not approach the man! At the word of Utu (= Shamash), bailiff of the gods, who defeats the sign that is evil for man, (who defeats) anything (evil) that approaches—though the man (lit. seed of man) himself may not be aware of it—it shall not approach him to his detriment! Like water—water poured into the canal—his punishment shall not approach him! His evil shall not hover about him! (These are) the words of Enki and Asalluḫi."[79]

These texts are cited in order to show that portents, rituals against certain portents, and ideas about the instability of the divine-human relationship present a complex picture. The *namburbi* against the *izbu* does not directly address divine anger, but does appeal to the sun-god to right the wrong signaled by the *izbu*. This may not be an argument against the idea of divine anger in the background of the *izbu* omens, but the ritual response to the *izbu* does not address it. The question of the relation of ominous signs to the wrath of gods can be posed with respect to other kinds of omens as well, such as those from lunar eclipses. The *namburbi* for the evil of an eclipse indeed offers another perspective.

A lunar eclipse was the manifestation of a disturbance of the moon-god, often expressed as that god's being in mourning or emotional distress (*lumun libbi*, literally "evil of the heart"). In some contexts, however, *lumun libbi* means "anger."[80] The afflicted person is required to set up an altar to the moon-god, present offerings, and, prostrated, recite a prayer three times before the moon/moon-god, as the celestial body and the god are, for the purpose of the ritual, one and the same: "May the great gods make you bright! May your heart be at rest! May Nannar of the heavenly gods, Sin the exorcist, look (hither)! May the evil of eclipse not approach me or my house, may it not come near or be close by, may it not affect me, that I may sing your praises and those who see me may forever sing your praises!"[81] The text adds for the *āšipu*: "You have him recite this, and you undo the offering arrangement. You perform the [ritual] for the evil of signs and portents, and the evil of eclipse will not approach him."[82] The exhortation in the prayer for the quieting of the moon-god's heart is a clear reference to *lumun libbi*. The connotation in astrological contexts is usually of the moon's grief, not his anger, and even more particularly, *lumun libbi* seems to be associated with an eclipsed moon that sets while still dark.[83] Marten Stol adduced the following passage in a late commentary to the incantation for Geme-Sin, the cow of the moon-god, Sin, who was giving birth:[84] "Darkened (*nandaru*) means carrying water means to embrace means pure of water means crown of splendor (i.e., the full moon), because É.LAM$_4$.MA means house of four." The word *nandaru* is elucidated in the commentary based on the homophony of the various roots from which it could be derived. As *nanduru* from *adāru* it could mean "darkened," but from *edēru* it means to be embraced, perhaps a reference to the sexual union of Sin and Geme-Sin leading to her pregnancy. It is also given an equivalent *e-lal*, perhaps Akkadian *elallu*, a poetic word for one who carries water, said of clouds, from the Sumerian E_4 "water" and LAL "to carry," perhaps a reference to the appearance of the moon in eclipse as covered by a cloud, hence carrying the water that was the source of his tears. Stol's interpretation leaned toward the image of the weeping moon in an eclipse, and suggested that "pure of water" (*ellammê*) referred to an eclipse.[85] The mourning of the celestial bodies sun and moon is attested as a metaphor for the state of being in eclipse, as in the omen where Shamash cries at the time of the decision of the Anunnaki, that is, at the end of the month, in what is a metaphorical—and anthropomorphic—expression of a solar eclipse.[86] What the commentary does not entertain is *nanduru* from *nadāru* "to go on a rampage," also spelled the same as *na'duru/nanduru* "to be apprehensive," or *na'duru/nanduru* "to become eclipsed." It is rather common to have the clause "the moon in his eclipse" expressed as "the god (meaning Sin) in his eclipse" (*ilu ina nandurišu*).

In any event the eclipse was construed as a sign of the moon-god's state of mind, which had to be restored to its normal state of brightness (= happiness) by an offering and the recitation of the prayer. Here in another prayer, the supplicant says: "I am afraid, I am upset and terrified. From the evil of the lunar eclipse, from the evil of the solar eclipse, from the evil of the stars of Ea, Anu, and Enlil (meaning the stars of the entire heaven)."[87] This explicit connection of evil with celestial signs of eclipse and with the divine fixed stars strikes a contrast to the *izbu*, which is not the manifestation of divinity in the same way as a lunar eclipse is the manifestation of the moon-god Sin, or a star may be the manifestation of a particular god.

The foregoing discussion aimed to show that *izbu*s, premature or malformed births, were not conceived of as monsters, and the word *izbu* does not signify a "monster," except in the classical sense of a portent. The evil portended by an *izbu* is not fixed in the *namburbi* ritual by appeasing divine anger. From the ritual evidence, therefore, the question of the conception of that portent as a result of divine wrath, or as punishment for human wrongdoing, is left open. An *izbu* was not an error of nature, or a deviation of nature from its own laws, but a portent in the same way as were other ominous signs in cuneiform, that is to say, whether a representation of or a deviation from a conceived norm, signs were the stuff of divine communication.

Variation with respect to a conceived norm made signs ominous, but not all omens in the cuneiform world were anomalies. Phenomena that fell within norms were also portentous. Again, where Canguilhem placed the conception of the monster, or the monstrous, in the context of living phenomena, the cuneiform material leveled the playing field for all ominous phenomena, not reserving "monstrosity" for the living, indeed, not expressing the notion at all.

CELESTIAL SIGNS AND ASTRAL PHENOMENA: REGULARITY AND ANOMALY

Norms for cyclical astral phenomena were defined differently from those in the biological realm, where the definition of health, or "normal," permits a good deal of variability before one begins to speak of a defect or an anomaly. Evidence of this kind of standard of measure by the healthy appearance of the liver or other organs is found in extispicy omens. In the *izbu* omens, the standard itself was anomalous, as just discussed.

Periodic phenomena in the heavens, on the other hand, are amenable to counting, or other arithmetical methods by which to construe regularity. As a result, the term meaning "normal" in astral omens was *minītu*, from the verb *manû* "to count." Thus the day when sun and moon were in opposition on an anomalous day of the lunar month was expressed as *ina la minâtišunu*, literally, "not according to their (calculated) norm," or a lunar eclipse might occur *ina la minâtišu* "not according to his (the moon's) (calculated) norm."[88] A similar expression was constructed with the word *simanu* "time," that is, "not according to its time," where the sense of the anomalous, unseasonal, or unexpected appearance of a celestial body was conveyed.[89] A third term, *adannu* "a period of time of predetermined length,"[90] also in the sense of "appointed time," or even "normal time," is also found in expressions for an astronomical phenomenon occurring at an irregular time. The first tablet of *Enūma Anu Enlil* has omens for the appearance of the moon using each of these expressions, with the unfortunate consequences of the decline of the market and the destruction (scattering) of the city.[91]

Such references relate to Canguilhem's first notion of normal as the measured or calculated mathematical mean. But the function of the norm in the various cuneiform divinatory contexts, astral, extispicy, and *izbu* omens alike, was to differentiate the meaning of those signs by deviations from "normal." Where Canguilhem focused on the development of ideas of pathology in life forms, he noted the problem with such a notion in physics and mechanics. While celestial divination was oriented to phenomena that would also be of interest to later physics and mechanics, the interest in them was as signs, in the same framework as the liver and the *izbu*. From the point of view of divinatory knowledge there was a unity between the signs in heaven and the signs on earth; all belonged to the category of signs. The other aspect of Canguilhem's investigation into the normal were his remarks about the ideal, which, as cited before, were described as "a positive principle of evaluation, in the sense of a prototype or a perfect form." David Brown has explicated this idea in early astronomical and celestial divinatory texts concerning cyclical astronomical phenomena and the arithmetical schemes devised for reckoning with them in a divinatory context.[92]

From the earliest periods an interrelated group of ideal units of time reckoning was devised for accounting purposes, and because those units came to undergird early Babylonian astronomy as well as the tradition of *Enūma Anu Enlil* omens, they would remain in the cuneiform scholastic tradition for two millennia.[93] This group of ideal units focused on the 360-day year of twelve thirty-day months. Of the local Sumerian calendars in the Ur III period (ca. 2100–2000 BCE), where real month lengths varied, calendar months in the city of Nippur became standard and were later taken over as the month names of the ideal calendar (twelve thirty-day months = one ideal year) common to the scholarly traditions of astrology (celestial and natal divination) and astronomy (*MUL.APIN*, Astrolabes) prior to circa 500 BCE and even later in some cases.

In the Old Babylonian period, the variation in length of daylight was understood as deviations from the ideal dates of the equinoxes. The earliest evidence for the quantitative model for daylight length is found in an Old Babylonian text (BM 17175+).[94] In four sections, one for each schematic season, the text gives the model as follows:

> [On the 15th of Addaru, 3 (minas, or 3,0 UŠ) are a wa]tch of the day, 3 (minas, or 3,0 UŠ) are a watch of night; [Day and night] are equal. [From the 15th of Addaru to] the 15th of Simanu is 3 months. [On the 15th of Simanu, the night] transfers 1 (mina, or 1,0 UŠ) of the watch to the day. [. . . 4 (minas, or 4,0 UŠ) is the wa]tch of the day, 2 (minas, or 2,0 UŠ) is the watch of the night.[95]

This model of the ideal year assigned the equinoxes and solstices to the midpoints, or fifteenth day, of months XII, III, VI, and IX. For each schematic season, or quadrant in the ideal year, the length of daylight shifted by 1 unit. Therefore, from vernal equinox to summer solstice, the length of day increased by "1," from summer solstice to autumnal equinox, daylight decreased by "1," and so on, producing a model for the change in the length of daylight in which the ratio of longest to shortest day length was 2:1. Thus:

3 (VE) + 1 = 4 (SS) − 1 = 3 (AE) − 1 = 2 (WS) + 1 = 3 (VE). This scheme is not practical for the geographical latitudes of Mesopotamia, but it is the simplest, indeed most elegant, way to model the experience of increasing and decreasing durations of daylight around two extremes (summer and winter solstices), provided the model is constructed on the ideal year. The mean value was expressed in sexagesimal notation as the number 3, that is, 180 (3 × 60), and represented one-half of the circle of the day (360 degrees) when daylight and night were of equal length (each 180 degrees).

The Astrolabes, discussed before, arrange three groups of heliacally rising stars month by month together with numerical values for the length of day in those months. This group of texts provided the full complement of numerical values that made up the model attested in the Old Babylonian example cited above. In table 4.1, "C" designates the value for length of daylight, taken as constant for the duration of the month. Again it is clear that the mean value of the table is 3 (= 3,0), representing the length of daylight (or night) at the equinoxes.

The text of *MUL.APIN* also shows that the Old Babylonian model for variation in daylight (table 4.1) was still being transmitted. The text says, for example: *ina Nisanni* UD.15 3 *mana maṣṣarti mūši* 12 UŠ *napāḫu ša Sin* "On the 15th of Month I, a nighttime watch is 3 minas; 12 UŠ the (daily retardation of the) rising of the moon.[96] This gives the same value for daylight length at the vernal equinox, but it occurs in the first month *Nisannu*, rather than the twelfth month *Addaru*. The remaining cardinal points of the

TABLE 4.1. Astrolabes' scheme for variation in daylight length

Month	*C (in mana)**	*C in UŠ**	*= Hours*	*Cardinal Points*
XII	3;0	3	= 12hr	
I	3;20	3,20	= 13hr 20′	Vernal Equinox
II	3;40	3,40	= 14hr 40′	
III	4;0	4	= 16hr	
IV	3;40	3,40	= 14hr 40′	Summer Solstice
V	3;20	3,20	= 13hr 20′	
VI	3;0	3	= 12hr	
VII	2;40	2,40	= 10hr 40′	Autumnal Equinox
VIII	2;20	2,20	= 9hr 20′	
IX	2;0	2	= 8hr	
X	2;20	2,20	= 9hr 20′	Winter Solstice
XI	2;40	2,40	= 10hr 40′	

*The relation between these measures is 1 *mana* = 1,0 UŠ. Mana was a unit of weight, for measuring water into the water clock. 1 UŠ = 4 time degrees.

year were also shifted up one month, from XII to I for the vernal equinox, as just seen, from III to IV for the summer solstice, and so on. This shift in the calendrical reckoning of the cardinal points did not alter the underlying schematic model for daylight length variation.

Another section of *MUL.APIN* clarified the scheme for an entire ideal year (*MUL.APIN* II ii 43–iii 12). The section not only spelled out the lengths of night for each ideal month but also included the value for the visibilities of the moon, whether from rising or before setting. *MUL.APIN*'s interest in the night lengths and visibilities of the moon is followed by a short passage explaining the calculations for the duration of lunar visibility using a "difference" coefficient (*nappaltu*), for example, 40 NINDA *nappalti ūmi u mūši ana* 4 *tanaššīma* 2,40 *nappalti tāmarti tammar* "multiply 40 NINDA, the 'difference' of daylight and night, by 4 and you will find 2,40, the 'difference' of the visibility of the moon" (*MUL.APIN* II iii 15).

The same numerical values for daylight lengths according to the model of *MUL.APIN* and the Astrolabes also underlie the calculation of the duration of visibility of the moon at night, found in Tablet 14 of the omen series *Enūma Anu Enlil*.[97] The duration of lunar visibility was of course related to the length of night, and the value given as the IGI.DU$_8$.A = *tāmartu* "visibility" of the moon is calculated as 1/15th of the length of night. For example, in Month I day 1, day = 3 and 1/6 and night = 2 and 5/6. On this day the IGI.DU$_8$.A of the moon is given as 11;20, the result of dividing the length of night by 15. For an equinoctial day, for example, Month I, day 15, the IGI. DU$_8$.A of the moon is given as 12. Night length at the equinox = 3,0 (moon rises at sunset and rises at sunrise and is visible the entire night, for 180°). The value for IGI.DU$_8$.A is 12, which is 1/15 of 3,0 (= 180°). The length of night was always the complement to the length of day, where on any given night of the schematic year, day + night = 12 *bēru* = 24 hours = 360 UŠ (360 degrees of time = 24 hours). Values were given in monthly intervals, but a statement from *MUL.APIN* confirms that these values could also be interpolated on the basis of semimonthly values for the length of the day: "The Sun which rose towards the North with the head of the Lion turns and keeps moving down towards the South as a rate of 40 NINDA per day."[98] Forty NINDA per day is the result of the regular increments or decrements of 10 units each 1/2 month, that is, in 15-day periods. Further interpolations could be made by dividing by 30 (the number of days in a schematic month) the semimonthly differences between values for the daily retardation of the moon throughout the schematic year, tabulated in *Enūma Anu Enlil* Tablet 14.[99] These quantitative descriptions were results of modeling, not measuring, the variation in length of daylight and the underlying structure was the ideal year, twelve thirty-day months or 360 days.

It seems to me that the numerical mean value in this ideal scheme, the value 3 (3,0 = 180), as the representation of one-half of the circle of the day, was not simply a derived mean value from the schemes for daylight length and duration of lunar visibility, but played a determining role in the construction of those schemes. Similarly, for use in later Babylonian mathematical astronomy, mean values for units of time such as the mean synodic month and the mean *tithi* (1/30 of a mean synodic month), as well as mean values for the computed periods of the synodic arc and synodic time, determined the structures of lunar and planetary theory.

According to Brown's understanding of ideal schemes, the numerical value assigned to the ideal, and deviations from the ideal, may have functioned as the basis for interpreting propitious and unpropitious signs. The following omen, repeated frequently in the reports to the king from the scholars, attests to the importance of the notion of a "normal" daylight length for a certain date in the lunar calendar, presumably derived from the scheme discussed before: "If the day becomes long in accordance with its count (i.e., reaches its normal length): a reign of long days."[100] Another report contains the comment: "[The normal length] of the month (means) it will complete [the 30th day]," showing that for the diviners the thirty-day month was indeed construed as a "norm."[101]

The principle of established norms was not only consistent among the various domains of signs, it was also instrumental to the entire system of divination through omens. This, no doubt, was one of its influential features as divination circulated within the geocultural sphere of the ancient Near Eastern and Mediterranean *oikoumene*.[102] Brown's insight that the categories by which the heavenly phenomenal world was structured in celestial omens, were, in his words, "devised in order to make the sky above interpretable," and that "it [the phenomenal world] was categorised in this manner *in order* that it could be encoded with signs,"[103] is much like that of Seneca concerning the Etruscans: "The difference between us and the Etruscans . . . is the following: that whereas we believe lightning to be released as a result of the collision of clouds, they believe that clouds collide so as to release lightning; for as they attribute all to the deity, they are led to believe not that things have a meaning in so far as they occur, but rather that they occur because they must have a meaning."[104]

That this is the case can be applied more widely within cuneiform knowledge corpora. As noted above, phenomena of the liver and exta that deviated from normal were studied as ominous signs. Those appearances that fell within the range of normal, were also counted as omens, signaling generally propitious events. Even some *izbu* omens had elements described as "normal." With respect to periodic celestial phenomena, Brown said, "the categorising of the temporal component of a phenomenon into an 'ideal period' and its subsequent elaboration into 'ideal schemes' made not just the infrequent or exceptional events in the sky open to interpretation, it ensured that the regular running of the universe could be deciphered. . . . The ideal schemes made possible a comparison between observed reality and an anticipated 'ideal.' They produced *anomalies* and *coherences* from the universe's regular and repeating behaviour, since on occasions the stars did *not* rise in their ideal months, the month was *not* 30 days long, and so forth."[105] This observation supports the main thesis here, that the investigation of what Brown called "observed reality," that is to say, the phenomena of the ancient scribes' perception, experience, and imagination, was undertaken not so as to understand "observed reality," that is, nature as such, or what we might call the workings of a structured world, but to interpret the perceived, experienced, or imagined phenomena for the purpose of divination, that is, to know what things meant. In the process of pursuing this goal, much was known of the world, including how parts of it "worked," such as heavenly phenomena, medicinal plants, or magical stones.

A substantial proportion of the cuneiform scholarly record represents the detailed study of what was found in accordance with Babylonian cultural (divinatory) and cognitive (rational) values to be signally important parts of the external world. Just how these two dimensions came together in that system by the use of associative reasoning will be the subject of the next chapter. The resulting system of knowledge about physical phenomena, to its fullest extent, included not only the calculation of periodic astral phenomena but also encompassed the other phenomena belonging within the scope of Assyro-Babylonian divination, as well as its supporting system of magic. Underpinning these various systems of knowledge were a variety of normative standards, that is to say, definitions of normal in the domains of the phenomena of interest, the ominous exta, births, and astral phenomena, that established an order of things.

From a hypothesis that the notion of a universal nature did not provide the foundational framework for cuneiform knowledge, it emerges that despite the "lack"—as we might conceive of it—of such a universal epistemic or a uniform physical framework, the sense of norms and anomaly, in other words, a way to establish order, characterized the systematic and analytic style of cuneiform divinatory knowledge. Such a focus on norms in terms of an ideal or a mean has an obvious kinship with natural inquiry seeking to find structures of order, regularity, and irregularity.

Classification of the domain of inquiry as well as the appropriate parts of that domain, however, responds to the particular needs of a community of knowers. Nature as such came to provide the domain of scientific inquiry in a context wholly outside of the cuneiform texts. But once that domain was conceived, it accommodated all sorts of objects of inquiry subject to history and culture, including the examination of qualities in Renaissance natural magic as well as of quantities in early modern mechanics. It gave rise to the idea of the "preternatural."[106] The point is, it was nature, consciously conceived, that provided the structure for those varied inquiries.

It seems critically important not to attribute the desire to come to an understanding of the order of things in the cuneiform texts to the desire to know or classify nature, or any of the kinds of phenomena defined in terms of a relation to nature, as would apply to the supernatural or the preternatural. Despite this discontinuity, the goal to find or create order in the world notably relates the culture of the cuneiform scribes to later intellectuals who formulated that same goal in terms of the conception of nature. This makes for an intellectual kinship while remaining two quite different responses to the world.

As the evidence in the foregoing chapter has illustrated, the multitudinous variability of the world can be successfully classified in ways not dependent upon the idea of nature as we currently see it, or as it was seen in Greek natural philosophy, or indeed as it came to be understood in the early modern and modern periods. The beginnings of such a definition of nature as a domain did not take full effect until the Hellenistic period within Greek philosophy, especially in the writings of Aristotle. The idea that science is oriented toward the natural world is the result of and entirely dependent upon that particular historical development. The cuneiform scribes occupied a world before that development took place.

The understanding of how the world works, what the world is, or how to find ways to bring parts of it within bounds of mathematical description can all be found in cuneiform knowledge. Its use of categories and principles of ordering is part of the way cuneiform knowledge participated in the epistemic process of science, perhaps even the essence of that participation. This might seem clearest when some of those ordering principles—namely, those found in the heavens—took mathematical form. But the evidence shows that divinatory knowledge and the practice of interpreting ominous signs were also based on ordering principles embodying norms and anomalies. Those principles, to be considered more directly in chapter 5, were put into operation in the system of divinatory knowledge by associations that were deemed to link events to events, particular to particular, in a way that could be used for making forecasts of future events.

If a relationship of cuneiform knowledge to what we understand as science can be argued on the basis of its concerns for order, norms, and anomaly, the possibility is thereby open that the history of science can accommodate further variations in the relationship of knowledge to its object(s). The fascination the cuneiform knowledge corpus holds for the modern reader is in the unique way the human subject sought to know select phenomena of the world as objects of inquiry, analysis, and interpretation. I do not see the necessity of pressing that relationship into the familiar subject/object axis of observer and nature. As the evidence presented here shows, the selection of objects of inquiry was based on other criteria than those that become available with the idea of nature. One of the anchors of cuneiform knowledge was the body of phenomena classified as signs, though ominous signs did not exhaust categories of knowledge within cuneiform texts. The relationship of interpreter to signs was, however, a significant dimension of cuneiform knowing. Signs and portents were thus indication of the structure—the scaffolding—of the scribes' world expressed as "the designs of heaven and earth" (*uṣurāt šamê u erṣeti*) or "the enduring designs of the gods" (*uṣurāt ilī kīnāti*). These expressions, however, do not correspond to, or present a substitute for, the word or the conceptual complex "nature."

Another significant relationship could be identified as the subject/object pole of human and divine. Knowledge of the gods' plans, "the designs of heaven and earth" (*uṣurāt šamê u erṣeti*), was certainly more than mere explanatory rhetoric, much as the relationship of God to nature was more than mere rhetoric in later natural philosophy. More than that, however, the fundamental coordinates of knowledge, as explored in the evidence presented here, as in science as we understand it, were formed around the axis of the human knower and the ordered phenomenal world, but specifically, a world in which phenomena were of interest for their potential to show change, deviation, malformation, or anomaly as against norms, or, to be serviceable in reestablishing norms in the realm of the divine as in the human.

PART III

Rationality, Analogy, and Law

CHAPTER FIVE

The Babylonians and the Rational

kullat ṭupšarrūti "Everything Pertaining to the Scribal Art"

Any talk of "the human mind" can only be presumptive, and will reflect no more than our partial knowledge of styles of rational discourse in the long register of traditions and forms of social life.

RODNEY NEEDHAM, *Belief, Language and Experience*

Irrationality, like rationality, is a normative concept. Someone who acts or reasons irrationally, or whose beliefs or emotions are irrational, has departed from a standard; but what standard, or whose, is to be the judge?

DONALD DAVIDSON, *Problems of Rationality*

ACROSS THE GREAT DIVIDE

In works from as long ago as Frankfort and colleagues' *Intellectual Adventure of Ancient Man* (first edition, 1946), or as recently as Robert Bellah's monumental *Religion in Human Evolution: From the Paleolithic to the Axial Age* (2011), Mesopotamian culture has not survived comparative analyses of rationality. For both accounts rationality is the marker of an evolutionary success story rather than being one element in the sometimes contradictory aspects of any culture's thinking (our own included) about the world in which it exists. In neither of these works has the thinking of "the Mesopotamians" crossed the great cognitive breach separating Us from "Them." "They" never matriculated.

The Frankforts' *Intellectual Adventure of Ancient Man* mounted an inquiry into the rational capacity of "ancient Mesopotamian man" and found it to be less developed than in later historical societies. In accordance with the standards of investigation of their time they gauged the difference in the kind of thought characteristic of ancient peoples in terms of criteria for rationality exemplified by Western logic and science, and, as well, in terms of whether they had separated nature from culture. They created a false dichotomy between the propensity either to reason one's way to an understanding of nature (science) on one hand, and on the other, to tell stories about gods (myths) as a substitute for such reasoned understanding—all that one can manage with a mythopoeic mind.

On the basis of this dichotomy the Frankforts did not find rationality in ancient Near Eastern society figuring prominently in its "intellectual adventure." Nor were they the only ones with such views of cuneiform sources. In the context of cuneiform medicine too, as Markham Geller pointed out, historian of medicine Henry Sigerist, writing in the middle of the twentieth century, assessed it to be "dominated by magic and religion."[1] No doubt the Frankforts and Sigerist would have readily conceded that ancient Near Eastern societies and states were managed on the most rational of bases, requiring the exercise of reason no less than in any other society or at any other time in history. But for scientific rationality the stakes were higher. This chapter is not so much a conspectus of but a response to some of the ways in which cuneiform texts have been evaluated with regard to the question of the rational, specifically, the rational and its relation to scientific knowledge.

The subject of the rational is of dauntingly vast proportions. There are many conceptions and aspects of rationality,[2] most of which will remain outside the bounds of the following discussion. Here I take as most useful for historical evidence the conception of rationality as conformity to normative standards.[3] While this is not an exhaustive criterion, it would be presumptuous to attempt anything more. Viewing rationality as conformity with a given normative system places it in the sphere of reasoning and cognitive style, rather than, say, decision making, and reasoning is what is at stake here.

Ever since Aristotle's *Posterior Analytics*, for example, I.2, 71b, reason by demonstration (*apodeixis*) has been the critical underpinning of scientific knowledge (*episteme*). Thus, in the *Nic. Ethics* VI.3.4, "Scientific knowledge is, then, a state of capacity to demonstrate, and has the other limiting characteristics which we specify in the *Analytics*, for it is when a man believes in a certain way and the starting-points are known to him that he has scientific knowledge, since if they are not better known to him than the conclusion, he will have his knowledge only incidentally. Let this, then, be taken as our account of scientific knowledge." From an Aristotelian standpoint, then, a particular form of reasoning by deductive demonstration produces that special form of knowledge that is science. The defining character of scientific knowledge as certain, justified, and true stems from its logical relationship to first principles, or premises. Logical consequence, therefore, became a key ingredient of rational knowledge.

While the Aristotelian model of knowledge was debated by Christian natural philosophers until the sixteenth century, to be mostly superseded in the seventeenth century, the logical criteria for knowledge set out in the *Posterior Analytics* still fundamentally define scientific rationality for us, serving to create the hierarchy that applies to knowledge as against belief. In 2010 a special issue of *Synthese* was devoted to "the classical model of science," described as "a millennia-old model of scientific rationality."[4] The papers there focused "on the role, the significance and the impact of the traditional axiomatic ideal of scientific knowledge in the history of philosophy from Aristotle to the twentieth century."[5] In their introduction, the editors, Arianna Betti and Willem R. de Jong, defined this model of science as a system of propositions and concepts satisfying seven conditions.[6] Their seven conditions systematized the ideal standards of scientific rationality, viewed as composing "the core of what a proper science should look like according to that ideal," and which, they said, had "remained remarkably constant for more than two millennia."[7] This classical model is inferred and reconstructed specifically from the history of Western philosophical thought about science, scientific method, and scientific explanation. Appropriate objects of knowledge, and the conditions for something to be an object of scientific knowledge, as established in this model of scientific rationality, belong to a discourse that sought to reason deductively from first principles.[8]

Aristotle classified all human beings as capable of reason, but on a sliding scale, from men to women to slaves.[9] And the capacity for reason, in a particularly Aristotelian philosophical sense, has frequently served as a comparative measure of intellectual development. Ernest Gellner said, "one persistent attempt to find a thread in the history of mankind focuses on the notion of Reason—Human history, on this view, is the unfolding of rationality. Human thought, institutions, social organization, become progressively more rational."[10] During the early to mid-nineteenth century, when this progressive view of human history was advancing, and with it racial hierarchies were becoming biologized, living, non-Western, nonindustrialized, nonwhite societies were judged to have remained behind in their rational capacities because they had not independently produced logic and science, favoring instead the irrational thinking of magic and ritual. Some of that research, inspired by scientific racism's founder, Samuel George Morton, took craniometry as its empirical basis, thus assuming a scientific veneer for what was in fact pseudoscientific and racist theory.[11] Dead, non-Western and premodern societies were later subjected to evaluation in terms of progressive ideals to similar effect.[12]

The question of how characteristic rationality is of the human species is long-standing and of interest from many perspectives. As Stephen Stich remarked:

> Aristotle thought man was a rational animal. From his time to ours, however, there has been a steady stream of writers who have dissented from this sanguine assessment. For Bacon, Hume, Freud, or D. H. Lawrence, rationality is at best a sometimes thing. On their view, episodes of rational inference and action are scattered beacons on the irrational coastline of human history. During the last decade or so these impressionistic chroniclers of man's cognitive foibles have been joined by a growing group of experimental psychologists who are subjecting human reasoning to careful empirical scrutiny. Much of what they found would appall Aristotle. Human subjects, it would appear, regularly and systematically invoke inferential and judgmental strategies ranging from the merely invalid to the genuinely bizarre.[13]

In view of Stich's clinical studies, the capacity for rational *and* irrational thought and action looks like the privilege of every normally functioning human being. Suddenly, rationality in human history does not appear to be inherently progressive, but simply variable according to contexts. Accordingly, just as it should not have surprised anyone, though it did, that E. R. Dodds would deliver his Sather lectures in Berkeley on the Greeks and the irrational in the late 1940s,[14] neither should it come as a surprise to speak now, though it might, about the Babylonians and the rational.

Two interrelated cognitive-historical projects bear on the use of evidence from the ancient Near East in a comparative analysis of rationality. The first is not explicitly, but certainly implicitly, progressive, and the second has great affinity for the progressive account of the history of humanity as one of unfolding and ever-advancing rationality. The first is the identification of cultures that either do or do not exhibit the kind of rational thought associated with science or philosophy. Those that do, by virtue of their cognitive style, join ranks with our own and are thus essentialized, and we with them, as scientific and rational. Those that do not, fail to consider the world in terms of nature and its laws. In such cultures the understanding of physical causality is undermined by an irrational belief in the supernatural. The boundary between rational and irrational thinking is judged by the rigorous yet limited criterion of whether knowledge is logically consequent (deductive) or not, thus appealing to Aristotelian methods of analysis. The second and related project looks for an explanation of the similarity between Others and Ourselves in terms of a watershed moment that cognitively changes everything forever. Such moments have been found in social behavior (as in the use of writing), spiritual attitude (as in being critical of an established authoritative worldview), or even cognitive ethos (as in the triumph of science over revelation).

References to a cognitive shift have been made by many, from Claude Lévi-Strauss's Savage Mind to Ernest Gellner's Big Ditch, and critiqued as well, from Jack Goody's Grand Dichotomy to Bruno Latour and Philippe Descola's Great Divide. The shift has sometimes been identified in the fifth century BCE, sometimes not until the seventeenth century CE. Most have been plainly critical of the dichotomizing strategy itself, specifically the division between archaic (whether in ancient or in modern ethnographic society) and modern, in terms of the presence or absence of rationality. Different solutions have been proposed. Gellner saw two kinds of rationality, one for each side of the ditch, described in terms of Durkheim and Weber, their sociological forefathers. Thus, "the two great sociologists of rationality were really concerned with radically different species of it . . . Durkheim meant by rationality the fact that men are bound, in thought and conduct, by the concepts shared by a culture. He found the solution to his problem in ritual. Weber's problem was the emergence of a distinctive style of rationality, rule-abiding, capable of instrumental efficiency in disregard of tradition."[15] Gellner and Lévi-Strauss each placed the changeover to scientific rationality in modernity as a result of shedding all forms of causality but the physical for an understanding of a fully autonomous realm of nature and its laws.[16] Latour argued that we are not, nor have we ever been, modern.[17] All the earmarks of the modern point of view, particularly the separation of nature from culture, he said, never really happened. We only *claim* "to mobilize nature,"[18] but our views are as much a function of a cultural filtration system as anyone's in the archaic past. Descola put the distinction between nature and culture as only one way of engaging with the world—namely, the modern Western way, and that we are mistaken to take naturalism as universal.[19] As he said, nature "has been constructed little by little as an ontological tool of a particular kind, designed to serve as the foundation of the cosmogenesis of modernity."[20]

Certainly, if rationality across the board historically is to be defined in terms only of Western epistemological, ontological, and methodological naturalism, in which, as David Papineau put it, "reality is exhausted by nature,"[21] then there is nothing to discuss here. This is clearly not the case.

By way of a final point of introduction, the question of positioning the ancient Near East on a cognitive historical map, with respect to a place called "Western" as well as to a time called "premodern," raises the specter of intellectual, or cognitive, imperialism. What Sheldon Pollock said in his critique of the history of politics in premodern India is relevant here: "Already a generation ago historians of Asia were attacking what they called 'intellectual imperialism' in the imposition of Euro-American models and presuppositions for studying non-Western polities. Yet the old critique was itself contradictory. At the same time as it challenged the epistemic domination of the West it . . . rejected as futile . . . and as pernicious any categorization that renders the non-West radically different. While the phrase 'intellectual imperialism' may have a dated ring today, the problem it flags has not vanished, and the contradictions of the critique are those we are still living with."[22] The cost of granting the ancient Assyrians and Babylonians their cultural and cognitive individuality, as against Western and modern counterparts, cannot be to typify their reasoning as irrational and deny them participation in the history of science. In other words, we should not have to make their thought like ours for it to be recognized as rational and for their ideas of knowledge to be participant in what we call science.

Here I propose, with respect to cuneiform texts, to look at contexts of rationality in scribal scholarship. My aim is to steer clear of the history of mind and turn instead to intellectual history, whereby the cuneiform world might be considered without the intervention of the Axial Age, the Great Divide, the Grand Dichotomy, or the Big Ditch. My way in, however, is by a short detour through the alleged irrational.

THE GREEKS AND THE DECLINE OF RATIONALITY

In his celebrated Sather lectures of 1949, E. R. Dodds debuted the irrational Greeks.[23] He explained the impetus for his study of the Greeks and the irrational as stemming from an encounter at the British Museum, when another visitor in the gallery confessed that the Elgin Marbles did not move him because they were "all so terribly rational."[24] Dodds admitted that compared with the art of some other cultures, ancient, anthropological, and even modern, "the art of the Greeks, and Greek culture in general, is apt to appear lacking in the awareness of mystery and in the ability to penetrate to the deeper, less conscious levels of human experience."[25] Dodds set out not to deal with Greek science, still for him exemplary of rational reasoning, but with the Greek experience of the divine, of possession, dreams, and the human psyche. For Dodds, science epitomized rational thought, making its appearance among the intellectuals of the late fifth century and again, briefly, in the Hellenistic period (third century BCE) among a wider social class, before things deteriorated.

According to Dodds, pseudoscientific (his term) writings of later authors, often attributed to divine revelation or fictive Eastern authorities (Chaldeans, Egyptians, Persians), injected irrational beliefs "of the Hellenistic masses"[26] into the thinking of educated Greco-Romans. The attribution of a decline in rational science to influence from "the Orient" was rooted much earlier. In the early twentieth century Lynn Thorndike pointed to Franz Cumont as being guilty of that "glib assumption," as he qualified his own position on magic in Greek religion, literature, and history:

> But they [the Romans], too, were supposed to have risen later under the influence of Hellenic culture to a more enlightened stage, only to relapse again into magic in the declining empire and middle ages under oriental influence. Incidentally let me add that this notion that in *the past* orientals were more superstitious and fond of marvels than westerners in the same stage of civilization and that the orient must needs be the source of every superstitious cult and romantic tale is a glib assumption which I do not intend to make and which our subsequent investigation will scarcely substantiate. But to return to the supposed immunity of the Hellenes from magic; so far has this hypothesis been carried that textual critics have repeatedly rejected passages as later interpolations or even called entire treatises spurious for no other reason than that they seemed to them too superstitious for a reputable classical author. Even so specialized and recent a student of ancient astrology, superstition, and religion as Cumont still clings to this dubious generalization and affirms that "the limpid Hellenic genius always turned away from the misty speculations of magic."[27]

Dodds's prime example of the Hellenistic trend toward the irrational was astrology. He quoted Gilbert Murray, who described astrology as having fallen "upon the Hellenistic mind as a new disease falls upon some remote island people."[28] While Dodds agreed with Murray on the nature of astrology, he noted that there was skepticism (that is, signs of rationality) in that period among Greek intellectuals about the precepts of astrology, for example, Eudoxus, as reported by Cicero in *De Divinatione* 2.87, and he (Dodds) also suggested that Berossus (280 BCE) failed to impress his audience. By Dodds's account it was later, in the second century, that Babylonian astronomy, and with it astrology, was taken up with enthusiasm, due to an intellectual climate for the acceptance of Babylonian astrology stemming from Platonic, Aristotelian, and Stoic ideas about the superiority and divinity of the heavens as well as of the intelligence and souls of the stars. Only the Epicureans seem to have demurred. Otherwise, Dodds assessed the popularity of astrology to be a manifestation of a great *trahison des clercs*, noting that, at a critical juncture, the Stoic Diogenes, called the Babylonian in Strabo, *Geography* Bk 16.1.16,[29] as the head of the Stoa in Athens (ca. 200–150 BCE), had the effect of slowing the progress of rational Greek science by his emphasis on religious cosmology, divination, and astrology.[30] To this, Dodds added that "besides astrology, the second century BC saw the development of another irrational doctrine which deeply influenced the thought of later antiquity and the whole Middle Ages—the theory of occult properties or forces immanent in certain animals, plants, and precious stones."[31]

The mid-twentieth century, in which Dodds's study of "the irrational" in Greek culture appeared, was a time that saw great successes in the scientific community and a renewed energy in the philosophical community to define scientific rationality. The polarization of science and nonscience, scientific rationality and lower forms of reason, was rooted in nineteenth-century arguments about the conflict and separation of religion and science, reverberations of Hume's aversion to religion and metaphysics.[32] This polarization affected the historical community and with it discussion of the Babylonians and the rational. Nor did Dodds challenge that polarity.

Nevertheless, Dodds drew a more nuanced picture of the Greek Enlightenment, both of its origins and of the reaction to it, particularly in Athens of the later fifth century, where, "disbelief in the supernatural and the teaching of astronomy were made indictable offences."[33] He said, "the evidence we have is more than enough to prove that the Great Age of Greek Enlightenment was also, like our own time, an Age of Persecution—banishment of scholars, blinkering of thought, and even (if we can believe the tradition about Protagoras) burning of books."[34] The complex coexistence of the rational and irrational (referring to dreams, magic, and astrological medicine) was manifested in a different way in the period of Galen, as Heinrich von Staden discussed.[35] Despite taking a dim view of the use of astrology and magic in the treatment of patients, calling these things *alogos*, that is, irrational, Galen subscribed to the healing power of the god Aesclepius and to therapeutic interventions revealed in cultically induced dreams, as Dodds pointed out as well.[36]

Otto Neugebauer, writing in the very period of *The Greeks and the Irrational*, said,

> To a modern scientist, an ancient astrological treatise appears as mere nonsense. But we should not forget that we must evaluate such doctrines against the contemporary background. To Greek philosophers and astronomers, the

> universe was a well defined structure of directly related bodies. The concept of predictable influence between these bodies is in principle not at all different from any modern mechanistic theory. And it stands in sharpest contrast to the ideas of either arbitrary rulership of deities or of the possibility of influencing events by magical operations. Compared with the background of religion, magic and mysticism, the fundamental doctrines of astrology are pure science.[37]

Setting astrology up against all that symbolized irrationality was, in effect, not saying much in its defense. Neugebauer sought to justify the status of astrology as an ancient science by saying that if we ignore the premise of stellar influence, the rest follows rationally, like any other mechanistic science. William Newman and Anthony Grafton, writing almost forty-five years later, found that Renaissance astrologers practiced a "rational art of living,"[38] but in the middle of last century, Neugebauer, who had studied closely many Greek astrological texts, indeed had edited (with H. B. van Hoesen) the Greek horoscopes, struggled to reconcile modern and ancient standards of rationality. Astrology belonged within the history of the ancient astral sciences, as Neugebauer saw them, yet, the distinction between science and "nonsense" (his word), that is, nonscience, was for him still defined along modern lines, as it was for Dodds.

Dodds's study was of immense importance for its attention to religious and psychological aspects of Greek culture that coexisted with the rise of philosophical explorations of rationality and reason, astronomy and physics. His subject was ritual life, possession, prophecy, dreams, madness, and the soul. It was a study of the human being in Greek antiquity, not of the Greeks' study of natural phenomena or formalization of methods for such study. As such, he did not detail criteria for determining the rationality of scientific knowledge. Nor did he have to. The rationality of Greek science was a given.

Dodds credited the Hellenistic period as one of great creativity and adventure, "as if the sudden widening of the spatial horizon that resulted from Alexander's conquests had widened at the same time all the horizons of the mind."[39] The shift in social values of the period seemed to correlate with a growth in the exact sciences (Dodds used the term *abstract sciences*), mathematics and astronomy, and in the systematic expression of human reason. This, he judged, was epitomized by Stoicism, in that, Diogenes of Babylon notwithstanding, "the Stoic contemplated the starry heavens, and read there the expression of the same rational and moral purpose which he discovered in his own breast . . . for both schools [Stoicism and Epicureanism], deity has ceased to be synonymous with arbitrary Power, and has become instead the embodiment of a rational ideal."[40]

THE BABYLONIANS AND THE RISE OF RATIONALITY

With respect to the cuneiform world the rationality of scribal knowledge has not been taken as a given. Quite the contrary. It is still up to Assyriologists to uncover and defend its rationality in the lingering afterglow of scientism and positivism from the last century.[41] As already indicated, Neugebauer was one of its early defenders. In reference to cuneiform mathematical texts, he said that "no trace of number mysticism has ever been found in these often highly sophisticated but perfectly rational mathematical texts which range from the twentieth to the first century BC. Nevertheless, the "Babylonian" origin of whatever is irrational or mystical (in fact or reconstructed) remains the inexhaustible resource for synthetic histories of science and philosophy, whether it concerns the Pythagoreans or the Ionians, Plato or Eudoxus, Nichomachus, Proclus, etc., etc."[42] And yet subsequent attempts to correct the misapprehension that the Greeks invented science and rationality, and to prove that rational reasoning, and with it science, does not have to be categorically excluded from the ancient Near East, has largely taken the form of showing that the contents, form, or methods of the cuneiform scholarly and technical writings satisfy criteria for rationality established in ancient Greek philosophy.

A number of studies in which a Babylonian rationality is or is not found in terms of the standards set by Aristotelian, or generally speaking, Greek, criteria have established a platform for further research. After the early formulation by the Frankforts, discussed in chapter 2, the next major statement was Mogens Trolle Larsen's 1987 investigation of the "Mesopotamian Lukewarm Mind."[43] There he concluded that despite the nascence of rational reason in cuneiform omen texts, a basic difference in mode of thought and an insufficient domestication of the savage mind prevented the "Greek Miracle" from occurring in ancient Mesopotamia. Appearing in the same year was Jean Bottéro's discussion of abstract thought and the "scientific spirit," again with respect to cuneiform divinatory texts, which he judged to have anticipated the miracle by "more than fifteen centuries."[44] Then in 2009 I analyzed the formal nature of Assyro-Babylonian omen statements in terms of the logic of indicative conditionals. I raised the question of whether omen statements were formally related to the inference rule known as material implication (*P* implies *Q*), athough I did not frame that question as one of progressive cognitive evolution.[45] Most recently, the study of the early Babylonian astronomical compendium *MUL.APIN* by Rita Watson and Wayne Horowitz in 2011 posited a cognitive shift toward science within cuneiform culture evidenced in a teleological progression from simple list making to more complex, abstract, even axiomatic reasoning.[46] Not invested in a comparison with Greek logic, and from a decidedly antiteleological standpoint, Markus Hilgert proposed for a different set of evidence from that discussed in these other studies a native epistemic style in the culture of the cuneiform scribes, taking as an artifact of that style the sign list Diri for compound logograms.[47] Hilgert defined within the structure and content of Diri speculative hermeneutic principles deployed in a highly nonlinear use of graphemic and semantic associations.[48] A consensus is not to be had among these studies, but they do provide a foundation from which to proceed.

Larsen's study was steeped in anthropological literature, and the concept of a "Mesopotamian lukewarm mind," pointed to Lévi-Strauss's "cold," that is, static, versus the progressive "warm" or "hot" societies, although Lévi-Strauss meant for the term to characterize an alternative orientation toward the world, not an evolutionary stage. The betwixt and between sort of society was called *lukewarm* by social anthropologist Luc de Heusch, to whom Larsen referred.[49] As a starting point for the argument, Larsen cited Jack Goody's critique of the rigid binarism and ethnocentrism that Goody had termed the Grand Dichotomy. But Larsen was particularly interested in the flaw of conflating contemporary ethnographic societies with ancient peoples, citing the *The Intellectual Adventure of Ancient Man* as an example from within the field of Assyriology. Nonetheless he suggested that Lévi-Strauss's notion of the "cold" "science of the concrete" was useful for understanding cuneiform casuistic texts, meaning those that compiled "If *P*, then *Q*" statements, or cases, such as compose omen texts. Larsen found Lévi-Strauss's "science of the concrete" relevant for its similarities in creating taxonomic order, system, and a particular kind of causation. He took his evidence both from the lexical and the casuistic literature, each of which he saw as fundamentally empirical and reflective of a desire to "present a systematic and ordered picture of the world."[50]

The emphasis on principles of ordering was important to Larsen as prerequisite for abstract reasoning. As the earliest evidence for the articulation of abstract generalizations, Larsen cited the commentary series to liver omens called *multābiltu*, which established principles for interpreting signs on the liver.[51] Thus: Thickness means peaceful dwelling as in: if the Increment is thick: the crops will thrive (Tablet I: 14), and, to bend downward means success as in: if the Presence (is bent) downward like a sickle: you will surround the enemy land in a siege (I: 17). Larsen saw that "in this respect the text appears to represent a step towards an abstract 'domesticated' science, away from the world of the 'savage mind.'"[52] It was not, however, a step that would be productive of science because, as he said, "there is no logical basis for the relationships which are postulated."[53] The absence of what he judged to be a logical basis for the relationships between sign and consequent reflected a lack of rationality. The complete realization of rationality was prevented because divination was a system that valued association over empirical relationships, and its goal was not the understanding of the effects of physical causality such as would be rewarded by Empiricism and experience. Omens made claims to the repeatable nature of events that for the most part do not relate to one another in physical causal terms, such as:

> If water secretes inside the gall bladder: The flood will come.[54]
>
> If the gall bladder is turned and has wrapped around the "finger": The king will seize the enemy country.[55]

These correlations have a semantic, or sometimes also an orthographic, analogous relationship, that serves to justify the connection. Occasionally, however, an omen will evince a different kind of connection between antecedent and consequent, something like Quine's "observation categoricals" (discussed in chapter 6), as this omen quoted in a report: "If the sun is surrounded by a halo: it will rain" (DIŠ 20 TÙR NIGÍN ŠÈG ŠUR).[56] It is the case that halos surrounding the sun or the moon can be an indication of coming rainstorms. The understanding that a halo around the sun will be associated with rain is not dependent upon the understanding of the high cirrus clouds containing ice crystals that create the illusion of a halo. For this omen, the correlation, not the cause, was of importance. On the basis of the overwhelmingly noncausal nature of Babylonian omen statements, Larsen's final assessment was that scientific rationality was not evinced in the cuneiform tradition, which remained conceptually equivalent to the "science of the concrete."

Although Larsen saw an incomplete development of a scientific rationality in cuneiform texts, he judged there to be a certain advance in cognition (a warming of the Mesopotamian mind) as a result of the technology of writing. Here he drew, as would Watson and Horowitz as well, on Jack Goody's *Domestication of the Savage Mind,* where he (Goody) said, "I have tried to take certain characteristics that Lévi-Strauss and others have regarded as marking the distinction between primitive and advanced, between wild and domesticated thinking, and to suggest that many of the valid aspects of these somewhat vague dichotomies can be related to changes in the mode of communication, especially the introduction of various forms of writing."[57] But Larsen was also critical of technological determinism (sometimes mistakenly associated, not by Larsen, with Goody[58]), and did not find the difference in cognition to be a direct or simple function of writing. Rather, he located the critical difference in the lack of explicit second-order thinking among cuneiform scribes, who were preoccupied more with the preservation of a tradition than with criticism or explication.

Also stemming from the late 1980s, and taking a different perspective on the question of the nature of Mesopotamian reason, Jean Bottéro argued for a transformation in the development of omen divination from what he identified as a posteriori to a priori knowledge. Thus, "from an observation of a sequence of events that do not have any apparent link between them, but were noticed to have followed each other once, it was thought that such events would always follow one another. That is what we would call empiricism."[59] From this he concluded:

> It is thus that divination passed from its primitive state of simple empirical observation to that of knowledge a priori, to "deductive" knowledge. From the moment they discovered that a lion is the sign, the ideogram, of violence or of power, it became useless to "wait for the events," which would have been indispensable in an empirical system. They could foresee without fail brutality, carnage, or domination from the moment that they noticed the presence of a lion in an ominous circumstance. This was a capital transformation and of considerable importance: because, in fact, a knowledge a priori, deductive knowledge, is already the essential element of science.[60]

Interestingly, sequential co-occurrences of the type that connects the sun setting and the cessation of bird song, for example, were rarely grist for Babylonian diviners' mills (as in the solar halo omen). Such phenomena can be explained by empirical experience, as they are reliably and inherently interconnected. But the omens do not unite such pairs of events.

Analogical relationships, on the other hand, are numerous, for example, from a Neo-Assyrian report: "If the Fish constellation stands next to the Raven constellation: fish and birds will become abundant."[61] Or, from *Enūma Anu Enlil*: If the King star (= Regulus) carries a sheen: the king of Akkad will achieve overpowering strength."[62] Other examples are not so obvious, but in such cases it is we who are hampered from "getting" the connections that are there. While we at this distance may not be capable of identifying the connective tissue between the ominous phenomenon and the correlated consequent in every case, empirical sightings of co-occurring events does not seem to me to be a plausible alternative to the question of the etiology of omens. The principle of analogy, or the use of the resemblance criterion, is paramount in all genres of omens, suggesting that argument by analogy, that is to say, argument from particular to particular, was essential in forging connections between phenomena in different domains. Argument by analogy can be classed either with inductive or deductive reasoning. It does not represent an inherently irrational cognitive process.[63]

What was operative in Bottéro's analysis was the change from a phase when the sequence of the events *P* and *Q* were observed, to a time when it was no longer necessary to observe such co-occurrences because it would be known by deduction that *Q* followed *P*. In the changeover from justification by Empiricism to justification by deduction, Bottéro identified the emergence of the scientific spirit, and therefore rationality, in cuneiform knowledge. But since we cannot show, or assume, that the Babylonians convinced themselves that *Q* was the result of *P* because every time they experienced *P* they experienced *Q*, we cannot reconstruct an imaginary inductive process by which the ancient diviners arrived at the idea that phenomena could therefore be ominous signs. I would argue that Babylonian divination was formally deductive, but it does not follow that such deduction was based in the inductive process, the "arch of knowledge," so to speak, that progresses from induction to deduction. This is a flawed reconstruction that puts uncritical faith in induction as the beginnings of the scientific process, a reconstruction that does not explain how divination can be construed as continuous with science. The etiology of cuneiform divination seems rather to lie in attention to particulars and their analogical relationships.

Bottéro judged deductive reasoning to be "what we call a science, in the proper and formal sense of the word, as it has been taught to us by the ancient Greek teachers, after Plato and Aristotle, and as it still in essence governs our own modern idea of science."[64] But he was willing only to grant to cuneiform texts the "spirit" of science, for the actual content of cuneiform science he deemed "empty of all meaning, outdated and 'superstitious.'"[65] The discovery of the "scientific spirit" in ancient Mesopotamia was meant to show that the Greeks had "learned and assimilated this 'scientific spirit,' born before them and in another country . . . it appeared in Mesopotamia more than fifteen centuries before Socrates, Plato, and Aristotle. Its birth and establishment cannot be observed better outside divination, a science which constitutes one of the most essential and typical characteristics of the ancient Mesopotamian civilization."[66] The Greeks, therefore, did not invent science, or at least the "scientific spirit," which was anticipated already in ancient Mesopotamian divination.

While I differ from Bottéro's position on Babylonian divination as superstition,[67] as well as from his thesis about the empirical origins of it, I concur with his assessment of the deductive character of Assyro-Babylonian omen statements, and in the idea that logical consequence was characteristic, even fundamental, to the formulation of omen statements. The objective of my earlier analysis was to show that a fundamental element of classical rationality, the argument form known as modus ponens, was present in Assyro-Babylonian omens statements.[68] Omens as given in the lists merely provide a series of cases that serve as a basis for knowing Q if P, or, put another way, what is indicated or implied by P. In effect, Q can be inferred from P on the basis of a written omen in a manner that, at least from a formal standpoint, follows the logic of indicative conditionals as follows: "If P, then Q. P: Therefore Q." Babylonian omens, therefore, employed the most common rule of inference, that of modus ponens, generally taken to be deductive. Although reasoning deductively to conclusions is not the only form of rationality, if we are in the presence of deductive reasoning, as Bottéro suggested, are we not by definition in the presence of the rational?

The way the antecedent and consequent are connected in written omen compendia undoubtedly rests on a long background of textual and ideological development. However, an argument for cognitive change on the basis that only in later omen series were all sorts of connections $P \rightarrow Q$ added on grounds of resemblance, paranomasia, orthographic similarities or etymographical word plays puts the horse behind the cart. Resemblance and analogy were active in our earliest examples of written omens, as illustrated in an Old Babyonian astral omen: "If the heavens are troubled, the year will be unlucky."[69] As already noted, a cognitive shift occurring somewhere between "Empiricism," based on the logical fallacy of post hoc, ergo propter hoc, as Bottéro suggested, and logical consequence, that is, rational prediction of Q from P, whether by analogy or not, is sheer supposition. I do not see evidence for post hoc, ergo propter hoc reasoning in an imaginary beginning stage of divination, therefore no cognitive difference between an earlier and a later understanding of why phenomena were ominous, and therefore no shift from the irrational to the rational. The decision as to whether divination is rational or irrational should depend on whether one deems the use of its forms of reasoning to be appropriate or inappropriate to its own cultural and epistemic norms.

I turn now to the most recent effort to identify and define the rational in Babylonian science, Watson and Horowitz's *Writing Science before the Greeks*. Theirs is a study informed by cognitive science and committed to the idea of literacy as a universal engine of cognitive change. In a return to Larsen's approach to the question of a Babylonian rationality, they asked, "does writing make us rational?" and "must one be literate in order to be logical?"[70] They posed the question of the cognitive effects of various kinds of texts and techniques of storing knowledge, such as in permanent archives. Resonant with Goody and Larsen, their model is of writing and conceptual change. In their formulation, writing "biases mental representation," which "may lead to the enhanced development of logic, or rationality."[71]

Although Watson and Horowitz repudiate the idea of a primitive mind, they allow for a progressive dimension in the effects writing can have to stimulate cognitive change. In reference to the astronomical text *MUL.APIN* they state that "the treatise has an historical-cumulative aspect,"[72] so that, like archaeological strata, *MUL.APIN* reveals increasingly complex and sophisticated modes of reasoning from the lowest and chronologically oldest level of simple lists to the appearance in the highest and chronologically latest level of what they read as an axiomatic statement. In their words, "the conceptual content of the later portions of MUL.APIN is of a higher order, and more abstract, than that of the earlier part."[73] Whether the inherent progressiveness of their thesis about *MUL.APIN* is derived from a commitment to the notion of human history being one of continuous cognitive transformation, along the lines of Gellner's "unfolding of rationality,"[74] is not explicit, but perhaps implicit, as the idea of writing as an agent of cognitive change "for the better," that is, toward logic and rationality, is central to their analysis.

Representing the oldest lowermost stratum is the first section of the text, *MUL.APIN* II 1–ii 35 (their section a). This is a catalog of seventy-one stars divided into the three celestial roads named for the three highest cosmic gods, Enlil, Anu, and Ea, as they are distributed from north to south marking arcs of rising and setting stars (and planets) along the eastern horizon. The names of the stars are given together with their divine names (such as the Great Twins, Lugalgirra and Meslamtaea, or the Panther, Nergal) or otherwise descriptive notes as to who they are ("messenger of Ninlil," "seeder of the plough," and the like) and where they stand in relation to nearby stars. As illustration, *MUL.APIN* I i 12–19:

ŠU.PA, Enlil, who decrees the fate of the land.
The star which stands in front of it: the Abundant One, the messenger of Ninlil.
The star which stands behind it: the Star of Dignity, the messenger of Tišpak.
The Wagon, Ninlil.
The star which stands in the cart-pole of the Wagon: the Fox, Erra, the strong one among the gods.
The star which stands in front of the Wagon: the Ewe, Aya.
The Hitched Yoke: the great Anu of heaven.

A wealth of information is built into the star catalog, most striking of which is the Sumero-Akkadian cultural underpinnings in the equivalences between stars and gods, and the importance of agriculture, seen in the references to the field, the plow, the yoke, the seeder of the plow, wagon, furrow, harrow, ear of grain, to the domesticated animals, pig, ewe, horse, bull, rooster, and the animals of the wild, lion, panther, stag, scorpion, eagle, raven, snake, and swallow. In accordance with Watson and Horowitz's archaeology of knowledge, as far as the names of celestial bodies are concerned, the tradition the names evoke seems to reach back to a much earlier period than that which saw the composition of the text.

At the other end, representing the uppermost stratum of the text in their analysis, is section L (*MUL.APIN* II ii 43–II iii 15), containing a list of values for the length of night and the derived duration of the moon's visibility at night throughout the year. Their assessment of section L is that it represents "the most purely abstract form of expression in the entire treatise."[75] Section L may be described as a condensed list of values corresponding to those found tabulated in the scheme for calculating lunar visibility in the omen series, EAE (*Enūma Anu Enlil*) Tablet 14. *MUL.APIN*, the Astrolabes, and EAE 14 all share the same scheme for calculating length of daylight (or night). Section L summarizes the important parts of the lunar visibility scheme, but whether the data can be described as abstract can be questioned; they are the same data tabulated in EAE 14. Watson and Horowitz have indirectly raised a fundamental question about how to understand such numbers, as abstract or concrete representations. In his study of astronomical procedure texts, Mathieu Ossendrijver classified this section of *MUL.APIN* with instructional texts, indeed as the first of such astronomical procedural texts that entail computation.[76] From that point of view, the numbers would seem to function as concrete representations of lunar visibility times.

Lastly, special rhetorical status is given to the conclusion of section L: "The summary statement," they said, "is distinguished from those of other sections, in that it opens with a formally-expressed axiom."[77] The line in question is *MUL.APIN* II iii 13: "4 is the coefficient for the visibility of the moon." The statement gives the coefficient, or multiplicative factor, for finding the duration of lunar visibility at night throughout the year, and is derived directly from the particular scheme that underlies the section as a whole. The scheme for lunar visibility assumes that the moon will be visible for 1/15 of the length of night (4 = 1/15 in the sexagesimal system, as 60 ÷ 4 = 15), and the length of night changes with the month. As an example: For the equinoctial months in which the length of night is equal to 3,0 (or 180 degrees, half of one day circle), the factor to calculate lunar visibility is 12 (12 = 1/15 × 180). Section L lists all the values for daylight length per month followed by the values to calculate duration of the moon's visibility, taking into account that in the first half of the month one sees the moon earlier each day, and for an increasing amount each day, from the thin crescent at sunset at the beginning of the month to the moon's visibility all night at full moon (day 15) when it rises with sunset and sets with sunrise, and then decreasingly from full moon to its final disappearance, being visible later and later and for a briefer length of time each night until one sees its final crescent appearance briefly just before sunrise. The values for these lengths of lunar visibility are all calculated by the coefficient 4, that is, 1/15 of night length, discussed before. Reference to the value 4 in line 13 extracts the factor used in the scheme. *MUL.APIN* II iii 13 is, I would suggest, not a self-evident or necessary truth, or some other kind of established or universally accepted principle, which is to say it cannot serve as an axiom or a first principle, but rather it is itself a derivative, thus automatically disqualifying the statement in II iii 13 as axiomatic.

Not only is section L consistent with Tablet 14 of the omen series *Enūma Anu Enlil*, but what follows, *MUL.APIN* II iii 16–iv 12 (section m), the closing section of the text, consists of a section of omens. The omens present a problematic endpoint for Watson and Horowitz's progressive historical-cumulative interpretation of the text as a whole. They suggest that the section was added by the astronomers "to justify the practical value of their work," the implication being that the astronomers viewed omens as "applied astronomy," that is, in a lesser light than the rest of *MUL.APIN*.[78]

Watson and Horowitz's interpretation of *MUL.APIN* as a stratigraphic map of cognitive-historical progress is an interesting one; however, the evolutionary scheme from list to axiom and thereby from a lower form of cognitive engagement with the world to a higher one is compromised once the axiomatic status of the section L's conclusion is questioned. Although the primitive stage they identify in the star catalog, in which the stars, constellations, and planets were named for gods in the context of activities for which those gods were known, doubtless reaches far back into preliterate times, the names of the stars were current and still practical throughout the first millennium despite such early origins. Given this, and the fact that the list remained a favored form in cuneiform scholarship (lexical lists, omens, mathematical and astronomical tables) it seems difficult to correlate the star names in *MUL.APIN*'s star list with a historical-cognitive stage.

A map-like structure and function can be seen in *MUL.APIN*'s use of celestial "roads," and their relative positions. As a descriptive "itinerary" of the heavenly region, *MUL.APIN* is functionally parallel to itineraries on the ground in the form of practical geographies.[79] Geographical itineraries described routes not out of the desire to render objective representations of lands and terrain, but because knowledge of places had an importance for political, economic, or military reasons. Similarly, the landscape of the sky was important to know but the elements of interest to *MUL.APIN* were those necessary for the practice of celestial divination. The ideal calendar of *MUL.APIN* is also consistent with the celestial divination series *Enūma Anu Enlil*, suggesting that the creation of *MUL.APIN* was not a disinterested description of "natural phenomena," but an account of the elements necessary for use of the series *Enūma Anu Enlil*. This has already been observed and well explicated by David Brown.[80] *MUL.APIN* presents a practical astronomy for celestial divination whose general attitude about heavenly phenomena it shared—namely, that these were the significant celestial bodies one needed to know, because the gods made them, or manifested themselves in them, to signal events in the future. It was moreover desirable from a divinatory standpoint to identify periodic phenomena and devise convenient ways of predicting them by numerical schemes, as seen in section L (*MUL.APIN* II ii 43–II iii 15).[81]

CORRESPONDENCES AND PROPERTIES

I turn now to Dodds's second form of irrational belief, mentioned briefly above—namely, the "theory of occult properties or forces immanent in certain animals, plants, and precious stones."[82] It is critically important to note that, as Nicolas Weill-Parot said, "the history of occult properties is . . . strictly linked to the history of the concept of nature."[83] If indeed the conception of nature is essential to the very idea of occult qualities, the occult becomes a category ill-fitted within the cuneiform system of knowledge.

Dodds attributed magic, divination, and medical practice to the paradoxographer Bolos (ca. 250–115 BCE), from Mendes in the Egyptian delta, judging his works "fatally attractive to the Stoics, who already conceived the cosmos as an organism whose parts had community of experience (*sympatheia*)."[84] The *Suda* called Bolos a Democritean, also the Mendesian Pythagorean, and ascribed to him, as enumerated by B. Hallum, "the following (lost) works: *Scientific Inquiry and Medical Art; . . . Concerning Wonders; Naturally Potent [drugs?]; On Sympathies and Antipathies* (the Souda [*sic*] adds *of Stones*, probably the vestige of another title); and *On Signs from the Sun, Moon, Ursa Maior, Lamps and the Rainbow*."[85] These objects of knowledge in the Greco-Egyptian milieu had a distinguished epistemic kinship with, if not ancestry in, *ṭupšarrūtu*, "the art of the (cuneiform) scribe," more specifically *āšipūtu* "the art of the medical conjuror/or incantation priest," which addressed itself in no small measure to the relational properties of phenomena. Cuneiform parallels that have to do with correspondences and properties testify to an important context of the rational, not the irrational, in cuneiform knowledge, because astrology, magic, and astral medicine employed analogical reasoning, among other cognitive strategies, which worked within normative standards for relating the particulars of phenomena in various meaningful ways.

The principal components of the cuneiform pharmacopeia were derived from animal products, plants, stones, and woods.[86] Explicit mention of plants, stones, and wood as categories of substances with healing properties is found in the following procedural statement from a Late Babylonian explanatory tablet: When you make (a medical treatment of) plant (medicinal herb), stone (bead), and wood (for fumigation), and the art of the healing profession for the sick man—one does (it) in accord with its explanatory comment(?).[87] Another procedural statement from a late natal astrological text is parallel: stone (bead), plant (medicinal herb), and wood (for fumigation), the animal(s) of 13 and 4,37 when you take each together, stone, plant and wood for the sick man, you smear (on him), feed him (the medicine), and fumigate him.[88] This second example contains reference to two numerical schemes employed in Late Babylonian astrology in order to increase the connections that could be made between dates in the year, positions in the zodiac, and *materia medica* appropriate to a person afflicted when those dates or positions were relevant. These are the *Dodecatemoria* scheme (to advance in steps of thirteen around the zodiacal signs) and the *Kalendertexte* scheme (to advance in steps of 277).

Exploration of the efficacy of animal, plant, and mineral matter was a long-standing tradition among Babylonian and Assyrian practitioners of medicine, the *ašīpus*, *mašmaššus*, and *asûs*. These scribes regarded the technical knowledge of their profession as of antediluvian origin, according to the following colophon from a text giving treatments for afflictions of the head, nose, and ears: "Proven and tested salves and poultices, excerpted from the lists, after an oral tradition from the ancient sages from before the Flood, which in the second year of Enlil-bani, King of Isin (1860–1837 BCE) Enlil-muballiṭ, sage of Nippur, deposited (lit. "left") in Shuruppak."[89] The practice of the *ašīpus*, *mašmaššus*, and *asûs* used several means for relief: preparation of salves to smear and rub on (ŠÉŠ = *pašāšu*) affected parts, mixtures to be ingested (KÚ = *akālu*), as well as the use of efficacious stone beads strung (*šakāku*) onto threads twined together of various substances (wool, gazelle tendon, goat hair, the stem-like leaves of the rush plant) to be bound onto painful body parts, as in the following example: "[If a man . . .] is seized and has chronic *sagkidabbû* (literally, "that seizes the temples," probably migraine): beads of silver, gold, carnelian, [. . .] SAG.DU-stone, *muššaru*-stone, *ḫulālu*-stone, *pappardilû*-stone, . . . male and female ŠU-stone, *saggilmut*-stone, [. . .] iron, *ajartu*-shell with 7 spots, *janibu*-stone, and *kapāṣu*-shell. These beads you string on (twine made of) virgin she goat hair, gazelle tendon, and male *ašlu*-rush that you wind together."[90] The use of strings of "stones," that is, beads made of substances including metals, shell, and semiprecious or other stones with properties to effect healing, was a staple of Babylonian medical procedure. Recitation of incantations was used together with administering salves for the removal or assuagement of the anger of an agent of pain, disease, or distress, meaning a god, demon, ghost, or sorcerer/sorceress, regarded as the cause. From the same prescription for migraine cited above, thus: "If *sagkidabbû* disease (lit. 'that seizes the forehead'), (which is) the hand of a ghost, persists in a man's body, and won't let go, and does not stop with treatments by compress or conjuration, slaughter a caged goose, take its blood, its windpipe, gullet, fat, hide, the outer part of its gizzard, and dry it in the fire, mix it into cedar balsam and recite the incantation 'Evil finger of man' three times. Apply the ointment to his head, his hands, and wherever it hurts him and he will be soothed."[91]

The quintessence of the Babylonian version of Dodds's second irrational theory wherein animals, plants, and stones were used for their inherent properties can be seen in a variety of late astrological texts from Babylon and Uruk dating roughly from the fifth to second centuries BCE. Some have been termed *Kalendertexte* because of the ideal 360-day calendrical schemes that enabled the association of positions and dates of the twelve zodiacal signs or twelve months, and 30 degrees or thirty days within the ideal calendar year of 360 days.[92] These texts make use of numerical schemes, mentioned before, from which one obtained a further position in the ecliptic from an initial position, one method projected positions separated by 13 degrees (*Dodecatemoria* scheme) and the other by 277 degrees (*Kalendertexte* scheme).[93] Characteristic of late astral medical texts is the association of maladies, or their treatments, with zodiacal signs, or their *Dodecatemoria* (BRM 4 19 and 20, LBAT 1626, SpTU V 243, and parts of BM 56605).[94] BM 56605 rev. col. i details the particular stone, wood, and plant that correspond to the zodiacal signs:[95]

(1) In Aries the stone is *zânu*-stone, the plant is *imḫur-līmu* (literally, "[the plant that] counteracts one thousand [diseases]), on the 20th day of Nisannu you should not eat fish and leeks.

(2) Pleiades: the stone (is) . . . , the wood is *e'ru*, the plant is *barirātu*, on the first day of Ajaru do not clean faeces.

(3) Capricorn: the stone (is) carnelian, the wood is *suādu*, the plant is *kamkadu*, you should not drink milk on the 15th day of Simānu.

(4) Cancer: the stone is *apsû*, the wood is šennur, the plant is pomegranate, you should not dry out latrine water.

The idea of a correspondence between human life and the stars was not new to the late period, nor was the idea of the efficacy of the stars in healing, or the idea of medical treatment in accordance with calendar days, such as is illustrated by the Neo-Assyrian period text STT 300,[96] or an explicit day in the lunar month. An entry in the seventh century Babylonian plant compendium *Šammu šikinšu* testifies to an early version of this latter idea, also attested at Sultantepe:

> The plant whose appearance is like (that of) the *kūru*-reed, whose leav[es] are like the leaves of the canebrake-fig, [whose x is like (that of) le]ek, whose pitch (lit. "blood") is as dark as (that of) the carob-tree . . . [that plant] is called [*sikillu* = plant?]; it is good for dispelling witchcraft. [(To be applied) by cle]aning the (bewitched) person's f[ac]e (with it) on the day of the moon's disappearance.[97]

But once the traditional schemata and relationships typical of celestial omens began to incorporate zodiacal signs and months, the human body, and other domains of phenomena, a system of correspondence between phenomena not previously correlated were developed in the late period.[98]

Similar connections were made to extispicy in the Late Babylonian period as well. A Late Uruk text associates traditionally ominous parts of the liver with a god, one of the twelve months, and a heliacally rising star, thus: "the path (of the liver) is Šamaš, Ajāru, Taurus; the gall bladder is Anu, Tašrītu, Libra," and so on.[99] The term "sign" or "ominous part" literally "flesh" UZU = *šīru*, used in liver omens, is found again in a Seleucid astrological context where the term refers not to ominous parts of the liver, but to ominous parts of zodiacal signs, thus 12 UZU.MEŠ ḪA.LA *šá* "the division of" sign such-and-such, in other words, the *Dodecatemoria*.

Magical practice also established new connections to the zodiac, such as in a list of spells with their correlated "regions" in the zodiacal signs, for example, "changing one's mind (is the) region of Leo; overturning a judgment (is the) region of Aquarius; loosening the grasp (is the) region of Virgo."[100] These are the first three lines of a long list of correspondences between signs of the zodiac and incantations to influence various people, or demons, or gods, though the purpose or implementation of these correspondences is unstated.

To aid in making correspondences among the elements of astral medicine were the numerical and calendrical schemes of the *Dodecatemoria*[101] and the *Kalendertexte*. Some examples of texts that correlated zodiacal signs with plants, wood, and stones (for amulets or beads), temples and place names, made use of these quantitative schemes, such as the much discussed texts of Iqīša, the *mašmaššu*, in fourth-century Uruk (quoted above). Related is LBAT 1593, a natal astrological text that assigns characteristic features to the native born in various zodiacal signs (line 2′ "Region of Libra . . . narrow of forehead"), and predicts the gender of the child born when a particular planet is present in the place of the moon, or predicts that twins will be born, as in line 9′: "Because Mars stands in Gemini with [or at the place of the] moon, twins will be born."[102] Iqīša's tablets, as Reiner said, "assign to each of the calendar dates an ointment whose ingredients are related to the zodiacal sign by a pun, either linguistic or purely orthographic, on the name of the sign."[103] SpTU 104, which covers Month IV, begins with the first day of that month and the first sign of the zodiac Aries with its position. This date/position pair is correlated with the blood, fat, and wool from a sheep, with which the patient will be anointed. Capricorn is next, correlated with the blood, fat, and hair of a goat.[104]

Compilations of the items belonging to the known elements of a corpus of *materia medica*, the stones and plants in particular, are known from long before the Hellenistic period, including the texts *Šammu šikinšu* and *Abnu šikinšu*. What emerges in the late period, as Reiner pointed out,[105] are the correspondences in some of the *Kalendertexte* between calendar dates and medical preparations based either in linguistic or purely orthographic associations between the names of the medical ingredients in the preparation and a zodiacal sign.

Such analogies used to forge a correspondence between entities in the world such as plants and stones and the signs of the zodiac, or their micro-zodiacal parts, manifest a continuation of the same kind of reasoning that assigned elements of signs and their consequents in omens and commented texts. The aim was to relate the particulars of phenomena and multiply correspondences among many elements of the world. In the astral medicine of the late period these particulars encompassed human beings, substances such as stone, wood, and plants, gods, zodiacal signs, *Dodecatemoria*, stars, planets, and the moon.

The *Kalendertexte* epitomize a method that relates traditional scholarly knowledge concerning the stones, plants, and animals of the Babylonian pharmacopeia with astronomical number schemata, the zodiac, and the ideal calendar. The number schemes seem to function as techniques for creating multiple correspondences. The particulars of these various parts of the world were interconnected, to be drawn together in a variety of correlations and correspondences, one essential component of which was correlation by analogy. Nils Heeßel not only established that the "stone, plant, and wood" system of astral medicine was an entirely new method of medical practice to be dated from the end of the fifth century (the reign of Xerxes, 485 BCE),[106] but he also read the development as an important marker of a fundamental epistemic and ontological change: "These new methods point to another, starkly altered perception of nature and its mode of operation. These innovations, which come to light in the system of 'stone, plant and wood,' can be explained if one assumes that the Babylonian scholars have conceived all the things of nature as an (invisible) web woven from sympathies and antipathies and interconnected. All things, be they stones, plants, wood, clay, metals or human constructions such as temples and streets thus have certain characteristics, qualities that attract or repel other things. Everything is permeated by a network of interdependencies, of sympathies and antipathies."[107]

The system of relating the component elements of medical preparations (plant, stone, wood) to zodiacal signs, planets, or dates in the schematic year was as potentially accessible in the second century BCE to Greek-speaking Hellenistic scholars as the rest of Babylonian *astrologia* (astronomy and astrology), though the specifics of transmission are as usual elusive. Dodds did not pin the new irrationality about stones, plants, and the stars directly onto the Babylonians, but the general sense is that this "magical" thinking, which had as its authority figures Pythagoras, Zoroaster, and Petosiris (the fictive Egyptian priest associated with Babylonian-style astral omens as well as knowledge of the healing properties of plants and stones), rather than logic, was part of the Eastern intellectual legacy.[108] That same Eastern intellectual legacy is also to be seen in early Arabic magic, exemplified by the *Ghāyat al-hakīm*, whose sources, like those of Abbasid astronomy, astrology, and alchemy, are traceable to Neo-Platonic, Syriac, and Greek texts circulating in pre-Islamic Syria.[109] Utimately, the indigenous Babylonian tradition connecting celestial entities with plants, woods, and stones lies at the foundation of the Arabic branch of celestial magic as well.

Ancient authors made the connection to the East. Pliny noted that "Zachalias of Babylon, in the volumes which he dedicates to King Mithridates, attributes man's destiny to the influence of precious stones."[110] Pliny's Zachalias might be connected to the scholar mentioned by the Greek physician Archigenes of Apameia in the late first to early second century CE as knowing about the properties of jasper. The attribution is preserved only in a tertiary Byzantine account by Alexander of Tralles, who said that, according to Archigenes, Zachalias commented on amulets for the treatment of epilepsy.[111] Pliny attributed knowledge of the properties of stones to another Babylonian called Sudines, who was also mentioned by Strabo as a Babylonian *mathematikos* alongside Kidinas (possibly the astronomer Kidinnu) and Naburianus (possibly the astronomer Nabû-rimanni). Pliny called Sudines a Chaldean astrologer and said he had special knowledge of stones (making reference to onyx, rock-crystal, amber, *nilios*, and pearls).[112] Polyaenus also cites Sudines as a diviner for King Attalos I of Pergamon.[113] These late attributions may not be historically verifiable from cuneiform texts, but they gain support from the general features of the late Babylonian astral medical culture.

Finally, it is interesting to note a further echo of the Babylonian practice of applying astral correspondences with medical practice in Greek sources—namely, as pointed out by Alan Bowen and Bernard Goldstein, by considering the parapegmata, whose premise was to establish correspondences between cyclically recurring astral and meteorological events in the same intellectual contexts as a number of Hippocratic texts. In reference to four late fifth century BCE Hippocratic texts, they noted that "these texts draw on the same conceptual framework as the parapegmata: both types of literature manifest that broadly Mediterranean view that order in Nature is not one of natures or natured things (as construed by Plato and Aristotle, for example) *but one of the periodic correlation or conjunction of events*. In other words, both rely on astral omens connecting stellar and meteorological events; and both suppose that these omens are useful in deciding what to do, since the connections and what follows are periodic."[114]

ANALOGICAL REASONING AND MAGIC

If we consider the internal standards of ancient cuneiform scholarship and its evident aims, it is clear that analogical reasoning, among other cognitive strategies, including deductive inference, worked within the framework of divination to relate and correlate the particulars of phenomena in various meaningful ways. While the content of cuneiform scholarship changed from the seventh to the third centuries BCE and later, analogical reasoning remained a consistent feature of the material over time. Analogical reasoning, as defined in the *Stanford Encyclopedia of Philosophy*,

> is any type of thinking that relies upon an analogy. An analogical argument is an explicit representation of a form of analogical reasoning that cites accepted similarities between two systems to support the conclusion that some further similarity exists. In general (but not always), such arguments belong in the category of inductive reasoning, since their conclusions do not follow with certainty but are only supported with varying degrees of strength. Here, "inductive reasoning" is used in a broad sense that includes all inferential processes that "expand knowledge in the face of uncertainty" (Holland et al. 1986: 1) . . . Analogical reasoning is fundamental to human thought . . . Historically, analogical reasoning has played an important, but sometimes mysterious, role in a wide range of problem-solving contexts. The explicit use of analogical arguments, since antiquity, has been a distinctive feature of scientific, philosophical and legal reasoning.

The roles of analogy and analogical reasoning have been slow to gain recognition as important parts of science. In counting analogy as one of the primary tools of the scientific imagination, Gerald Holton said that "philosophers have long warned that such a technique of thought can have no good purpose in science. The *Dictionary of Modern Thought* declares that analogy 'is a form of reasoning that is peculiarly liable to yield false conclusions from true premises.' Indeed, analogy, and its close cousin, metaphor, have been called the essence of poetry."[115] The poetic, with the fictive, has occupied a place opposite science, as associated with the irrational as science is with the rational. But analogy plays an important part in both the literary and the scientific imagination, as it did in the scholarly imagination of the scribes. Assyrian and Babylonian divinatory, astrological, magical, and medical texts that characteristically deal in correspondences and properties testify to one important context for the use of analogy in cuneiform scholarship.

Of all the things that happen in sequence, why two particular events might be seen or experienced as connected can be explained in different ways. Such connections may have nothing to do with reasoning by empirical induction or by physical causality, yet in analyzing Babylonian divination and its relation to science, the tendency has been to try to find an empirical connection at some initial point in its development, and thereby some process of empirical induction, as did Bottéro, discussed before.

Connections were made between many ominous signs and their anticipated events where the connective tissue between them could be based on orthographies, homophony, or analogy between key words in the protasis and apodosis. Polarities were also effective, such as the positive and negative associations attributed to right and left, dark and light, and so on. As suggested by the use of the 13x and the 277x schemes, additional techniques could propagate other values, positions in the zodiac, or dates in the ideal year to be combined and correlated with various elements in new systems of astrology and astral medicine. Signs of a physical causal relationship—such as the omen predicting rain from a solar halo[116]—were rarely a part of divinatory texts, whose focus was not commensurate with our search for physical causal relationships. And yet, as evidenced by the consistent application of analogies not only to structure omens but also to explain the meanings of omens in commented texts, observations aimed at finding relationships between many kinds of phenomena in heaven and on earth created a different normative standard as to what constituted knowledge and understanding of phenomena.

Stich noted that "a common theme in the research on human inference is that people are inclined to overextend the domain of an inferential strategy, applying it to cases where it is normatively inappropriate."[117] Normative inappropriateness is, of course, the definition of the irrational. But deciding what is normatively appropriate or not depends on one's point of view. Stich was interested in inference strategies used to make a judgment or a decision about the likelihood of an event. Where similarity between events is used as a way to judge the likelihood of occurrence, psychologists Daniel Kahneman and Amos Tversky introduced the term "representativeness," and its use in decision making or prediction as the "representativeness heuristic."[118]

Although in the following passage Richard E. Nisbett and Lee Ross appealed too much, in my view, to the ideas of primitive otherness, magical thinking, and prescience, using Zande culture as the default and exemplary Other, it is nonetheless interesting to consider in the light of Babylonian divination and medicine what they had to say:[119]

> People have strong *a priori* notions of the types of causes that ought to be linked to particular types of events, and the simple "resemblance criterion" often figures heavily in such notions. Thus, people believe that great events ought to have great causes, complex events ought to have complex causes, and emotionally relevant events ought to have emotionally relevant causes. . . . The resemblance criterion is transparently operative in the magical thinking of prescientific cultures. For example Evans-Prichard [*sic*] . . . reported such Azande beliefs as the theory that fowl excrement was a cure for ringworm and the theory that burnt skull of red bush-monkey was an effective treatment for epilepsy. Westerners unacquainted with Azande ecology might be tempted to guess that such treatments were the product of trial and error or laboriously accumulated folk wisdom. Unfortunately, the truth is probably less flattering to Azande medical science. Fowl excrement resembles ringworm infection; the jerky frenetic movements of the bush-monkey resemble the convulsive movements that occur during an epileptic seizure.[120]

Referring to Nisbett and Ross's statement, Stich takes the Azande's use of the representativeness heuristic to be inappropriate to the domain in which it is applied. It is by his reckoning irrational, because a treatment for illness based on the resemblance between the affliction and the cure is "an extreme example of the overextension of an inferential strategy."[121] It seems to me, however, that if the grounds for an inference scheme are not equivalent to those of modern science, as in Zande medicine, or Babylonian divination (and medicine), and do not operate with or require the assumptions or methods of naturalism, then the charge of irrationality is categorically wrong and historically anachronistic.

What the psychologists refer to as the *resemblance criterion* is in fact the crux of analogy, which draws a meaningful connection by associating two things, the analogue and the target, that are not alike in other respects, such as in the Old Babylonian gall bladder omens cited above, where the analogy is made between water in the gall bladder and the flooding of the river, or between the arrangement of the gall bladder and the "finger" (of the liver) and that between the king and his enemy. Indeed, the extispicy corpus provides vivid illustration of the fact that analogy is built solidly into its foundation, where the features of the liver are made to relate to various areas of royal concern and thus become suitable for prognosticating about those concerns. As Heeßel explained, "the Akkadian designations of these parts [of the liver] refer to the different areas of interpretation of the observed signs and also had a symbolic value. Thus a groove with the name 'presence' refers to the presence of the gods in this extispicy . . . while the umbilical fissure named 'palace gate' refers to events and income of the palace; the groove 'path' refers to campaigns and expeditions and the area called 'throne dais' to the king's private life, his court and the stability of the dynasty."[122]

Connections in Akkadian omen texts based in analogy can be multiplied and many examples tallied, as has been done elsewhere.[123] Analogy seems to me a fair description of the kind of reasoning at the heart of Babylonian divination, as well as some of the lexical lists built up through semantic and phonological associations within the cuneiform writing system and which provide imaginative spellings of words, as in the lexical lists Antagal and Diri, or in commentaries such as *iNAMgišḫurankia*. Alasdair Livingstone already noted the propensity of the cuneiform scribes for juxtapositions of related words or ideas as a fundamental of scribal methodology: "The explanatory works exhibit characteristics typical of many other genres of Babylonian scholarly literature. It was usual for almost every type of information to be summarized and recorded by listing pairs of associated items, arranged in columns. This technique acquired specialized conventions appropriate to the particular subject matter involved. The principle of expressing information by simple juxtaposition is so universal in the literature that it is sometimes necessary to raise the question of the extent to which the actual thinking of the ancient scholars was influenced by this aspect of their practical methodology."[124]

It is also possible to see the listing of juxtaposing pairs not as a technique giving rise to but rather the outcome of analogical reasoning. Listed pairs of associated items are certainly descriptive of omen texts, thus providing a ready structure for the juxtaposition of analogous (or opposing, or other) relationships. Analogical reasoning, the construction of meaningful similarities and correspondences, is not unique to the cuneiform world. Descola has found in Amerindian cosmology a particular way of resolving aspects of ontology predicated upon the importance of analogical reasoning, which he termed "analogism." He has said that "analogism was the dominant ontology in Europe from Antiquity to the Renaissance, and is still extremely common elsewhere: in China and India, in Western Africa or among native cultures of Mexico and the Andes."[125] It is also associated widely with what we call magic, which is supposed to differ from science by virtue of its acceptance of the causal workings of the supernatural. Magic, from the point of view of modern scientific naturalism, operates on the basis of an irrational belief in the supernatural, hence Dodds's identification of the two principal irrationalities of Hellenistic Greek culture with astrology and the occult qualities of plants and stones. But cuneiform divinatory or ritual texts that deal with the agencies of gods, demons, ghosts, or sorcerers were not concerned with the supernatural. They did, however, make widespread use of associations and analogy.

There are reasons for rejecting as well as for maintaining the use of the term *magic* in the context of cuneiform texts. Arguing for keeping the term, Daniel Schwemer said, "in many respects the texts . . . will match our provisional notions about magic but a number of aspects will take us into the spheres of religion and (premodern) science and reveal the limited applicability of our traditional compartmentalizations."[126] He further commented on the "powerful heuristic framework" still wielded by J. G. Frazer's triad of magic, science, and religion, one taken up in detail by S. J. Tambiah with particular emphasis on the history of anthropology.[127] Schwemer offered that "we should approach the extant sources bearing in mind a general notion of our own concept of magic, which then can be modified by the study of the sources themselves, leading towards an adequate understanding of what can be subsumed under the heading 'Mesopotamian magic.'"[128] The *āšipu* had a corpus of technical literature:[129] the apotropaic rituals (*namburbû*); purification rituals (e.g., *mīs pî* "washing of the mouth," the rituals of *bīt rimki* "bath house," and *bīt mēseri* "ritual enclosure"); and the texts with medical preparations and incantations for healing. For reasons of similarity and even continuity in some cases with magic in other cultural contexts, as well as a lack of a suitable alternative term, therefore, Schwemer takes the position, and I would agree, that the term *magic* seems still to have some life left in it for cuneiform texts.

On the other hand, the use of the category magic as an ahistorical tool for demarcating between science and nonscience in this period, or as a priori evidence of irrational thinking and misbegotten causality is to be clearly rejected. Here I would agree with Eleanor Robson's comment that "the term 'magic' categorises a set of thoughts and activities as alien to the mindset and lifestyles of those of us moderns who study the ancient world, and this risks belittling, trivialising, or even denigrating those ideas and practices, while ritual—like activities, and concerns about bodies' and spaces' cleanliness or pollution, are as much a part of modern, Western society as they were of Mesopotamian culture. That is not to say we should somehow domesticate or over-identify with ancient healing practitioners, but rather that historians should have techniques of both familiarisation and de-familiarisation in their repertoire."[130] The ahistorical and essentially demarcationist usage of the term *magic* to mean the irrational communing with the supernatural has also been problematic in the historiography of later European medieval and Renaissance science, but for rather different reasons. As John Henry explained:

> One major reason for the prevailing mistaken conception (by positivist historians and others) of the nature of magic in the Renaissance is the lack of any understanding of what was known as natural magic. Lack of awareness of the natural magic tradition is due to the fact that it was to a large extent completely absorbed into what we now think of as science, while other, lesser, aspects of the tradition have remained in what should be regarded as merely a rump of the magical tradition—what was left over after parts of the tradition had been absorbed into natural philosophy. Today, we tend to identify magic with the supernatural (if we leave aside the stage trickery of "show-business" magic), but in the period we are looking at, to describe an event or a phenomenon as supernatural was to say that it had been brought about miraculously by God—only God was above nature, and only God could perform a supernatural act.[131]

From a sixteenth-century handbook on natural magic, Henry cited Giovanni Battista Della Porta as saying, "Magic is nothing else but the knowledge of the whole course of Nature."[132] In this context, then, magic belonged to nature, not the supernatural, while, I would maintain, in the cuneiform context it belonged to neither.

The case of later European magic underscores the necessity of being clear about what magic represents in the cuneiform world, for while the later literature may provide a number of parallels and similarities, the conception of what we would otherwise call the magical properties of phenomena cannot be defined in a matrix with nature and God as conceived in that later period in European culture. The Babylonian scribes' investment in magical ritual and incantation was predicated on the idea of divine communication, to appeal to the divine to effect desired changes in the world. Their knowledge of what we call magic was not conceived of as the knowledge of "the whole course of nature," as later conceived, but rather as the knowledge of particular associations between elements of plants, animals, stones, and human beings.

Even though in both ancient and later traditions of magical texts things in the celestial and terrestrial worlds were relatable, the framework within which those relations were conceived was, in my view, fundamentally different in those traditions. What is potentially distorting about classifying the systematic knowledge of *materia medica*, incantations, and ritual with magic is not so much that as magic it is made to oppose science, but that as magic it is taken as a system for influencing or controlling supernatural agents. That definition can have no purchase whatsoever with cuneiform knowledge. Descola quoted the following relevant insight of Durkheim: "In order to call certain phenomena supernatural, one must already have a sense that there is *a natural order of things*, in other words that the phenomena of the universe are connected to one another according to certain necessary relationships called laws. Once this principle is established, anything that violates these laws necessarily appears to be beyond nature, and so beyond reason."[133] And he added: "As Durkheim stresses, such clarifications become possible only late in the history of humanity, since they resulted from the development of the positive sciences undertaken by the Moderns. Far from indicating an incomplete determinism, the supernatural is an invention of naturalism, which casts a complacent glance at its mythical genesis, a sort of imaginary receptacle into which one can dump all the excessive significations produced by minds said to be attentive to the regularities of the physical world but, without the help of the exact sciences, not yet capable of forming an accurate idea of them."[134] What the cuneiform divinatory and magical texts represent are not examples of the irrational, but rather contexts of the rational, given their epistemological and ontological assumptions. Heeßel, I think, was addressing precisely this when he proposed, in the context of the use of analogical reasoning in late astro-medical texts to adopt from Antoine Faivre the concept of "animated nature" (*belebte Natur, nature vivante*), a concept taken from various forms of what Faivre called Western esotericism (including many "occult" sciences). Heeßel remarked: "It is important in this context to bring to mind the fundamental, qualitative difference between the new idea of a Nature conceived as animated, and analogical thinking, long since known in Mesopotamia. Thinking in analogies assumes that open and hidden correspondences exist between certain parts of the cosmos, which those who are knowledgeable can read and decipher. . . . From here it is, however, a huge step, if not a leap, to the conception of an 'animated Nature' and the related idea that everything is not only connected to everything but also mutually influential."[135] Analogy, and perhaps also some form of ontological analogism, therefore, provides a continuity from the reasoning typical of earlier cuneiform scholarship to that of the later period, though something new is certainly reflected in late astro-medical texts, as Heeßel rightly described. The discontinuity was in the method of correspondences and properties that integrated traditional celestial divinatory and medical practice with zodiacal astrology and its numerical schemata.

I remain, however, somewhat skeptical about Faivre's Western esotericisms as comparanda, consequently of his notion of a living nature as descriptive of the Babylonian scribes' conception of the external world. I question whether the conception of nature, that universal order of things constitutive of various naturalisms (methodological, ontological, epistemological) without a counterpart in the cuneiform world, is helpful for describing the way in which the scribes saw and understood that external world. For the present discussion, though, the more important aspect is that analogical and associative reasoning functioned as a way to make rational inferences about the meaning of phenomena and did not exclude other kinds of rational reasoning attested in other areas of cuneiform scholarship and in ancient Mesopotamian life in general.

Even though continued use of the descriptor *magic,* with its historical freight of irrationality, can prevent us from seeing how cuneiform knowledge provided a context for rational reasoning, I am sure that we are not finished with the term. Schwemer recognized the crux of the matter when he quoted Versnel's remark that "you cannot talk about magic without using the term magic."[136] However, in setting aside ditches, divides, and dichotomies, however big or small, a reconstruction of the intellectual contexts of rationality, here divination, astronomy, astral medicine, and magic, and the reasoning strategies employed within them, such as the analogical and the deductive, will further our understanding not of the imagined and imaginary "Mesopotamian mind," but of cuneiform intellectual history.

CHAPTER SIX

Causality and World Order

uṣurāt ilī kīnāti "The Enduring Designs of the Gods"

> The only immediate utility of all sciences is to teach us how to control and regulate future events by their causes.
>
> HUME, *An Enquiry concerning Human Understanding*

> This is the difference between us and the Etruscans, who have consummate skill in interpreting lightning: we think that because clouds collide lighting is emitted; they believe that clouds collide in order that lightning be emitted. Since they attribute everything to divine agency they are of the opinion that things do not reveal the future because they have occurred, but that they occur because they are meant to reveal the future.
>
> SENECA, *Questiones Naturales*

CAUSALITY AND LAWS

In the entry "Naturalism" in the *Stanford Encyclopedia of Philosophy*, David Papineau pointed out that,

> there is an interesting history to modern science's views about the kinds of things that can produce physical effects. The mechanistic physics of the seventeenth century allowed only a very narrow range of such causes. Early Newtonian physics was more liberal, and indeed did not impose any real restrictions on possible causes of physical effects. However, the discovery of the conservation of energy in the middle of the nineteenth century limited the range of possible causes once more. Moreover, twentieth-century physiological research has arguably provided evidence for yet further restrictions. It will be worth briefly rehearsing this history, if only to forestall a common reaction to ontological naturalism. It is sometimes suggested that ontological naturalism rests, not on reasoned argument, but on some kind of unargued commitment, some ultimate decision to nail one's philosophical colours to the naturalist mast. And this diagnosis seems to be supported by the historical contingency of naturalist doctrines, and in particular by the fact that they have become widely popular only in the past few decades. However, familiarity with the relevant scientific history

> casts the matter in a different light. It turns out that naturalist doctrines, far from varying with ephemeral fashion, are closely responsive to received scientific opinion about the range of causes that can have physical effects.[1]

As Papineau said, even in the "heyday of Newtonian physics, science raised no objections to non-physical causes of physical effects."[2] At the culmination of a long tradition, ontological naturalism only recently provided the exclusive causal underpinnings for science, together with physicalism, according to which doctrine "any state that has physical effects must itself be physical."[3]

Far from any notion of a unifying universal nature, or a concern to discover physical causes, what underpinned the system developed by the cuneiform scribes for understanding visible phenomena was the capacity for relation and correspondence. Such correspondence was based frequently upon analogies and associations between words for, or spellings of words for, things, as discussed in the last chapter. The correspondences were sometimes derived from schematic arrangements, such as the four quadrants of the visible lunar disk as a means of schematizing the meanings of lunar eclipses and how to relate them to geographical regions on earth.[4] Because patterns in divinatory interpretation were applied in general across many unrelated phenomena, as though regulated by established principles, they bear discussion in the context of ideas about law in the cosmos. There seems, however, to be a gap between the application of such schemes by various semiotic relationships and the more mechanistic system of correspondences and relations that ultimately underpinned astrology in later antiquity. The gap is to be attributed to a difference in cosmology defined by the presence or absence of the idea of nature. Lynn Thorndike viewed astrological influence as tantamount to a universal natural law. As he put it, "During the long period of scientific development before Sir Isaac Newton promulgated the universal law of gravitation, there had been generally recognized and accepted another and different universal natural law, which his supplanted. And that universal natural law was astrology."[5] It is clear that such an idea requires the framework and structure of nature within which to theorize that the eternal heavens controlled and dominated the changeable earth, or as Thorndike said, "that celestial influence was the general and universal cause of all inferior nature."[6]

In the cuneiform world, making correspondences, connections, and relations through semantic and orthographic association, number schemes, and so on, were the expression of a different orientation to the world, not physicalistic (or materialistic) in nature. The plurality of things, and the ways in which they could be seen as interconnected, was not expressed in terms of a singular causal framework meant to function in the way that nature has served for us since the modern era to universalize investigation of physical phenomena. Indeed, the very term *physical phenomena* implies, via etymology of the word *physical* from Greek *phusis*, its identity within nature. In the cuneiform world it was the effect of divine will manifested in omens that pervaded the two visible realms of the cosmos, the terrestrial and the celestial.

CONNECTIONS BETWEEN THINGS

When Cicero, in his treatise *De Fato*, focused on the nature of propositions, particularly those that "make a statement about a future event and about something that may happen or may not" (*De Fato* I.1), a two-thousand-year-long tradition of such propositional statements about future events in the form of Babylonian omen texts stood in the background and was certainly known to him (*De Divinatione* I.1, cf. 19). We are fortunate to have Cicero's works on divination because of his engagement with other thinkers, especially Stoics such as Diogenes of Babylon[7] and Chrysippus, whose works are not otherwise preserved. Consequently, we know the terms of a certain portion of the ancient discourse on divination, the Greco-Roman portion, some of which was contemporary with the late, that is, third century and later, continuation of Babylonian divination, especially astrology. Because some of the philosophical argument among Greeks and Romans interested in prognostication from signs had an underlying polemic attached, that is, does divination really work, and if so, how, their discussion touched upon the subject of causality and how signs and their portents were thought to be connected. This chapter redirects the question of causality to the cuneiform portion of ancient divination. Of central interest are the questions of how Assyro-Babylonian divinatory texts constructed connections between signs and portents and whether inference can be made to a sense of causality in those connections.

The difficulty of accounting for causality has had a long and distinguished career, made particularly acute in the eighteenth century by David Hume. "The problem," quoting Wesley Salmon,

> is that we seem unable to identify the *connection* between cause and effect, or to find the *secret power* by which the cause brings about the effect. Hume is able

to find certain *constant conjunctions*—for instance, between fire and heat—but he is unable to find the connection. He is able to see the spatial contiguity of events we identify as cause and effect, and the temporal priority of the cause to the effect—as in collisions of billiard balls, for instance—but still no *necessary connection*. In the end he locates the connection in the human imagination—in the psychological expectation we feel with regard to the effect when we observe the cause.[8]

The way connections between things are conceived in cuneiform omen texts is unlike that of fire and heat or the collisions of billiard balls. Yet we can read them as a reflection of the imagination, of culturally particular ways of imagining the connections between events and how phenomena bear meaning for human beings who observe or know them. Omen texts therefore seem to be a prime body of evidence for our consideration of Assyro-Babylonian thinking about causality, particularly among the intellectuals of the Neo-Assyrian period, from which most of the sources come.[9]

As long as we seek from our own physical perspective the connection between sign and portent, it remains difficult to meet the Assyrian and Babylonian scholars on their own terms. We are fortunate to have a rich body of evidence in the many series of conditional statements "If *P* then *Q*" that refer to ominous phenomena and their consequents. Unfortunately, as compared with later Greco-Roman texts concerning divination and astrology, no explication of the relation between signs and what they signified, such as was attributed to Stoic philosophers, specifically interested in conditionals,[10] is given by the cuneiform scribes. On the surface, the "If *P* then *Q*" statements in cuneiform omen texts express the expectation of something (the consequent or apodosis) on the condition of something else (the antecedent or protasis). But what *is* the connection between *P* and *Q* in Babylonian omens? It is an interpretive challenge to penetrate below the surface level of the Akkadian omens to how the ancients thought *P* and *Q* were related, hence to the level of the ancient theory of divination.

DIVINATORY SIGNS AND CAUSALITY

In terms of form, as mentioned briefly in the previous chapter, Babylonian omens conform to the most common of all inference schemes, modus ponens, the mode that affirms the consequent: If *P* then *Q*. *P*, therefore *Q*. This rule of inference was first identified, or defined and expressed as such, by Stoic philosophers of the Hellenistic period. In Stoic logic this formulation was called the "first undemonstrated inference scheme." One can view the entire body of knowledge codified in cuneiform omen texts as representing the first proposition "If *P* then *Q*" of the Stoic inference scheme. Divinatory practice extrapolated from these propositions by way of inference, we would say, if we are analyzing the divinatory process in terms of its logical structure. In the vast corpus of such omens, inference from *P* to *Q* is made on the basis of a variety of relations construed between them.

Among the possible relationships that would establish a connection between *P* and *Q*, as discussed before, an important one is analogy, analyzable in terms of the cultural norms of the scholars. In some cases antecedent and consequent are linked by a phonetic or semantic relation between a word in the protasis and one in the apodosis: "If the coils of the intestine look like the face of Huwawa (written logographically ^{d}ḪUM.ḪUM): it is the omen of the usurper king (*ḫammā'u*) who ruled all the lands."[11] Here the antecedent is related to the consequent by a word play based on the phonic echo of ḪUM.ḪUM in *ḫammā'u*, not by any empirical connection between intestines coiled that way and a usurpation. Another example is from the Old Babylonian *Izbu* series has it that "if the anomaly has the face of a lion, there will be a harsh king and he will weaken that country," where *nēšu* "lion" and *enēšu* "to weaken" provide the link.[12] In the case of the *Izbu* omen the relation is embedded in orthography, as *nēšu* is written logographically (UR.MAḪ). Such relations of sound attraction between words in the antecedent and consequent are found throughout the omen series.

Conceptual rather than phonic connections were also frequently made, as in the Jupiter omen: [If Jupiter becomes steady in the morning]: enemy kings will be reconciled.[13] Here Jupiter's steadiness is connected to reconciliation between enemy rulers on the basis of an analogy drawn between the bright planet Jupiter's association with the god Marduk, king of the gods. Jupiter thus connotes rulership, and its "steadiness" connotes rectitude and stability, thereby related to peace between enemy kings. Another omen correlates an unnamed star trapped inside the moon's halo with the situation of the king and his troops being besieged (SAA 8 376: 4–5), in what seems to be a visual metaphor. Consistent with this style of making connections, a Late Uruk omen text sets planetary latitude in relation to market prices, as follows: "If Jupiter is faint, or attains minimum latitude, or disappears, and Mars is bright or attains maximum latitude, or Mars and Jupiter are in conjunction: Business will greatly decrease and the people will experience severe famine."[14] As these examples suggest, resemblance on the levels of the conceptual, visual, and the phonetic is the essence of the connection between antecedent and consequent, from the beginning to the end of the tradition of celestial divination, Old Babylonian to Late Babylonian.

Connections made by resemblance between some aspect of the sign *P* and its consequent *Q* lack the dimension of necessity that connects cause to effect according to our way of thinking. We therefore would rather say the omens reflect a system of correlation not causation. But this is because we define a cause as something that directly and necessarily produces an effect, that is, that the antecedent should be directly, physically, and necessarily responsible for the consequent. Elizabeth Anscombe said, "It is often declared or evidently assumed that causality is some kind of necessary connection, or alternatively, that being caused is—non-trivially—instancing some exceptionless generalization saying that such an event always follows such antecedents."[15] This, however, is clearly not the kind of sign relationship the cuneiform scribes used in constructing omen texts.

In the realm of cuneiform divination, the idea of "exceptionless generalization," from a mechanical-causal standpoint, is clearly not found, but in consideration of the schemata and patterns of association, such as in the analogies just illustrated, a certain resonance can be noted. Inherent in analogies is a generalizing force, in the sense that similar things behave in similar ways. And although the aim of interpreting signs, seemingly, was not to generalize but to relate a sign of a particular character to a particular portent, the connections established in the omen texts were in fact held generally. If they were not, the utility of the omen list for future reference and guide to the interpretation of signs would have been nil, and the fact that the lists continued to be copied and cited by the scribes in letters and reports concerning observed signs plainly testifies to their general application.

If we are not concerned with exceptionless general physical causality, or with causality in a mechanical way, is there a place for causal thinking within the Mesopotamian divinatory system? Perhaps the signification of phenomena and the causality of phenomena are not mutually exclusive, but operate on different and not incompatible levels. That we can identify the techniques by which an omen signifies should not mean that the system is reducible to the mere manipulation of words, symbols, analogies, or any of the other linguistic devices that create meaning. Divination is fundamentally a technique of communication with divinities. It is perhaps in terms of this that we can see the emergence of a causal language, revealing where causality is located in the framework of thought and experience to which Assyro-Babylonian divination belongs.

DETERMINISM AND THE GODS

That the gods speak to human beings within the Mesopotamian divinatory system is expressed in a variety of ways, for example, in constructions with verbs performed by gods such as *parāsu* "to decide (a decision)" or "make a judgment (as in a legal case)"; *dânu* "to make a judgment (in court)"; *šâmu* "to decree (as in fate)"; and *wamā'u* or *amû/awû* "to speak," *tamû* "to swear an oath," or *ṭêmu* "to inform, give orders, command"; and their related nouns, the objects of the gods' performative speech: *purussû* "decision, judgment, or verdict"; *dīnu* "judgment"; *šīmtu* "fate"; *tamītu* "speech, or oath"; *tāmītu* "oracle query"; and *ṭēmu* "report, decision, or counsel." Tzvi Abusch has shown the parallelism between *ṭēmu* and *alaktu* in the meaning "decision," "decree," or "oracle,"[16] suggesting specifically that this oracular decision is conveyed through ominous celestial signs, as the request for such a divine pronouncement is addressed to the gods of the night, or to specifically named astral deities. In the opening lines of *Maqlû* ("Burning"), Abusch translates *dīni dīna alaktī limdā*: judge my case, grant me an (oracular) decision."[17] The synonymy of *alaktu* with *dīnu*, *ṭēmu*, and *purussû* also extends the legal metaphor in play within the semantics of divination.

Omen divination therefore evinces a fundamental anthropomorphism, where what we call nature is perceived as divine speech, matter turned expressive, meaning materialized in the world of phenomena. In an omen, celestial bodies (or other phenomena) function as parts of divine speech, elements of meaning that can be "read" and interpreted in accordance with a grammar of repeating structures of sense. Focusing on the ancient metaphor of divine speech is one way in which we might understand how divination by omens from the phenomena, surely one of the primary vehicles for what we think of as Babylonian knowledge, fits within a broader view of the connections between observed or imagined phenomena and human social life.

The crux of omen divination lies in the relation between the antecedent *P* and the consequent *Q*. The grammar of the conditionals that form omen statements, with the verb of the *šumma*-clause in the preterite and the verb of the apodosis in the durative, suggests that *P* is temporally antecedent to *Q*. This implies that whatever is given in the second clause is expected to occur after or in some sense "as a result of" the phenomenon in the first clause. That magic was thought to be efficacious by virtue of divine-human communication further suggests that Babylonian omens functioned as defeasible, or contingent, conditionals, that is, *P* implies *Q* unless something else obtains—namely, the apotropaic ritual appealing to the divine to undo the consequent.[18] Divine agency, therefore, not mechanical causality, explains the relation between antecedent and consequent in omen statements.

One of the more telling indications of the force of divine agency in the antecedent/consequent relation is that the Akkadian term for the consequent is *purussû* "(divine) decision"or "verdict."[19] *Enūma Anu Enlil* Tablet 20 includes an instruction to the diviner at the close of each omen to observe the moon-god's eclipse and to "hold wind such-and-such in your hand"; thereby a decision (*purussû*) is given for such-and-such land and the king of that land.[20] Indeed, many comparable references to the apodosis clause, the consequent, as a divine decision or verdict (*purussû*) are to be found in omen commentaries, reports, and letters from scholars concerning omen texts, as well as other genres. Conjurations and rituals for the purpose of "seeing a divine decision" are found in the subscripts to the prayers ending with the statement that "if you do such-and-such, you will see a divine decision" (*purussâ tammar*). It is still found in this usage in the subscript to a late astronomical text from Uruk, which reads: "In order for you to see a divine decision concerning the king" (BE-*ma* EŠ.BAR 3,20 *ana* IGI-*ka* TU 11 rev. 37).[21]

The depiction of Mesopotamian gods as judges, who issue decrees that establish the way things are, is well known. Divine epithets such as "the ones who judge the law of the land," "who draw the cosmic designs," "who decree the destinies" all exemplify the gods' ultimate power to decide, control, and command. Considering that written omens and written laws in cuneiform culture share the same casuistic "if . . . then" formulation, this suggests that in dealing with omens we are not only in the realm of signification but, conceptually speaking, of case judgment as well, and that, in the manner of the so-called law codes, the omen compendia represent a kind of codification of divine judgments. In this sense the omens in the cuneiform texts reflect a written record of what has been promulgated through divine speech. But before the omens were codified on cuneiform tablets, the signs themselves had an explicitly written character, being an inscriptional record of divine will in the observable world. Thus communication between god and human was made possible. Phenomena, by this reasoning, not only represented, or manifested, the divine word but also embodied the further notion that the future is written by the gods on the world.

The conception of ominous phenomena as a written language is well attested in first millennium scholarship. Whether it goes back to the third millennium to the description of the goddess Nisaba's tablet, the *dub mul-an*, attested both in the Gudea Cylinder[22] and in the Sumerian composition "The Blessing of Nisaba," is another question.[23] The meaning of *dub mul-an*, literally "tablet of the 'star of heaven,'" can be read as employing a complex metaphor. The reading of *mul* not simply as "star" but also "script" can be made on the basis of an Old Babylonian lexical correspondence between *mul* and *šiṭirtum*, thus "the tablet of heavenly writing,"[24] or "the tablet of the star [which is] the writing of heaven." It is tempting to read the metaphor back even further and extend it to the "tablet of the 'star of heaven'" itself, in which case *dub* is metaphorical for the sky as its script is for the stars. The metaphor of the celestial bodies as signs written on the heavens extends itself as far as Neo-Babylonian royal inscriptions,[25] and beyond the cultural boundaries of Mesopotamia in the magical and religious-philosophical literature of Late Hellenism. The divinity of script, in Greek and Late Antiquity, seen as the letters of the alphabet (*stoicheia*), was projected onto the heavenly cosmos in much the same way. Notably, in a passage arguing for a linguistic signification through analogy, reminiscent of what has been defined above as the antecedent-consequent relation in Babylonian omen texts, rather than for a causal nature of the heavenly phenomena, Plotinus said: "those who know how to read this sort of writing can, by looking at them as if they were letters, read the future from their patterns, discovering what is signified by the systematic use of analogy."[26]

Omens and divinatory texts support an even more general conceptualization within scribal scholarship of the cosmic order of things as being the result of divine command and utterance. Illustrative of this broader conception, that the world order is produced by the creative power of divine word, is the passage cited earlier from *Enūma Eliš*: "At your (Marduk's) word the constellation shall be destroyed, 'Command again, the constellation shall be intact.'"[27] And in a prayer accompanying an interrogatory divination, a request for an oracular consultation in a Sultantepe text, the diviner says, "The gods, your fathers, listen to your sublime words. . . . Since you have been so kind (before) as to let me know your divine decision (*ṭēmu*), so (again) send me your decision (*ṭēmu*) and let my mouth pronounce it!"[28] Further reference to divine speech is found in the oracle queries, called *tāmītu*. In a prayer to Sin the great gods ask the moon-god to "give the divine answers to the oracular questions (*tāmītu*)," specifying the day of the disappearance of the moon on the twenty-ninth as the day of Sin's responses. As W. G. Lambert pointed out, the various usages of the term *tāmītu* have in common the basic root meaning of "formal speech or judicial utterance."[29]

The linguistic underpinning of what we might call the Babylonian theory of divination seems also to be reflected in the use of the term *pišru* (from *pašāru* "to release, undo, or solve") by the scholars to refer to interpretive elements of an omen, including the quotation of an omen (preceded by the phrase *anniu piširšu* "this is its interpretation") for the purpose of elucidating an observed sign.[30] The verb has a number of usages, including "to recount" as of a dream, and "to explain or report" what someone said.[31] The root has the force of releasing or revealing, in this case, one form of speech by means of another. The noun *pišru,* used in the scholars' reports overwhelmingly with reference to celestial omens, is defined in *The Assyrian Dictionary* as "interpretation" or "hidden meaning." In a letter from the scholar Balasî, the meaning, or interpretation (*pišru*), of monthly omens—*šume ša urḫāni* (ITI.MEŠ) are said to be "not comparable" (*la mušul*), and that each goes its own way. Each omen having its own interpretation, if that is what this means, affirms the importance not of generality but of particularity in the system. How causality figures in the scribes' language with regard to divination is supplied in the statement from the same letter that "the one who 'made' (*epēšu* 'caused') the earthquake also made the *namburbi* against it." The responsible party is Ea: "Ea has done/caused (*epēšu*), Ea has undone/solved (*pašāru*)." Both the sign and the ritual are thought to come from the same divine source. The connection between the sign *P* and its portent *Q*, so we may then infer, is attributable to divine intent, presumably actualized by divine verbal pronouncement.

There seems to be little meeting ground between what we can extract from Balasî's letter or other evidence of Assyro-Babylonian divination and more familiar parts of the history of the investigation of causality. In Greco-Roman antiquity the discussion about causality was often tied to various commitments regarding necessity.[32] But the classical philosophical tradition is rife with ambiguity on the question of whether the relation between the antecedent and consequent of astrological divination involved necessity and its implied determinism. Cicero, in his literary dialogue with the Stoic Chrysippus, discusses the deterministic implication of the omen about Fabius's birth at the rising of the Dogstar—namely, that he will not die at sea. Having already occurred in the past, Cicero argues, Fabius's birth at the rising of the Dogstar cannot be changed and so is necessary (on the grounds that "what is past cannot turn from true into false" and "all things true in the past are necessary" [*De Fato* vii.14]). But the necessity of the occurrence of the second proposition, that he will not die at sea, depends on whether one thinks (as Cicero claims the Stoic Diodorus does) that there is a correspondence between the necessity of the antecedent and the inevitability of the consequent.[33]

These are not concerns of the Assyro-Babylonian scholars. In the absence of grounds for seeing any interest in determinism or necessity with respect to the signs and portents, the connections between the sign and portent in cuneiform texts is closer to the Humean idea of constant conjunction, though Hume referred to conjunctions between physical things (Hume 2.3 1739 book I part III sections IV and XIV), not as in the omen texts, between observed phenomena and social events.[34] In Richard Sorabji's words, the Humean idea is that "If A causes B on a particular occasion, this implies that events like A are constantly conjoined with events like B,"[35] and he draws the distinction between deterministic laws and constant conjunctions, where *A*'s causing *B* implies that a deterministic law relates *A* to *B*.

The deterministic law is the equivalent of the "covering law model of explanation," associated with Carl Hempel.[36] This model consists of three elements, the universal generalization (law statement) in the form of "whenever an event of type *b* happens, an event of type *a* happens,"[37] the initial conditions or occurrence of event of type *b*, and the consequent, which is the occurrence of *a*.[38] On the surface this model bears formal resemblance to the reconstruction of how an omen works, though one would be reluctant to see the pairing of the events in the omen statement as the equivalent to the "law statement," if the basis of the universally general law is physical. But it also bears resemblance to the Stoic "first undemonstrated inference scheme." It is interesting, however, given the Assyro-Babylonian juridical terminology of divine "verdict" to think about the resemblance to a "law" given by divine command. Sorabji concluded, in his study of causality in Greek philosophy, that the idea of a cause as a necessitating condition, or part of one, is original to the Stoics, but had substantial longevity after Hellenistic antiquity.[39] Babylonian omens may bear a formal resemblance to later Greek formulations of the relation of two events by logic, determinism, necessity, or law, but do not share the philosophy, physics, or cosmology that underpin them. Both, however, share a drive to create a rational system that can apply generally to a great many and various particulars.

The questions both of fate and causality in cuneiform intellectual culture seem to have been attributed to the gods, specifically through their judicial decision-making role. Even the omens could be changed by the gods, as in this line from a prayer to Nabû: "You (Nabû) are able to turn an untoward physiognomic omen into (one that is) propitious."[40] The scholars were equally unambiguous about the role of the gods in changing the interpretation of a sign, as in this statement from Nabû-nadin-šumi: "the king, my lord, should not be worried about this omen (*ittu*). Bēl and Nabû can make a portent (GISKIM) pass by. They will cause it to bypass the king, my lord. The king, my lord, should not be afraid."[41]

Divine decisions were conceived of as being inscribed in or on the world, and the scholars who were trained to read and interpret the divine script thereby had access to knowledge of future events. In his work on divination, Cicero (*De Divinatione* I.127) said that "since everything happens by fate . . . if there were a human being who could discern the connection of all causes with his mind, surely he would never err. For someone who grasps the causes of future things necessarily grasps what the future thing will be. But as nobody can do this except god, it is left to human beings to gain their foreknowledge by means of signs which announce what will follow."[42] I read this as fitting comfortably with the Babylonian omens and the cuneiform scholars' language of divine causality. Indeed, it is the idea that signs are not themselves causes but rather convey divine decisions about what will happen that makes for common ground.

Babylonian divination, particularly celestial divination and astrology, came within the ambit of Hellenistic intellectuals. Although the texts are mostly no longer extant, titles of Greek works concerning the problem of signs, fate, and causality attest to the continuing interest in these subjects for at least four hundred years, beginning with Chrysippus in the third century BCE.[43] Evidence for the transmission to and influence on this tradition from Babylonian ideas about divination, fate, and divine causality is indirect but compelling. The idea of the linguistic character of ominous signs persisted into much later antiquity, as for example when Augustine, influenced by and echoing Plotinus, said, the positions of stars are "some kind of speech which foretells the future." This comment is found in Augustine's argument against stellar influence, that is, against the idea of astrological signs as causes.[44]

It is significant that Babylonian scholars sought to formalize their understanding of the gods' judicial role in the cosmos in a vast system of conditional statements. The meaning of conditional statements can vary widely, from co-occurrences without an understood causal connection (Hume's "constant conjunctions," such as fire and heat), to certain events in the future necessarily following certain events in the past (Cicero's analysis of the omen of Fabius being born at the rising of the Dogstar[45]), to certain kinds of causal relations (where "If *P*, then *Q*" can mean *Q* occurred because *P* occurred, often taken as an explanation of the form "*Q* would not have occurred but for *P*," for example, if he trips, then he will fall). We can look at the omen statements and say that *P* and *Q* are not causally related, since they bear no physical or mechanical relationship. But a different causal language can be derived, both from the omens and from related literature that draws another picture altogether, one of divine causality effected not only through speech in the form of judicial verdicts but also through a kind of writing on the world, as on a clay tablet, in the form of the ominous phenomena themselves. If divine causation, described this way, subsumes all other forms of causal links between events, then Babylonian ideas about causality are the result of a focus on the role and effect of the divine in the world, and distinctions between god, causality, and fate are not sharply drawn, but make possible the statements about future events and about what may happen that we find preserved in the textual record of cuneiform divination.[46]

THE LAWS OF NATURE BEFORE THERE WAS "NATURE"

Missing from the discussion of causality thus far is a trope that is frequently invoked as an important and decisive concept in the development of causal explanation of phenomena: the laws of nature. In the historical discourse about nature, especially about nature's relationship to gods, or God, the use of the conception of law as a way to describe perceived order and regularity in the world of physical phenomena shows nearly continuously from Greek and Greco-Roman antiquity down to the seventeenth century. Even today the several related conceptions, the laws of nature, laws of physics, and laws of science, perpetuate a metaphor of law to refer to structures purportedly embedded in a world apart from human thought or intervention, and in the view that the aim of science is the discovery of these laws. It is the power of the metaphor of the "laws of nature" to connect what is seemingly outside the sphere of human culture with what is human and cultural. Real and independent as we may think nature and its orderliness are, the very notion of physical phenomena being subject to laws is a profoundly cultural claim, one that imparts a human value to the world external to human society. In so doing, that noble part of civilization, law, is further dignified by being written into the very substance of the world, and, in turn, the world is made intelligible and even predictable by its law-like behavior.

Three conceptions drawing on the legal metaphor are endemic to the development of Western science. They are divine law, the order imposed upon nature from a transcendent source; the laws of nature, seen as a property of the physical world; and the natural law, an ethical theory grounded in a commitment to a universal human reason. Scholarship on the origins of the conception of laws of nature or physical law, as well as of the natural law, has traditionally focused on the classical and Greco-Roman periods, with a view to tracing that history down to the fully nomological and secularized laws of nature in early modern Europe.[47] So closely associated with the West is the trope "laws of nature" that Joseph Needham cited it as a key point of difference with science in China. "In Western civilization," Needham wrote, "ideas of natural law (in the juristic sense) and the laws of Nature (in the sense of the natural sciences) go back to a common root. . . . For without doubt one of the oldest notions of western civilization was that just as earthly imperial lawgivers enacted codes of positive law, to be obeyed by men, so also the celestial and supreme rational creator deity had laid down a series of laws which must be obeyed by minerals, crystals, plants, animals and the stars in their courses."[48] In his article from the *Journal of the History of Ideas* of 1951, Needham was probably the first to assert that "there can be little doubt that the conception of a celestial lawgiver 'legislating' for non-human natural phenomena has its first origin among the Babylonians."[49] He also proposed that the Stoic universal law (*koinos nomos*) was "derived from Babylonian influences," specifically by means of the transmission of the astral sciences after circa 300 BCE.[50] But the "celestial lawgiver" of Needham, by which he meant Marduk, was not the exclusive or even the principal source for the legal metaphor in ancient Mesopotamian texts reflecting on the connection between divine and world, though it must be counted toward a statement of the idea.[51]

In the Western tradition, the metaphor of nature's lawhood was ultimately realized in two separate but related doctrines, one of the so-called natural law and the other of the laws of nature, each one predicated on the universality of nature. Although distinct in the domain of their effect, natural law applying to human social life and the laws of nature to physical phenomena, they interconnect historically in Stoicism and its philosophical heirs by virtue of the identification between the universe of nature and a divine universal creator who established both the law-like nature of nature and a universal ethical order (*koinos nomos*). The context of such ideas about physical order as a function of theocratic order, implied in the terms *ius divinum* "divine law" and *deus legislator* "divine lawgiver," has thus far been looked for in earlier Greek materials and in the various relationships assumed between the terms *nomos* "law" and *phusis* "nature."

The history of the laws of nature belongs properly to the mostly Greco-Roman period and later Greek and Latin sources that speak explicitly in those terms. Whereas the cuneiform corpus altogether lacks a lexical counterpart to the word or the conception "nature," and thus, strictly speaking, belongs prior to and outside the bounds of the Western discourse about nature, description of the relation between the divine and the world in the Akkadian language employed legal and juridical terms. Having a lexical counterpart to "nature" is obviously not a requirement for the application of the legal metaphor in cosmological thought, as Derk Bodde's study of the Chinese evidence demonstrates.[52]

Whether, or to what degree, Stoic natural law theory actually emanated from a Babylonian background is a question that would be significant for understanding the very roots of an important point of Western jurisprudential and theological thought. As Malcolm Schofield said, with Stoic natural law "the stage is set for *ius naturale* as it appears in Cicero's *de officiis* and the *Digest*—and in Grotius, Pufendorf and beyond."[53] This, however, is a goal far beyond the scope of the present work or my competence.[54] Moreover, the interest here is not in tracing the transmission of ideas from the Near East to the classical West, but only to review the cuneiform sources relevant to the connection between "divine law" and "cosmic order," leaving specialists to draw their own conclusions about its later Western legacy. To that end, first the Assyro-Babylonian trope of the divine judiciary; second its extension to the phenomenal world; and lastly, and in all brevity, the question of the casuistic, or case-law, formulation of omen statements are reviewed in the following discussion.

Cuneiform sources reflecting on the conception of law run the gamut of chronological periods, geographical locations, and genres. None are philosophical in nature and so do not afford second-order articulations on the nature of cuneiform law. Their terminology is juridical, referring to elements of the enactment of law, the administration and function of the law court, and consists of words for "legal cases" (*amatu, dīnu*), "verdict" (*purussû*), "judge" (*dajānu*), "to render judgment" (*dīna dânu*), and "decide a legal decision" (*purussâ parāsu*). Yet terms for the moral principles of truth and equity, or justice, are included within the juridical framework, such as the "equitable laws of Hammurabi" (*dīnāt mīšarim ša Hammurabi*),[55] and the hendiadys *kittu u mīšaru* "truth and right," which expresses ethical principles that effect the righteous administration of justice.[56]

From as early as there are texts that bear on the administration of justice, "law" was attributed to a divine source and was legitimated by a claim to divine foundations. The divine legitimation of royal dispensations of justice is found in one of the earliest attested of historical documents, Enmetena's account of the border dispute between Lagaš and Umma in the Early Dynastic period (ca. 2500–2400 BCE). There, it is reported that Mesalim of Kiš once decided a border dispute between those two cities, but that judgment was projected onto the god Enlil who decided the case as though it were a matter between the gods Ningirsu of Lagaš and Šara of Umma. The divine judgment was conveyed to king Mesalim through Ištaran, god of justice.[57] This early manifestation of the idea of the divine warrant for law becomes foundational for later Babylonian legal thought. As Raymond Westbrook emphasized, the gods stood "behind and above" the entire judicial system. Moreover, oaths and the ordeal[58] effectively moved the court of appeal directly to the gods, whose judicial authority was final.[59]

A number of gods have epithets calling them judges, or they appear in contexts where they function as judges,[60] but in Babylonian mythology, Shamash, lord of truth (*bēl kittim*[61]) and lord of law cases or judgment (*bēl dīnim*), is the cosmic judge par excellence. The large iconographic register atop Hammurabi's monumental law stele depicts the enthroned Shamash, patron of the law and universal judge, in the act of bestowing upon the king "truth," or "right," by extending toward him the symbols of divine rectitude, the rod (*šibirru*) and ring (possibly *kippatu*).[62] The depiction of this handing over of judicial precepts is the pictorial image of the metaphor, that is, of the underpinning and validating of law by the transfer of "right" or "justice" from divine to human.[63] The law collection confirmed the king as the recipient of "truths" (*kinātu*) from Shamash, legitimating his own promulgation of law by divine command. Although the meaning and purpose of Hammurabi's stele in all its complexity is beyond the present scope,[64] the firm rooting of the conception of justice in divine truth or rectitude is of central interest here.

In the scholarship on the *Code of Hammurabi*, discussion of the ideology of the Code has emphasized its religious and ethical context, indeed, that Shamash did not give the king laws but rather "truth" or "truths," in the plural (*kinātum*), as it is stated in the Epilogue, line 90. On this basis, Norman Yoffee defined the notion of law underpinning the code as "law in the sense of 'natural law,' i.e., law that *transcends* human creativity but whose suprahistorical ideals are those toward which lawmakers should strive."[65] E. A. Speiser had also stated that in the cuneiform tradition, "law is an aspect of cosmic order and hence ultimately the gift of the forces of the universe,"[66] seeing *kittum* as "an immutable aspect of the cosmic order" and the principle of cuneiform civil law expressed by *kittu u mīšaru* as a reflection of an ethical law in the universe. The metaphysics of *kittu u mīšaru*, whether they were thought to be "of" the divine or to transcend even the divine, is not a feature of the texts in which the terms are used, and it is difficult to posit anything transcendent of the gods in Mesopotamian terms, as any consideration of their creation accounts will suggest. What is clear in the case of legal principles of *kittu u mīšaru* is that they denote not only what was equitable and right by human law, but also later, in the Neo-Assyrian period, were used to refer to regularity in heavenly phenomena, thus extending the legal metaphor into a cosmological context.

Contemporary with Hammurabi's law collection are Old Babylonian petitionary prayers (*ikribu*) to Shamash and Adad, gods of judgment and the divinatory inspection of the exta, respectively. These prayers are a particularly evocative source for the juridical metaphor in its Mesopotamian form. Spoken in preparation for the performance of an extispicy, they ask the gods directly for judgment and a "true" (that is, just, or, reliable) verdict. At daybreak, presumably, the diviner, having made himself ritually clean by means of cedar, addressed Shamash and Adad, saying: "Cleansed now, to the assembly of the gods draw I near for judgment (*ana dīnim*). O Shamash, lord of judgment, O Adad, lord of prayers and acts of divination (*bēl ikribi u bīri*), in the ritual I perform, in the extispicy I perform, place the truth! (*ina ikrib akarrabu ina têrti eppušu kittam šuknam*)."[67] The same entreaty to "put truth (for me)" (*kittam šuknam*) is made in the Old Babylonian nocturnal prayer to the celestial deities, the "gods of night," before the lamb is offered and the extispicy is performed in the morning. The *bārû*-diviner says, "in the extispicy which I am performing (or will perform), in the lamb/ritual blessing which I am offering (or will offer), put truth for me!" (*ina têrti eppušu ina puḫād/ikribi akarrabu kittam šuknam*). Another Old Babylonian *ikribu* makes clear that the divination on behalf of the client is, metaphorically, a legal case and the sun-god is asked to put truth (or a true verdict) in the offered lamb: "(Just as) you (Shamash) judge the case of the great gods, (just as) you judge the case of the beasts of the field, (just as) you judge the case of mankind, judge today the case of so-and-so, son of so-and-so. On the right of this lamb (place) a true verdict, and on the left of this lamb place a true verdict."[68] The plea to the god Shamash, or the gods Shamash and Adad, to "put truth" in the entrails of the lamb is a formula that occurs as well in the later extispicy literature, and Lambert suggests, including *tāmītu* inquiries in which questions are posed to Shamash and Adad, who are, in turn, as in the *ikribu*s, expected to make a judicial deliberation and answer (that is, give the verdict) in the form of "written" signs upon the liver.[69]

The function of Babylonian divination was therefore to give the diviner a "hearing" with the gods and to receive "truth," as Jean Bottéro was the first to observe.[70] Extispicy was the direct process, requiring that an offering and a formal petition be made before the divine judges. Celestial (and other terrestrial) divination, on the other hand, were indirect, happening by virtue of the diviner's perspicacity, although still part of the juridical complex of thought, as evidenced by the use of the term *purussû* "verdict" for the consequent (apodosis) of the unprovoked omen statements.[71] That the celestial deities gave judgments in the form of their own appearances as signs and portents is also implied in the aforementioned nocturnal prayer (*ikrib mušīti*) to the "gods of night," where it is said, "the gods of the land, the goddesses of the land, Shamash, Sin, Adad and Ishtar have gone off into the lap (*utlu*) of heaven.[72] (There) they give no judgment, nor decide cases."[73] Not being visible, therefore, these celestial deities as manifested in the planets or weather cannot produce ominous signs.

While various gods are named as judges who convene before an extispicy, the sun-god presides as supreme judicial authority in the cosmos, both in heaven and in the netherworld. The conception of his universal jurisdiction is expressed in an incantation from the *bīt rimki* ritual: "(Utu) judge of the land, worthy of lordship, [Ut]u, great lord, king of the totality of heaven and earth." That the sun-god held sway above and below the horizon, over the heavens and into the netherworld, is further reflected in other epithets, showing him to be lord and ruler in the Great City, in Arali, and over the Anunnaki and the spirits.[74]

The ritual *Maqlû* is a nocturnal procedure in which Shamash is called upon for judgment and sentencing. In the first part, the images of the witches are raised up to the divine judge before being placed in the brazier for burning. The incantation:[75] "O Shamash, these are the images of my sorcerer; These are the images of my sorceress; . . . You, Shamash, the judge, know them, but I do not know them. . . . You, Shamash, the judge, overwhelm them so I not be wronged! Judge my case, render my verdict! Burn the warlock and the witch!"[76] The text of *Maqlû* is replete with the imagery of the divine judge. Here, judgment is specifically tied to the decision of the divine court, and to the juridical terminology already noted, an additional collocation from this text may be added—namely, "render a divine decision" (*alakta lamādu*), as explicated by Abusch.[77] In the prayers to the divine judge(s), the phrase *alakta lamādu* occurs as a complement to the expression "decree fate" (*šīmta šâmu*), as in the entreaty "Decree my destiny, render for me an (oracular) decision (*šīmti šīmā alaktī limdā*)."[78] The expression *alakta lamādu*, literally, "to make known/reveal the course," has specific reference to the course of the stars, which provided yet another means of reading the divine will through divination. Prayers to gods as celestial bodies, as for example, the prayer to Ninurta as Sirius, similarly request of the god/star that the future should be revealed by them: "To reveal the future I stand before you, to give a right judgment I pray with uplifted hands to you."[79]

Complementary to the idea of the course of the stars (*alaktu*) representing the decision of a divine judiciary is the use of the cosmic *kittu* to mean the established visible order in the cosmos, thus "right" in the sense of "regularity." An inscription of Esarhaddon states: "The stars of heaven stood in their positions and took the correct (*kitti*, "regular") path (and) left the incorrect (*la kitti*, "irregular") path."[80] In a similar passage, the moon-god and sun-god "took the road of truth and justice (*ḫarrān kitte u mišāri*)," in order to give "verdicts of truth and justice," that is, through their regular celestial signs.[81] Finally, a letter to Assurbanipal from his son makes the same references to the regularity, in this case, of Jupiter, as an indication of the propitious and auspicious nature of Assurbanipal's reign: "In your days, which Marduk loves, Jupiter has taken the right paths in the heavens (ii 21′ dSAG.ME.GAR *ina šamê ḫarrāni kittu iṣṣa*[*bat*]), (while) Mars, your star, is clothed with a glitter in the heavens. As for you, total power [. . .]. Your star is clothed with a glitter like UR.GAL; the true king of the lands, Aššur, has shined forth with truthful judgement (ii 26′ *dīn kitti*). You have [. . . ed] the closed gate of heavens [. . .].[82]

The locus classicus for the establishment of order in the cosmos is in the Babylonian poem, "When Above" (*Enūma Eliš*) Tablet V. There, Marduk fixes the heavenly bodies in specified areas of the sky, the Paths of Ea, Anu, and Enlil. Their regularity is coordinated with the marking of time in a manner that correlates with the Astrolabe.[83] The moon-god is assigned the regular marking of the month by his changing appearance from bearing horns for six days (the crescent moon) to wearing a crown (*agû*[84]) when he stands opposite the sun on the fifteenth of each month. The moon is instructed to draw near (*šutaqrubu*) to the "path of the sun" (line 21, *ḫarrān* dUtu) and to assume the same position (*šutamḫuru*) as the sun on the thirtieth day (conjunction). No juristic terminology is found in this passage, nor is Marduk referred to as judge. But, as new creator and ruler of the cosmos at this stage of the narrative, the positions and courses of the heavenly bodies are determined by Marduk's verbal command, delivered in direct discourse and with imperatives addressed to the moon and sun as the gods Sin and Shamash. In Tablet VII Marduk's role as the god who regulates the celestial universe is reiterated with the line: "May Marduk make the courses (*alkātu*) of the stars of heaven constant and enduring."[85] Upon completing his orders to the heavenly bodies, Marduk states, "I de[signated] the celestial sign,"[86] and he commands Sin and Shamash to "proceed on its path (that of the sign), . . . approach each other and render (oracular) judgment."[87] The juridical metaphor for the appearances of heavenly bodies as divine judgments is therefore present in the passage.

A variant account of the formation and ordering of the celestial cosmos is found in the several introductions to the celestial omen series *Enūma Anu Enlil*. In the bilingual exemplar, the Sumerian version attributes to Anu, Enlil, and Ea the fashioning of the "great divine powers of above and below" (*me gal-gal-la an-ki-a*), rendered in the parallel Akkadian as *uṣurātu*, the "designs" of heaven and earth (GIŠ.HUR.MEŠ AN-*e u* KI-*tim*). In the Sumerian, the moon's course is next established as an indicator of the month (*iti*) as well as of a "sign" (*giskim*). The Akkadian version describes the celestial markers of time in both sun and moon, referring to the creation of the day and the renewal of the month. What is of interest in the present context is that these acts of cosmological ordering are, as elsewhere, effected by "divine decision," expressed in Akkadian, here, with the word *milku* "instruction," or "order," taken here to mean "(divine) decision,"[88] and in Sumerian, with the word *galga* "(fore)thought, plan(ning)," or "instruction."[89] But another lexical equivalent of *milku* is *a-rá* "way" or "course," and can be written logographically as A.RÁ, which is also the writing for *alaktu* "course" or "divine oracular decision," even "omen."[90] Despite slight differences in terminology, a unified conceptual frame of reference in which the regularity of heavenly appearances is the result of divine legislative decree appears consistently in Babylonian cosmological texts.

One last example for the legal metaphor for divine cosmic order must be adduced. It is a *namburbû* incantation that employs explicit juridical terminology with respect to the gods Ea, Shamash, and Marduk. It not only has these gods handing down the verdicts of heaven and earth, but also determining the fates and, as was attributed to Anu, Enlil, and Ea in the introduction to *Enūma Anu Enlil*, organizing, or designing, the cosmic plans (*uṣurātu*):

> [Incantation.] Ea, Shamash, and Asalluḫi, great gods, who hand down the verdicts of heaven and earth, who determine the fates, who make decisions . . . , who confirm the lots, who fashion the designs, who apportion the lots. . . . To determine the fates, to fashion the designs (is) in your hands. The fates (destinies) of life you alone determine. The designs of life you alone fashion. The decisions of life you alone make. . . . You alone are the great gods who direct the decisions of heaven and earth, (and of) the depths of the seas.[91]

The *namburbi* positions the deities Shamash, Ea, and Asalluḫi/Marduk at the highest level of cosmic creative and authoritative power, where they hold court and enact decisions in no less extensive a jurisdiction than all the world, administering judicial power throughout the universe and all life. Shamash, Ea, and Asalluhi/Marduk's decreeing of fate recalls the mythology both of the Sumerian creator god Enlil, whose holy command decided everlasting fate,[92] and also *Enūma Eliš,* in which Marduk decrees destinies in his newfound capacity as creator and king of the universe. In the incantation to this divine triad, however, divinatory judgment (*dīnu,* the law case, and *purussû,* the verdict itself) and the decreeing of fate (*šīmta šâmu*) are explicitly and repeatedly brought together within a single frame of reference.

The evidence thus far adduced for the deployment of legal metaphors represents only a small selection of a much greater wealth of sources, but already it offers a multifaceted picture. First, legitimation of civil law is secured by the transmission to the king of the principles of right and justice from Shamash, chief justice over the whole of the cosmos. This, and the prayer petitions to deities for truth in oracular judgment, make it clear that *kittu* and *mīšaru* are legal and ethical principles that are the gods' to give, ideas that lie at the center of what we would call "divine law."[93] Second, access to knowledge of the future was described by the metaphor of obtaining divine judgment through divination from extispicy or from celestial signs. Again, this places "divine law" at the center of the interaction between human beings and the gods, where human stands humble before that law awaiting judgment. Third, the idea that the celestial phenomena are signs of divine judgment was written into the very creation of cosmic order, as attested in Tablet V of *Enūma Eliš* and in the introduction to *Enūma Anu Enlil.* Creation and arrangement of a celestial order was conceived of as the product of divine decree, whether it was Marduk who assumed the role of organizer of the cosmos and, in that capacity, fixed the regular positions of the stars, thereby creating the signs both as visible indications of time and as celestial omens, or, indeed, Anu, Enlil, and Ea, who produced the celestial positions and their meaning as signs. The nature of heavenly appearances as conveying written judgments of the gods, the so-called heavenly writing, is in the background of the prayer literature when the plea is made to "reveal, or make known the course (of the stars)," meaning "to deliver a favorable judgment."[94] Finally, the divine triad, Shamash, Ea, and Asalluhi/Marduk, is placed at the center of the administration of divine law, addressed as the judges of cosmic judgments (*dā'inū dīnī ša šamê u erṣetim,* literally, "who judge the cases of heaven and earth"), who decree all fate (*mušimmū šīmāti,* literally, "who decree decrees") and establish all things (*muṣṣirū uṣurāti,* literally, "who design designs"). The Mesopotamian legal metaphor for cosmic order is therefore based in a conception of a divine law with universal legitimacy, whether it is the divine judge Shamash and his cosmic jurisdiction; Marduk as the cosmic regulator of heavenly phenomena; Anu, Enlil, and Ea's decision to lay down a cosmic plan; or the association of Shamash, Ea, and Marduk with cosmic law.

In view of this evidence it is difficult to argue against some connection between the Babylonian legal metaphorical language used to express divine cosmic order and the later history of the topoi of the natural law and laws of nature, particularly inasmuch as they too were deeply rooted in the idea of divine law. Cicero stated that "one eternal and unchangeable law will be valid for all nations and all times, and there will be one master and ruler, that is, God, over us all, for he is the author of this law, its promulgator, and its enforcing judge."[95] And, even more definitively: "Nothing, moreover, is so completely in accordance with the principles of justice and the demands of Nature (and when I use these expressions, I wish it understood that I mean Law) as is government, without which existence is impossible for a household, a city, a nation, the human race, physical nature, and the universe itself. For the universe obeys God; seas and land obey the universe, and human life is subject to the decrees of supreme Law."[96] Philo, similarly, put it, in one of the earliest statements of natural law theory, that "the same being was the father and creator of the world, and likewise the lawgiver of truth; secondly, that the man who adhered to these laws . . . would live in a manner corresponding to the arrangement of the universe with a perfect harmony and union."[97] Even as the two doctrines separated and became more fully articulated in the European Middle Ages and into the seventeenth century, Jean-Robert Armogathe showed how the term *deus legislator* "divine lawgiver" continued to figure prominently in the medieval and early modern scientific discourse, where the theological, the juridical, and the physical were intertwined. The omnipotence of God, in Armogathe's words, "manifests itself by its decrees, which constitute a juridic apparatus of govenment (*gubernatio*)."[98] Thus, as he said, "one cannot understand the transfer of concepts which have a juridical origin onto the physical world unless one understands the *theological matrix* that undergirds them both."[99] By the seventeenth century, therefore, according to J. R. Milton, "the idea of nature being governed by laws had become widely acceptable. Francis Bacon quotes James I as saying that 'kings ruled by their laws as God did by the laws of nature.'"[100]

Although strikingly similar to the use of a legal metaphor in the Western tradition of natural philosophy to describe regularity in the physical world, and the similarity found in the later Western foundations of both tropes—the laws of nature and natural law—in the idea of divine law and the expression *deus legislator*, the frame of reference and the aim of the cuneiform use of the metaphor is of a different order. The target domain for the legal metaphor in Mesopotamia was not nature, that is to say, physical phenomena qua phenomena, but rather phenomena qua signs of divine will and intent. The divine-human relation, whether effected by means of divinatory techniques to obtain knowledge directly or indirectly from the gods, or by means of ritual acts of entreaty to gain a response from a divinity, is what was described juridically, not the phenomena themselves (i.e., not nature itself). However, insofar as phenomena were taken as signs of divine communication, the legal metaphor was extended to them as well, as in Esarhaddon's use of the word *kittu* to denote the regular path of the stars, and in the formulation of omen statements as "laws."

OMENS, CASE RULINGS, AND LAWS

In view of the question about a connection between a legal metaphor describing cosmic order in the ancient Near East and the later Western idea of laws of nature, I want to consider further the formulation of omen statements in the series, *Enūma Anu Enlil*, *Šumma ālu*, *Izbu*, and so on, as laws, or case rulings, in parallelism with the civil laws of royal law collections. What it is that makes something a "law" can be a matter of locution, for example, "All *P*s are *Q*s," or, indeed, "If *P*, then *Q*." The set of phenomena deemed appropriate to such statements will vary across history and culture, as will the warrant for something qualifying as a "law," such as by empiricism, authority, or logic.

Quine saw as "our first faltering laws" what he termed *observation categoricals*, defined as the "generalized expression of expectation" in the form of "whenever *P*, *Q*," or (his example), "when it snows, it's cold."[101] He viewed such generalized associations as the "direct expression of inductive expectation, which underlies all learning."[102] The difference between observation statements (or sentences) and observation categoricals is that, while the observation statement refers to a single contemporary observation and does not contain any predictive content ("inductive expectation"), observation categoricals do represent such inductive expectations and can be passed on as cases with their own integrity. Indeed, Quine viewed them as a "sketch of a causal chain" and "a rudimentary theory of the world."[103] More explicitly, "an observation categorical is a miniature scientific theory that we can test experimentally by waiting for an occasion where the first component of the categorical is fulfilled, or even by bringing about its fulfillment, and then watching for fulfillment of the second component. An unfavorable outcome refutes the theory—the categorical. A favorable outcome leaves the theory for further consideration."[104] The causal chain underpinning the relationship defined for Quine's categoricals excludes the vast majority of Babylonian omens from qualification as such categoricals. The problem, however, is not in the relationship between the phenomena constituting such "categoricals," but the aspect of physical causality. The lists of omen cases "If *P*, *Q*" may be formally like observation categoricals but do not share the kind of relational content Quine had in mind, where physical causality obtains. There are exceptions worth noting, as mentioned briefly in chapter 5, for example, "If the sun is surrounded by a halo: it will rain." In the Diaries too, many observations of solar and lunar halos were recorded together with the rain showers or overcast skies and wind that followed them, thus casting a somewhat more empirical light on that relationship. The question remains as to whether the omen concerning the halo and rain represents a "sketch of a causal chain" or an association in the manner of other omen statements.

To return to the idea of law, what separates law-like physical phenomena from other kinds of regularities is a quality of necessity and universality in all cases. Helen Beebee described the difference between physical laws and non-law-like, or accidental, regularities in the following way: "We tend to suppose that laws (as opposed to accidental regularities) 'govern' what goes on in the universe. We tend to suppose that laws are rather like pieces of divine legislation; decrees that the universe must obey certain rules rather than mere general description of what in fact happens."[105] For her to have put it in that way is testament not only to the deeply ingrained nature of the legal metaphor but also to its theological formulation.

Indeed, Edgar Zilsel, in what is still one of the principal essays on the history of the concept of physical law, said, "the concept of physical law, as it is used in modern natural science, does not contain any ideas of command and obedience. Yet it obviously originates in a juridical metaphor."[106] Zilsel saw what he called the embryonic stage of the laws of nature in a nonempirical theologically grounded conception, where divine law and physical law were coeval. He stated that "the roots of our concept go back to antiquity. They consist in a few passages of the Bible and the *Corpus Iuris*. A few other ancient ideas are of less importance."[107] In 1942 when Zilsel wrote, the cuneiform evidence was perhaps not accessible enough to be taken into account, or it was ignored for other reasons, and it would be nearly ten years before Needham would see the relevance of Babylonian tradition.

As has been noted elsewhere, omen statements were formulated in just the same way as the "laws" in law collections, that is, in conditional sentences introduced by the subordinating conjunction "if" (*šumma*).[108] The rulings in the civil law collections are of the nature of precedents, the bases of future judicial decisions given the same material facts in a case. Despite discussion and debate over the precise nature of the civil law collections and the degree to which they represent legislative codifications, at a basic level, legislative or not, simply by virtue of being systematically arranged rulings, they can be considered codifications of those rulings, or cases.

Similarly, omen series are also codifications, in the sense of systematic arrangements of rulings in accordance with various criteria or subject matter. The crux, as far as the historical relation between omen collections, in which phenomena are arranged into codifications of "divine law" and laws of nature as recognized in later Western tradition, lies in the respective criteria by which the collected statements "If *P*, then *Q*" are taken as law-like. As omen statements, the law-like (or case-like) nature of phenomena is a function of their being correlated with other, mostly social, phenomena, rather than there being a conception of lawhood intrinsic to the phenomena themselves.

In *Enūma Anu Enlil* the systematic structure imposed upon the phenomena for the purpose of their inclusion within the series—such structures as groupings of like phenomena (for example, heliacal risings of various stars), or arrangement by variations of a single phenomenon (for example, colors of a particular planet, or the number of halos around the moon)—does not seem to argue for an interest in the inherent lawhood of those particular phenomena, in any way similar to our thinking about laws of nature. The "If *P*, then *Q*" form of the omen statements represents their status as cases in all their particularity.

Still, the case law formulation of omen collections constitutes another aspect of the legal metaphor as it was projected onto divination. As precedents, omen statements fit within the juridical metaphor applied elsewhere in cuneiform texts. Just as in Hammurabi's or the other law codes where case rulings represent what was (theoretically) decided by the judge in the case of *P*, so the omen statements refer to what was "ruled/decided" by the gods, in the event of *P*, where *P* is some possible ominous phenomenon. And just as in the use of precedent by judges to make the same decision as in a prior case where the material facts are the same, so the diviner would find the ruling in his case in the same way each time the same ominous phenomenon occurs. There need not be (physical or efficient) causality to connect the event with its consequent. Indeed, the Babylonian omen statements do not represent causally connected events in that way. The juridical metaphor for the gods' decisions, literally "verdicts," about the future through signs in the liver or in the heavens is in fact a way of expressing divine causality, that is to say, the action of an omnipotent divine will in the universe.

Omens are law-like statements because they were meaningful within a juridical frame of reference, and where judicial acts and legitimation through divine sanction were of primary significance. Because law statements and nonlaw statements can both be formulated in the same way, the formulation of omen statements as case rulings, "If *P*, then *Q*," alone does not qualify them as law statements by any other standards outside Babylonian culture. Although formulated in a law-like way, criteria such as "causal necessity," "necessary condition," "physical possibility," and "real connections between matters of fact" will not be met by the elements of Babylonian omen statements.[109] Within the Babylonian framework, however, omens and civil case rulings have a parallel formulation for the reason that the parallel structure of the universal and the human was a conscious understanding; further that the understanding of great world events (invasions, plagues, crop failure) as consequences of the appearance of ominous phenomena was itself an extension of normative social practice as seen in the establishment of codes of consequences for civil society. We would call this extension metaphorical, though it is not possible to know if such a conception makes, as it were, Babylonian sense.

The goal here was not to approach the subject of the legal language used in the context of universal order as a question of the transmission of ideas from the Near East to the West, or to trace the origins of the concepts of laws of nature or natural law to Mesopotamia, certainly not to propose a linear evolution of these ideas from the ancient Near East to Athens or to the Stoics and on to the Christian West. There are common elements to be found throughout this complex history. Perhaps the commonalities are a function of even more basic human impulses, to implement metaphors from human social mechanisms of control as a conceptual tool for the description of things manifestly not within human control.

Although the subject of legal or juridical metaphorical language is by no means restricted to the history of science, it has direct relevance for it, in that science, in its desire to understand natural phenomena, is still susceptible to the use of metaphor, the legal metaphor being particularly attractive and entrenched.[110] Examples from cuneiform texts demonstrate that the legal metaphor has had a continuous presence in the history of our engagement with phenomena. By means of the metaphor, recurring cyclical phenomena were described as being regulated by some outside agency (gods, or God), only much later transferring this same notion to one of internal regulation through (physical) law, altering but not relinquishing the metaphor. The commitment to the view of nature as having laws has had a long-standing claim on our conception and representation of nature, and even though, strictly speaking, ancient Mesopotamian cosmological texts antedate that development, their juridical language has a place in the history of that claim.

PART IV

The Cuneiform World of Observation, Prediction, and Explanation

CHAPTER SEVEN

Observation of Astral Phenomena

ina muḫḫi maṣṣarti "Concerning the Watch"

On the other hand, Epigenes, a writer of very great authority, informs us that the Babylonians have a series of observations on the stars, for a period of seven hundred and twenty thousand years, inscribed on baked bricks.

PLINY THE ELDER, *The Natural History*

OBSERVATION AND EMPIRICISM

The opening to Lorraine Daston and Elizabeth Lunbeck's 2011 *Histories of Scientific Observation* states that "observation is the most pervasive and fundamental practice of all the modern sciences, both natural and human. . . . Yet scientific observation lacks its own history: why?"[1] They point out that many kinds and contexts of observation have been the focus of study in the history of science, but that "observation itself is rarely the focus of attention and almost never as an object of historical inquiry in its own right . . . one might well wonder whether a history of observation wouldn't simply be the history of science in its vast entirety—or the still more vast history of experience."[2] Cuneiform sources for observation of celestial phenomena in particular warrant examination as part of the vast entirety of which they spoke, whether taken as the history of science or, indeed, the history of experience.

As a response to his logical-empiricist predecessors, who saw a direct relation between sensory inputs and knowledge of the world, Norwood Russell Hanson argued for the act of seeing as "an experience,"[3] invariably drawing upon other dimensions of life—one's culture, one's worldview, one's reasons for looking and watching. Accordingly, there is in seeing, and consequently observation, many facets, including perception, knowledge, language, and expectation. More than fifty years later, Hanson's idea of the theory-laden nature of observation is widely taken for granted, and with it the rejection of an epistemological given. In a well-known statement, he called attention to the importance of the particular epistemic framework of observation, as well as its historical, linguistic, and notational influences on the ways phenomena are seen and recorded: "There is a sense, then, in which seeing is a 'theory-laden' undertaking. Observation of x is shaped by prior knowledge of x. Another influence on observations rests in the language or notation used to express what we know, and without which there would be little we could recognize as knowledge."[4]

Since Hanson, it is nearly everywhere conceded that the influence of theory is at work in the very act of observation, determining what one sees. Such influence can take the form of a set of nonempirical metaphysical beliefs about how the world is structured, the understanding of how a phenomenon behaves, or an idea that motivates an observational program. Systematic observations, brought within an organized presentation designed for collective use, embody and depend upon cultural orientations to data within which the observations are believed to fit. This view raises the possibility of considering the framework, or frameworks, within which Babylonian celestial observation operated, as a significant dimension of the bodies of recorded observation.

Daston and Lunbeck's *Histories of Scientific Observation* brought fifteen hundred years of the history of scientific observation together from the sixth to the late nineteenth centuries CE. With cuneiform texts we have an equal duration of fifteen hundred years of recorded observations in records long antedating those of Western and European traditions. The observation of ominous celestial phenomena by the scribes of the Neo-Assyrian royal court and the more long-lived observation of the positions of the moon and planets archived in the city of Babylon every night for a period of some eight hundred years not only deserve a place in the history of observation but also consideration in relation to the later stream of what we value as scientific observation. A full exposition of that relation is well beyond the present scope, but if observation is what we want to understand, a look at some of the ways observations of the phenomenal world, especially those of the heavenly phenomena, were systematized and recorded in Assyrian and Babylonian scholarly texts will contribute to the big picture. And if we are also interested in how else the phenomenal world might be construed against our own deepest assumptions about nature, these same sources are again a good place to start.

Cuneiform evidence of systematic engagement with astronomical phenomena is first attested in Old Babylonian versions of *Enūma Anu Enlil* dated roughly to the eighteenth century BCE, and in related texts containing astronomical knowledge that supported celestial divination, such as is found in the Astrolabes, *MUL.APIN*, and the so-called Great Star List.[5] These texts contain a wealth of detailed schematization of the starry heavens, identifying groups of stars, the appearances and disappearances of planets, variant names for a particular star or planet, or the correlations between stars and gods. What is called the Great Star List is especially rich in providing alternative names for the planetary gods. Mars (ᵈ*Ṣalbatānu*) is the most fully preserved section for planetary names in extant exemplars, where names such as the Fiery Red, the Red, the Yellow, the Sinister, the Strange, the Hostile, and so on are given.[6] Where *MUL.APIN* classifies stars in accordance with their positions with respect to the horizon (the Paths of Ea, Anu, and Enlil), the Great Star List has a variety of classifications relevant for the divinatory interpretation of celestial signs, such as the twelve stars each of Elam, Akkad, and Amurru, or the identification of the phases of the moon during the first half of a lunar month with Anu, Ea, and Enlil. Thus:

> The moon in its appearance from the first day to the fifth, five days, is a sickle, it is Anu;
> From the sixth day to the tenth day, five days, is a kidney, it is Ea;
> From the eleventh day to the fifteenth day, five days, it wears the crown of splendour, it is Enlil.[7]

The phenomena compiled in such early texts were based presumably on long-standing knowledge derived from naked-eye experience, though in no way do these compilations represent observational texts. Such compilations of the names of stars, their seasonal risings, associations with deities, and schematic theories of the appearances and disappearances of the moon and planets are the standard content of early Babylonian astronomical texts, which in turn reflect the interest in the heavenly bodies as counterparts or correspondences (in Akkadian, "likenesses" *tamšīlātu*, from the verb *mašālu* "to be similar," or "equal") of deities, and thus could appear as signs produced by those deities. The opening of *Enūma Eliš* Tablet V 1–2 has Marduk create "the stations for the great gods" and position "the stars which correspond to them." Similarly, the passage at the close of *Enūma Anu Enlil* Tablet 22 has Anu, Enlil, and Ea "draw" the constellations as likenesses of the gods.[8] These passages point to the framework for the reference texts (Astrolabes, *MUL.APIN*, Great Star List, *Enūma Anu Enlil*) as partly the visual map of the sky and partly a divinatory structure with its codes of correspondences tied to writing conventions, that is, sign forms and spellings of the celestial bodies and their phenomena, such as the number 30 for the moon-god. Late Babylonian astronomy makes use of the logogram ÍR/*bikītu* "grief" (literally, "weeping" or "lamentation") for the maximal phase of a lunar eclipse, which is not used in omen texts but reflects a similar use of the script to convey meaning through association.[9]

Strictly speaking, ominous phenomena were visible phenomena. In celestial omens the phenomena considered ominous were expected to "appear" or be "observed," as is clear from the frequent use of the verb "to see" (*amāru* in active and passive stems) in the protases of celestial omens. In this way, the omens of all the divinatory fields embodied an observational element, in the sense that the things deemed portentous of the future were watched for only in visible domains of the world, the terrestrial or celestial. Other cosmic regions, such as the sweet subterranean watery *Abzu*, or the "Heaven of Anu," known more from literary and magical texts than from divination, were not sites for seeing omens.[10]

On the basis of the perception of patterns, regularities and irregularities, ominous phenomena were constructed from conceptions of how variation is possible. Compilations of omens in series then aimed to organize phenomena around schematic structures, which were built upon perceived and conceived phenomena so that they were amenable to the formation of meaningful correlations and associations between things celestial and terrestrial. One of many examples is the division of the lunar disk into quadrants by means of which the direction of the lunar eclipse shadow from onset to clearing could be correlated with geographical regions on earth in accordance with the four cardinal directions, and the four winds.

An important element of the structure of the antecedents ("If *P*") of all types of omen texts is the use of just such thematic schemata based on binaries (up and down, right and left), symmetries (the four directions), and other standard sequences (the five colors black, white, red, yellow/green, and variegated, or the three watches of the night). The theory of ominous phenomena to be extrapolated from the series integrates the observable with the unobservable, the occurring with the nonoccurring, the possible with the impossible. Examples of such impossible phenomena are the appearance of the sun at midnight (in the middle night watch) or the lunar eclipse shadow that travels from west to east across the lunar disk. Indeed, most of the extant Jupiter omens of *Enūma Anu Enlil* have the planet "entering," "passing," "coming close to," or "being in the middle of" fixed stars whose latitudes with respect to Jupiter's path in the ecliptic preclude these Jupiter events. In fact, as David Pingree pointed out, "this choice of constellations far removed from the path of Jupiter seems to be deliberate,"[11] because when the planet is north of the equator (between the spring and fall equinoxes) the constellations it is associated with in these omens are to the south and vice versa.

Particularly if we are interested in the observational dimension of celestial divination, such inventions in the omen series raise interesting questions, because, if we know that P cannot occur, we wonder what purpose statements beginning "If *P* . . ." serve. If we know, for example, that Venus can never be seen at the zenith, a statement beginning "If Venus is seen at the zenith" to us seems pointless and wrong. What of the eclipse omens for days of the month when eclipses do not occur? The Babylonians' attention to eclipses as omens is directly related to the early determination of their periodicity and eventual ability to predict them.[12] We therefore hesitate to say that the scribes were unaware of the impossibility of a solar eclipse, say, on the tenth day of the month, or a lunar eclipse on the twentieth. The scribes' understanding of the behavior of the phenomena suggests that such phenomena were included in the omen series not because they were thought to be usual or likely occurrences, but because they represented the limits of the conceivable, within the framework of omen schemata.[13] It also shows the limits of observation, or observational "reality," for divinatory speculation.

Exploration of such limits is certainly not unknown in later scientific tradition. John E. Murdoch and Edward Grant have described the way in which this was manifest in medieval natural philosophy. In the context of the many questions asked of the nature and structure of Aristotle's natural world, Grant said, "scholastic natural philosophers and theologians asked questions not only about what is but also about what could be, but probably wasn't."[14] And on observation specifically, he said,

> Medieval observations were not introduced for their own sake, namely, to learn more about the world. They were intended rather to uphold an a priori view of the world or to serve as an example or illustration. The idea of observation was important in the Middle Ages because it was the basis of the medieval Aristotelian epistemology, which was founded on sense perception. But it was clearly not enough . . . during the late Middle Ages, empiricism was, and remained, the servant of the analytic and a priori, which provided the "why" of things to explain and interpret the empirical world. John Murdoch has perceptively argued that although it is true that "empiricist *epistemology* was dominant in the fourteenth century this did not mean that natural philosophy then proceeded by a dramatic increase in attention being paid to experience and observation. . . . On the contrary, its procedures were increasingly *secundum imaginationem* (to use an increasingly frequently occurring phrase) and when some 'natural confirmation' of a result is brought forth, more often than not it too was an 'imaginative construct.'"[15]

These remarks on medieval science are not mentioned in order to explain cuneiform omens in terms of the epistemic practices of medieval Scholasticism, but simply to note that the history of science has accommodated a wide range of speculative imagination stemming from the contemporary or local background of ideas about the phenomena. In the cuneiform world, the choice of what to observe of celestial phenomena was guided by perception and by the potential for constructing meaning through correspondences between phenomena in the sky (and on earth) and those in the world of political and economic events. Constructions of meaning were given form in the lists of conditional omen statements, the "ifs."[16]

Not only were imagined celestial phenomena a feature of the omen series, the consequents of ominous phenomena were equally constructed. Illustrative of the constructed, as opposed to the observational, nature of omen statements are the lunar eclipse omens of *Enūma Anu Enlil* Tablets 20 and 21, in which lunar eclipses of a given description are paired with events set in various city-states well known from early Mesopotamian history. In naming the city-states of Eridu, Agade, Ur, Subartu, Gutium, Ešnunna, Emutbal, Babylon, Amurru, as well as the enemy known as the "Ummanmanda," for example, the events of the apodoses of Tablet 20 point to second and even third millennium dynasties, such as the Early Dynastic, Ur III, and Old Babylonian. Tablet 21 makes reference to areas of political significance to the eastern borders during the Old Babylonian period, such as Dēr, Elam, and Anšan.[17]

The Neo-Assyrian period *ummânū* who copied and used these texts were well educated in the dynastic past, although the exemplary meaning of that past for them, and how it was deployed for the interpretation of phenomena, renders these passages difficult for us to deal with from a modern historical perspective, certainly as the basis for dating these texts. Tablet 20 §III, for example, describes an eclipse occurring on the fourteenth day of Simānu (Month III), beginning on the east above and clearing on the west below, and lasting from the first night watch until the middle watch. Following the description of the eclipse, the text gives the "prediction," that is, the consequent ("*Q*"), thus: "The prediction (literally, "decision") is given for Ur. The king of Ur will experience famine. There will be much pestilence. The king of Ur, his son will wrong him, and the son who wronged his father, Shamash will catch him. He will die in the mourning place of his father. The son of the king who was not named for kingship will seize the throne."[18] An eclipse for the fourteenth day of Abu (Month V), beginning on the west above and clearing on the north below during the middle watch, is associated with the following events in Ešnunna: "The servant, beloved of his lord, will kill his lord. The land will become fearful. The son of the king, because of the murder, will flee from his beloved city and avenge his father. He will cut off the hand of the servant, the murderer of his father. The country, not his own, will turn to him, and he will not return to his beloved city, and the servant (variant: noble) will be captured."[19]

The apodoses of *Enūma Anu Enlil* Tablets 20 and 21 in particular have an exemplary character similar to the so-called historical omens that mention the names of historical figures (Gilgamesh, Sargon, Rimush, Naram-Sin, Ibbi-Sin) of much earlier periods. Whereas the historical omens name royal names but provide little in the way of a narrative of events, the lunar eclipse omens do not name names but have a narrative character. In a similar way to the historical omens, they are, in Beate Pongratz-Leisten's terms, "paradigmatic in nature."[20] These little stories were not meant to record all the details of historical events, but to exemplify them. They are no more empirical than the descriptions of phenomena with which they were paired as protasis and apodosis. The descriptions of eclipses in the omen protases are correspondingly paradigmatic, schematic, and exemplary, as well as sometimes purely imaginary, as in the example given above for the fourteenth day of Abu, in which an eclipse begins on the west side of the lunar disk. As a consequence, the dating of the phenomena codified in those omens, neither those of the protases nor of the apodoses, is virtually impossible.[21]

Contemporary, sometimes datable, observations of lunar and planetary phenomena, on the other hand, are found in two groups of sources. The first and earlier group, in which the phenomena are to be interpreted as omens, was produced on a special type of tablet called *u'iltu* "report," distinguished from letters or omens by being wider than they are long and written parallel to the long side.[22] Reports were written by scholars throughout the Assyrian Empire, in cities such as Assur, Kalhu, Arba'il and Kalizi in Assyria; and Babylon, Borsippa, Dilbat, Cutha, and Uruk in Babylonia. The bulk of datable reports fall within a short fifteen-year period between -679 and -665, though outliers include one from the eighth century (SAA 8 501).

Chronological distribution is difficult to know as most of the texts are not datable. They were a medium for communicating to the Assyrian king on observed astronomical phenomena that had ominous significance. A letter from Nabû-ahhē-erība makes mention of the "reports of the scribes of *Enūma Anu Enlil*": "Concerning the report on the lunar eclipse about which the king, my lord, wrote to me—they used to receive and introduce all astrological reports (literally, 'reports of the scribes of *Enūma Anu Enlil*') into the presence of the father of the king, my lord. Afterwards, a man whom the father of the king, my lord, knew, used to read them to the king in a *qirsu* (some kind of ritual enclosure) on the river bank. Nowadays it should be done as it (best) suits the king, my lord."[23] This letter shows the importance of having reports come from cities throughout the Assyrian Empire, to make sure that nothing was missed, in this case, the observation of a lunar eclipse:

> Now there are clouds everywhere; we do not know whether the eclipse took place or not.
>
> Let the lord of kings write to Assur and all the cities, to Babylon, to Nippur, to Uruk and Borsippa; maybe they observed it in these cities. The king should constantly listen.[24]

A number of reports by the scholar Nabû'a from Assur begin with the statement, "We kept watch (*maṣṣarta nittaṣar*) on the nth day," followed by short observations of the moon, or of the date of the equinox, without interpretation by citing omens from the series, as is common in other reports.[25]

Some reports sought to render omens from the series that were either obscure or invented by reinterpreting their elements, for example, references to fixed stars "reaching" or "coming close" to other fixed stars are reinterpreted as planets.[26] The reports correlated a wide range of actual observations of the moon, sun, weather, planets, and fixed stars with omens from the series *Enūma Anu Enlil* and made recommendations for the performance of ritual action in response to portended events.

While during the eighth and seventh centuries BCE the Assyrian imperial court emerged as a significant center for the observation of celestial signs, simultaneously, an observational program in Babylon called the "regular watch" (*naṣāru ša ginê*) was initiated, which produced the second group of observational records—namely, the astronomical Diaries. After the fall of the Assyrian Empire the astronomical reports to the Assyrian kings ceased, but in southern Mesopotamia the collection of observational material that would eventually become Diaries continued. Extant Diaries are datable into the first century BCE, but we have reason to assume they might have been produced until the end of the cuneiform tradition of writing in the early CE. Numbering over fifteen hundred texts, comparable to the combined number of extant Neo-Assyrian reports, letters, and queries concerning divination, the Diaries corpus is of incomparable importance for any understanding of the role of observation in cuneiform astronomy. With texts preserved from the middle of the seventh century (–651 is the earliest extant Diary, or an early form of a Diary) to the first century BCE (–60 is the latest extant Diary), the archive represents the longest running celestial observational program in history.

Indirectly supporting the idea of an eighth-century beginning for the Diaries archive is the use of the so-called Nabonassar era in Ptolemy's *Almagest* as an epoch by which Babylonian data are dated. This era was not used in contemporary Babylonian texts, but cuneiform eclipse reports also seem to begin in −747/746 (LBAT 1413–57), preserving records of 269 observed or predicted lunar eclipses and ninety observed or predicted solar eclipses (for further discussion of eclipse reports, see below). The existence of the eclipse reports suggests strongly that the Diaries, or something very like them, with their reports of eclipses, were available in this early period as a source of data. A connection between astronomy and the reign of King Nabonassar (Nabû-nāṣir 747–734 BCE) is furthermore indirectly supported by the first century BCE text known as the Saros Tablet, which lists regnal years in eighteen-year intervals, beginning in the seventh year of Nabû-nāṣir (−548) up to SE 213 (−98).[27]

Finally, the reign of Nabû-nāṣir is also associated with the inception of the Babylonian Chronicle series, which begins with Nabû-nāṣir and ends in the third century BCE, making its duration comparable to that of the Diaries.[28] Chronicles and Diaries also have overlapping interests in political, military, and religious events. The chronicles include mention of royal accessions and assassinations, rebellions and strained Assyro-Babylonian relations, while Diaries mention administrative activities in regions outside Babylonia, for example, Susa, Bactria, Transpotamia, and Sardis. Battles, plundering, massacres, and the capturing of cities are found in Chronicles, while Diaries mention the movement and fighting of troops. Religious and cultic events worthy of note in the Chronicles include sacrifices, the interruption of the *Akītu* festival, and divine statues. Diaries too are concerned with divine statues, divine accoutrements, conducting divine rites and sacrifices, clearing debris from Temple Esagil, and the kettle-drum ritual. Otherwise notable events appearing in both genres include the appearances of wild animals in the city.[29]

The political cultures of Assyria and Babylonia were distinct in many ways, but the value of divination in the scribal culture of each state was a foundation for the celestial watching throughout Mesopotamian cities where scribes were active. The fact that observational texts come almost exclusively from Babylon (few texts from other southern cities) during the period after the fall of Assyria may be an accident of discovery. Uruk was an important center for astronomical activity as well, though similar observational records are not common from there.[30]

The scribes who began the regular nightly watch of the moon and planets and the recording of their positions in the form of the Diaries were the contemporaries and peers of the Assyrian celestial diviners. The astronomical substance of a Diary, however, has little in common with omens, unless one speaks in the most general of terms. Those general terms, however, bear some significance. It cannot be entirely discounted that the continued attention to visible celestial phenomena had to do with ominous signs. At the same time, it is not possible to conclude from the shared interest in the positions of the moon and planets, or the weather, on certain dates that the Diaries' observations were solely for the practice of celestial divination. Differences in observational content, and the quantitative data—positions of the moon and planets given in numbers of cubits from the counting stars, Lunar Six data given in time degrees—distinguish this corpus from the omen tradition. The relation between the two corpora is complex, signaling neither a rejection by the Diaries' authors that certain celestial phenomena were ominous nor a singular unity of purpose with omen texts.

As mentioned before, the knowledge for which the Assyrian and Babylonian scholars developed observational practices, that is, what they were observing and why, was first motivated by the practice of divination. Apart from such nonempirical dimensions of cuneiform knowledge, however, observation, in the sense of "the watch" (*maṣṣartu*) or the "regular watching" (*naṣāru ša ginê*), played a fundamental role in the way cuneiform astral knowledge continued to be shaped. Despite the continued preservation of the omen texts until late in the Seleucid period, and further developments in nativity omens, or birth astrology, the Diaries introduced a new element into the landscape of cuneiform knowledge and a new idea into the practice of celestial observation that cannot be reduced to celestial divination.

It must be said that the Diaries did not serve to record exclusively observed phenomena. It is well known that some phenomena were calculated, either when weather did not permit any observation or just as a matter of course (equinox and solstice dates, for example). The question of the meaning of visible phenomena would become almost moot in birth astrology, as attested in the cuneiform horoscopes, because the planetary phenomena were calculated, not observed, in horoscopes. However, celestial observation continued to play a role in astronomical practices and in the kinds of records those practices produced. The meaning of celestial signs also changed with the changes in astronomical practices and the creation of various text types (omens, horoscopes, horoscopic or zodiacal omens, medical-astrological texts). It is therefore not only useful to isolate observation, that is, the systematic act of watching, for discussion in this context, but also to differentiate between observation and Empiricism. The two are obviously implicated in each other but can be separated both historically and philosophically. It seems to me arguable that both the reports and the Diaries reflect an empirical attitude ranking highly among epistemic values in cuneiform scribal culture. This attitude is also brought to bear in other intellectual contexts—in terrestrial divination and extispicy, predictive astronomy, and medicine. Here, however, the focus is on celestial divination and astronomy, leaving the question of observation and Empiricism in Assyro-Babylonian medicine or extispicy to specialists in those corpora.

In light of the idea of the interdependence of observation and theory, Alan Chalmers's notion of an observation statement can be of some use for approaching the two principal bodies of cuneiform sources that relate to the observation of phenomena in Mesopotamian antiquity. An observation statement, as coined by Chalmers,[31] is a statement of direct observation that stands in relation to experience and to its conceptual framework. He said, "observation statements, then, are always made in the language of some theory."[32] The antipositivism of this definition traces back to the highly positivistic protocol sentences, or *Protokolsätze*, of Rudolf Carnap, Otto Neurath, and other members of the Vienna Circle in the 1930s, who debated the problem of objectivity and its relation to individual experience and sensory inputs. For them protocol sentences were as stripped of inference, presupposition, and judgment as was possible in a linguistic expression. Neurath emphasized that such a statement about an observable object be distinguished by that statement's being analyzable or subject to evaluation, offered up for public assessment. An observation statement thus had a rhetorical character.[33]

Subsequently, in the aftermath of Hanson and Kuhn, the observation statement, or observation sentence, as Quine had it, moved further away from the physicalism of the positivists and represented more of a learned rhetorical statement about experience.[34] Observation statements, no longer representing a supposed raw a-theoretical perception of the world, but, being linguistic expressions, embodied cultural knowledge and understanding, thus "carving" the world in particular cultural ways. Quine called the observation sentence "Janus faced," in that "it faces outward to the corroborating witness and inward to the speaker."[35] If we are looking for observation statements (or sentences) in cuneiform, they are to be found in the *u'ilāti* "reports" and in the Diaries.

Empiricism, on the other hand, relates to a set of commitments about what science is, or what science can or cannot do with respect to describing and explaining the world. Strictly speaking, therefore, Empiricism can only be indirectly embedded within the aims of cuneiform observation statements, though this is in itself a feature worthy of consideration.[36] Empiricism has been of central importance in philosophical considerations of the foundations of knowledge from the Greek medical Empiricists to twentieth-century debates about the nature and purpose of science. Some have variously hailed and critiqued Empiricism for being antitheoretical and antimetaphysical. It has also been used to drive a wedge between Western scientific rationality and that of non-Western systems of thought.

Marshall Sahlins commented that, "historically, the West has seen an epistemological union of the empirical with the instrumental, which together make up the rational, also known as the real or objective, in contrast to the fictionality of the irrational."[37] The "irrational," or perhaps the sense of another rationality, was at the center of his objection to anthropologist Gananath Obeyesekere's position on the thinking of the Hawaiians of the eighteenth century in Obeyesekere's book *The Apotheosis of Captain Cook*.[38] Sahlins took issue and charged Obeyesekere with turning the Hawaiians "into Bourgeois Realists."[39] In his discussion of Sahlins's position, Steven Lukes found Empiricism to be decisive in defining the terms of this debate:

> This Western account of rationality is a package combining several constituent elements. First, it involves a series of dualisms: of "logos and mythos, empirical reason and mental illusion," the practical and the mystical, the observable and the fictional. Second, it involves an empiricist view of knowledge, ever since Bacon saw this "a redemption from the error of inclining before false idols, such as custom and tradition." Third, the "native Western praxis theory of knowledge . . . is not simply that we know things through their use but *as* their utilities" . . . And fourth, and most interestingly, there is the idea, clearly expressed by St Augustine, but deriving from ancient Judaism, that the world is purely material and without spiritual presence. This "cosmology is the metaphysical ground of . . . instrumental rationality . . . with the same implication of a suffering humanity alienated from an impersonal nature."[40]

Even if we set aside Western dualisms, each of which is at odds with what pertained in the cuneiform world, we are left with Sahlins's cautionary remark that "one cannot simply posit another people's judgments of 'reality' a priori, by means of common sense or common humanity, without taking the trouble of an ethnographic investigation."[41]

For the dead civilization of ancient Mesopotamia, plentiful textual evidence attests to what the Assyrian and Babylonian scribes observed—the observation statements identified in celestial omen reports and Diaries—and in what contexts the observations in those texts had meaning. This evidence runs counter to each of the four constituent elements of the "package," in particular the element sometimes most closely affiliated with Empiricism, that is, materialism, called "a sub-species of naturalism" by David Armstrong.[42]

In taking the contexts of astronomical observation into account, the relationship of observation to Empiricism can be parsed with reference to cuneiform material without embroiling the Babylonians in the same conflict about rationality that was focused on the eighteenth-century Hawaiians, unless, however, we maintain that observing the world can only be a self-conscious observation of nature. Similar to its impact on our understanding of rationality and causality in the cuneiform systems of knowledge, the conception of nature also has the potential for clouding the question of Empiricism in these traditions by introducing anachronistic ontological stakes.

Cuneiform knowledge can take its place in the history of observation without naturalism. Given the high epistemic value of "the watch" and the observation statements produced by the practice of "watching," observation played an important role in the culture of the scribes. As a primary source of knowledge for the diviners and astronomers alike, reports and Diaries testify to certain commitments about how to describe the world, one element, at least, of the way we define Empiricism. The following sections will consider the nature of the two observational programs reflected in the collections of reports and Diaries.

REPORTS ON OMINOUS CELESTIAL PHENOMENA

Observation of celestial phenomena requires the use of reference systems for assigning the positions of stars, planets, sun, and moon. Over the course of the history of cuneiform knowledge of the heavens, four such reference tools for systematic reckoning of the positions of heavenly bodies were developed: (1) the Paths of Ea, Anu, and Enlil defined within the *tarbaṣu* "cattle pen,"[43] interpreted either as the circle of heaven, or as the eastern horizon; (2) the Path of the Moon; (3) the counting, or norming, Stars, and; (4) the twelve signs of the zodiac.[44] For the omens and reports, only systems (1) the Paths of Ea, Anu, and Enlil and (2) the Path of the Moon are relevant, and frequently statements about the phenomena in those sources are given without such specifications. The earlier attested of these two celestial reference systems, the three Paths, or "roads" (KASKAL = *ḫarrānu*), described arcs across the heavens from the eastern to the western horizon.

The roads described the parts of the sky, named for the cosmic gods Enlil, Anu, and Ea, within which risings and settings of stars, constellations, and planets were observed. In our terms, the Path of Anu was reckoned as the arc over the horizon where stars relatively close to the celestial equator, within approximately 15° ± declination, are seen to rise. The Path of Enlil was to the north, and included the circumpolar stars. That of Ea was to the south of the celestial equator. In *MUL.APIN* the Path of Enlil is called the head of the cattle pen, the Path of Ea the foot. The same terminology is found in *Enūma Anu Enlil* Tablet 50, where it states that "the road of the sun at the foot of the cattle pen (belongs to) the stars of Ea; the road of the sun in the middle of the cattle pen (belongs to) the stars of Anu; the road of the sun at the head of the cattle pen (belongs to) the stars of Enlil."[45] Use of the Paths of Enlil, Anu, and Ea is attested in the Middle Babylonian period (KUB 4 47: 43–44), the Neo-Assyrian scholarly corpus, and very occasionally in later Diaries.

This letter from Mār-Issar to Esarhaddon illustrates how the Paths determined the positions of planets:

> Concerning Jupiter about which I previously wrote to the king, my lord: "It has appeared in the Path of the Anu stars, in the area of Orion (The True Shepherd of Anu)." It was low, and indistinct in the haze, (but) they said: "It is in the Path of the Anu stars," and I sent the relevant interpretation to the king, my lord. Now it has risen and become clear; it is standing under the constellation Chariot in the Path of the Enlil stars. Its interpretation will remain the same, even if it . . . to the Chariot. But the interpretation pertaining to Jupiter in the Path of the Anu stars, which I previously sent to the king, my lord, is not valid. The king, my lord, should know this.[46]

The Paths were also employed for giving the position of a comet (*ṣallummû*), as reflected in the quotation of an omen in the following report: "If a comet becomes visible in the path of the stars of Anu: there will be a fall of Elam in battle."[47] The practice was still in use in the second century BCE, as seen in this passage from a Diary from 120 BCE: "Night of the 26th, beginning of the night, the comet which [had appeared?] in the east? (in) Month I on the 29th in Aries in the path of Anu."[48]

The second reference system, attested in *MUL.APIN*, *Enūma Anu Enlil*, and the reports was a group of seventeen constellations ("gods") describing the "Path of the Moon" (*ḫarrān* ^{d}*Sin*):

> The gods who stand in the Path of the Moon, through whose regions the Moon in the course of a month passes and whom he touches: the Stars, the Bull of Heaven, the True Shepherd of Anu, the Old Man, the Crook, the Great Twins, the Crab, the Lion, the Furrow, the Scales, the Scorpion, Pabilsag, the Goat-Fish, the Great One, the Tails of the Swallow, Anunītu, and the Hired Man. All these

> are the gods who stand in the Path of the Moon, through whose regions the Moon in the course of a month passes and whom he touches.[49]

All the planets, including sun and moon, were understood to travel this same path: "The Sun travels the (same) path the Moon travels. Jupiter travels the (same) path the Moon travels. Venus travels the (same) path the Moon travels. Mars travels the (same) path the Moon travels. Mercury whose name is Ninurta travels the (same) path the Moon travels. Saturn travels the (same) path the Moon travels. Together six gods who have the same positions, (and) who touch the stars of the sky and keep changing their positions."[50] The Paths of Enlil, Anu, and Ea functioned as a qualitative expression of what we would call declination, while the Path of the Moon demarcated, again qualitatively, what we call sidereal longitude. Twelve of the stars in the Path of the Moon would come later in the fifth century to demarcate the subdivisions of the continuous path of the zodiacal circle conceived through the stars, with respect to which the sun, moon, and planets' positions were then given in degrees of arc, 30° per each of twelve signs of the zodiac. Degree of arc, or angular distance in longitude, is first used with reference to the zodiac after the fifth century BCE.

Before circa 500 BCE, the Path of the Moon was the preferred reference for the moon and planets in such texts as *MUL.APIN* as well as in the over five hundred *u'ilāti* reports and nearly as many letters of the Neo-Assyrian scholars.[51] Nearly all the fixed stars mentioned in the reports belong to the Path of the Moon. The practice of observing the moon or a planet with respect to a fixed star, later made quantitative in the Diaries with numbers of cubits or fingers below or above, in front of or behind a more extensive set of ecliptical stars (Normal Stars, see below), is found in the following report from an unknown writer: "Mars stan[ds] below the right foot of the Old Man (= Perseus). It has not ente[red] it, but stan[ds] in the area."[52] Other fixed stars, such as the culminating or *ziqpu* stars, were used to indicate time at night in the reporting of lunar eclipses: "A lunar eclipse took place on the 14th of Simanu, [during] the morning watch. It started in the south and cleared up in the south. Its right side was eclipsed. It was eclipsed in the area of the Scorpion. The Shoulder of the Panther was culminating. An eclipse of two fingers took place."[53]

Reports were written so that relevant omens from codified omen texts could be extracted to determine the meaning of observed astral and related phenomena. A report by the scholar and *rab ṭupšarri* "chief scribe" under Esarhaddon and Assurbanipal, Issar-šumu-ēreš,[54] concerning the increase in daylight length makes reference to such excerpting: "It is exceedingly good to excerpt these omens now considering the fact that the first days [lengthen] after each other regularly."[55] Another report from Issar-šumu-ereš concerns a solar eclipse: "[Mannu-k]î-Ḫarrān [wrote me] today: 'The sun was [eclips]ed on the 29th; [what day do you have] today? We have [the . . .] [. . .] We reject this [date . . .]. (Break) The messenger who went to Marduk-[. . .] came (back and) reported: "We did not see the moon, there were clouds." They did not see, we did not see; (consequently) we do not reject [the (30th) day].'"[56] This broken text attests to the importance of the contemporary observation, in this case, establishing the date of an eclipse. Reports, and later, Diaries, frequently stated that observation was not possible due to clouds.

Many reports and letters were prompted by an inquiry from the king: "Concerning what the king, my lord, wrote to me: 'You must certainly have observed something in the sky!'—I keep a close eye on it (but) I must say, I have seen nobody and nothing, (therefore) I have not written to the king. Nothing has risen, I have seen nothing. Concerning the watch of the sun about which the king, my lord, wrote to me, it is (indeed) the month for a watch of the sun."[57]

The authors often referred to the king's inquiry, and either excused the fact that there was nothing to report, as in the previous example, or followed it up with citations of appropriate omens from either *Enūma Anu Enlil*, or possibly from its commented digest, "If the moon at its appearance" (*Šumma Sin ina tāmartišu*).[58] In this next example, the same Issar-šumu-ēreš began his report with two quoted omens concerning the favorable appearance of the moon on the first day (of the lunar month) and the lengthening of days. This he followed with a report on Mars, as follows: "Twice or thrice we watched for Mars today (but) we did not see (it), it has set. Maybe the king, my lord will say as follows: 'Is there any (ominous) sign in (the fact) that it set?' (I answer): 'There is not.'"[59]

The passage underscores the fact that for a phenomenon to be ominous it must be visible, an idea that might seem counter to the existence of many nonoccurring phenomena in omen compendia, but points to the fact that the two corpora, omen compendium and reports, had different aims. Where the compendium focused on the schematic and the conceivable, the reports dealt with perceptual experience, thus offering observation statements. But the authors of the reports were watching specifically for celestial signs, and so the material content of their observation statements corresponds to phenomena of equal interest to the omens of *Enūma Anu Enlil*. A comparative look at selected phenomena in the omens will illustrate.

In *Enūma Anu Enlil* the lunar omens, produced by the moon-god Sin, fall into two parts, the appearance of the moon in its first crescent (Tablets 1–13, termed *tāmarāti*/IGI.DU$_8$.MEŠ *ša Sin* "visibilities of the moon"), and when full and in eclipse (Tablets 15–22).[60] These omens describe the configuration of the crescent (the "horns"), conjunctions of the moon with fixed stars (the ecliptical stars "in the Path of the Moon"), dates of opposition (when "the gods see one another," that is, on either side of the horizon), and the dates and characteristics of the eclipsed moon (direction of the shadow, the reddish color of the moon, and so on). The moon at its syzygies, at conjunction and opposition with the sun, was of paramount importance astrologically, a feature that continued in cuneiform horoscopes.[61]

Because the Babylonian calendar was based on the lunar month, the days of the months correlated with the phases of the moon. Whether the moon's significant visibilities occurred on the appropriate days of the month was taken into account for omens. Additionally, the calendar days, in accordance with hemerological tradition, were assigned as lucky and unlucky together with proscriptions and rituals to be followed on those days.[62] The day of the new moon, for example, was propitious for the performance of "witchcraft-releasing" (*ušburrudû*) incantation-prayer, and the seventh day for rituals against the consequences of a broken oath (*māmītu*). Daniel Schwemer cites an *ušburrudû* that called upon the personified new moon: "New Moon, who purifies the heavens and the Subterranean Ocean, who undoes witchcraft (and) magic: the messages of the night and of the whole day which you (2nd pl.) keep sending against me: New Moon, may your day of wrath overpower them!"[63] The lucky/unlucky character of days, and their association with festivals, goes back to second millennium calendars, where, for example, a festival devoted to cleansing was held on the seventh day of the seventh month.[64] Positive or negative consequences of the appearance of the moon on certain days is not, however, always consistent with the hemerological tradition, but first visibility on the first day generally signaled a good omen, for example, "if the moon becomes visible on the 1st day: reliable speech; the land will become happy."[65]

As the lunar syzygies at the beginning and middle of the month were of particular interest throughout the entirety of the lunar section of *Enūma Anu Enlil*, it is perhaps not surprising that one (undatable) report focuses on the setting of the moon near opposition (*rabû ša Sin*) during the morning watch, from the eleventh to the fourteenth of the month, paying particular attention to its visibility or invisibility due to clouds:

> [The night] of the 11th day [was cloudy(?); in the] morning [watch] the Moon came out [In the daytime of] the 11th day there was much [. . .]; the Moonset.

> [The n]ight of the 12th day [was cloudy(?); in] the morning [watch] the Moon came out. [The daytime of the 1]2th was cloudy; the setting of the Moon was not visible. The night of the 13th day [was cloudy(?); in] the morning [watch] the Moon came out. [The daytime of the 1]3th day was cloudy; (the Moon) was not visible. [. . . an eclipse(?) Of] the Moon passed by. The night of the 14th day [was cloudy(?); in the mor]ning [watch] the Moon came out; the disk did not wane. [—On the] «15»th day the setting of [the Moon] was not visible; solstice(?).[66]

This is a particularly significant report in light of the Diaries, which adopt much the same format of reporting night by night, giving the moon's intervals of visibility around conjunction (*na*) and opposition (the Lunar Four), and noting when the desired phenomena were not visible due to clouds or weather. In this report, the watch for the moon during the morning watch, or third watch of the night ending at sunrise, is certainly for moonset (ŠÚ-*u ša Sin* in line rev. 1).

The Diaries recorded the intervals of visibility around full moon known as the Lunar Four, one of which, called ŠÚ, is the interval between moonset and sunrise when the moon is visible (in its setting) for the last time before sunrise. The Lunar Four intervals were often given with quantitative values, measured by some sort of device (a "water clock" *mašqû*, or a more accurate descendant?) in UŠ "(time) degree" and its subdivision the NINDA (1 UŠ = 60 NINDA).[67] It is not impossible that the reference to the "setting of the moon" in the report is to be understood as the interval of ŠÚ known from the Lunar Four, but it would be the earliest such attestation, predating the text from Cambyses year 7 by something on the order of one hundred years. If not the interval ŠÚ, the report might simply be recording the last sighting of moonset during the morning watch. The similarity between this report and the Diaries, however, is notable.

The following is a line from the earliest preserved Diary, from −651 (year 16 of Šamaš-šuma-ukīn, King of Babylon, who ruled simultaneously with his brother Assurbanipal of Assyria until 648 BCE): "Night of the 12th, overcast. The 12th, the 13th, the 14th, overcast. The 14th, one god was seen with the other."[68] The statement that "one god was seen with the other" retains the idiom of the reports—namely, that the moon and sun were in opposition. This phraseology drops out of the later Diaries.

The next earliest comparable passage comes in the Diary for −381, where the interval ŠÚ is given a measured quantitative value: "The 11th very overcast. Night of the 12th, overcast. The 12th, moonset to sunrise: 4° 30′, measured (despite) clouds. Night of the 14th [. . .]"[69] Apart from the question of the identity of the phenomenon called ŠÚ *ša Sin* in the seventh-century report, both the report and the Diaries were concerned with visibility, and with invisibility due to clouds or overcast skies.

In the context of the importance of the syzygies in the omens, and therefore also in the reports, *Enūma Anu Enlil* Tablet 20 attached special significance to the observation of the day of the moon's last visibility (related to the later phenomenon KUR, the interval between moonrise and sunrise when the moon was visible for the last time before conjunction).[70] A procedural instruction was added to the end of each omen in Tablet 20 addressing the reader in the second person. Two such clauses follow:

> EAE 20 §I. (8–9) In *Kislīmu* the 28th (or 29th) day, observe his last visibility (that of) the [god who in his eclipse] beg[an the last watch], delayed 1/3 of the watch, and set while eclipsed, and Venus entered within him . . . Observe his last visibility [on the 28th, variant: 29th of *Kislīmu,* and you will predict an eclipse. The day of last visibility will show you the eclipse.[71]
>
> EAE 20 §II. (8–9) [Observe his] last [visi]bility on the 28th of Nisannu, and [on] the 14th of *Ajāru* [you will predict] an eclipse. [The day of last visibility will] show you [the eclipse.][72]

The Late Uruk text, TU 11, devotes a section to the calculation of the day of last visibility of the moon by means of a lunar visibility scheme and its associated scheme for the length of night (as duration of the moon is a function of the length of night) known from the much earlier schemes of *MUL.APIN* and *Enūma Anu Enlil.* In Lis Brack-Bernsen and Hermann Hunger's description, TU 11 §19 "reveals to us how the Babylonians surveyed the movement of the moon during the days of invisibility around conjunction, i.e., through extrapolation from the time between moonrise and sunrise, observed on the last morning before the moon became invisible."[73] The passage explains the procedure of calculating the day of invisibility on the basis of values of KUR on the twenty-seventh day. The procedure therefore determined the position of the moon in relation to the sun around conjunction, and found the day when the moon would again become visible, thereby marking the first day of the new month. The result is a method of determining the length of the month, either twenty-nine or thirty days.

As the most dramatic of events that occur at syzygy, eclipses were certainly among the most significant for which the scholars kept the watch. The reports echo the omens by observing the elements found in *Enūma Anu Enlil* eclipses and citing the corresponding omens. Munnabitu explains: "The evil of an eclipse affects the one identified by the month, the one identified by the day, the one identified by the watch, the one identified by the beginning, where (the eclipse) begins and where the moon pulls off its eclipse and drops it; these (people) receive its evil."[74]

Lunar eclipses and solar eclipses alike affected the king in particular. The death of the king was the most serious of the consequences of an eclipse, and apotropaic measures following a lunar eclipse were taken to ensure his protection.[75] The "passing by" of an eclipse meant that the eclipse would not occur, or would not be visible, and therefore, would not affect the king. The lunar eclipse omens mention this situation in *Enūma Anu Enlil* Tablet 17 §VI.6, Tablet 21 §I.1 and §VI.1, as follows:

> (Tablet 17 §VI.6) If an eclipse occurs in *Ulūlu* on the 14th day, there will be rebellion against the king; the extensive army will fall; If the ec[lipse passes by the king], brother will consume [his] brother; variant: a brother will murder the king; [there will be] depletion of da[tes].[76]
>
> (Tablet 21 §I.1) If an eclipse occurs on the 14th day of *Nisannu* and it begins in the south and [clears in the . . .]; it begins in the evening watch and clears in the middle watch. You observe his (the god's) eclipse and [you bear in mind the south]. The prediction is for the king of Akkad: The king of Akkad will die. If the eclipse passes by the king: There will be destruction and famine. The people will send their children out to the market (to be sold). The great country will go to the small country for food.[77]
>
> (Tablet 21 §VI.1) If an eclipse occurs on the 14th day of *Ulūlu*, and it begins in the north and clears in the south, variant: east; it begins in the evening watch and clears in the middle watch. You observe his (the god's) eclipse and bear in mind the north. The prediction is given for the king of Akkad: Revolt for the king. If the eclipse passes by the king: Rains in the sky, floods in source will cease; there will be famine in the land. The people will sell their children for money.[78]

Particular concern, then, was devoted to how the eclipse would affect the king, or if it would not affect him, in which case it was said to "pass by." Nabû-ahhē-erība reported: "The moon will compl[ete] the day in Tammuz (IV); on the 14th day, it will b[e seen] with the sun. It will let the eclipse pass by; it will not mak[e it]."[79] And Munnabitu reported: "The eclipse will pass by, (the moon) will not make it. Should the king say: 'What sign did you see?'—[. . . the god]s did not see each other [. . .] for the night [The eclipse] will pass by (and) [the moon] will be seen [together with] the sun."[80] The reports indicate that the appearance of the benefic planets, Jupiter and Venus, during a lunar eclipse was a signal of good fortune for the king: "In the eclipse [of the moon] Jupiter stood there: well-being for the king, a famous important person will die in his stead. The king should have much trust in this omen."[81] The following two letters attest to the reassurances made to the king on the basis of the presence of Jupiter and Venus:

> The king sho[uld not] be afraid of this eclipse! The planets Jupiter, Venus, and [Sa]turn were present [during the eclipse].[82]

> The eclipse swept from the east(ern quadrant) and settled over the entire west(ern quadrant of the moon). The planets Jupiter and Venus were present during the eclipse, until it cleared. With the king, my lord, (all) is well; it is evil for the Westland. Tomorrow I shall send the king, my lord, a (full) report on (this) eclipse of the moon.[83]

Not every ominous phenomenon was periodic, but in the case of the lunar halo, it could be associated with the full moon, and even in some cases was observed to precede an eclipse. The lunar halo is normally seen around a full moon for optical reasons, but here it is associated with the moon's first visibility: "If the moon in its first appearance appears as on Day 1, (or) Day 27 and is surrounded by a halo: an eclipse will occur."[84] Interest in the stars or planets seen within the lunar halo was another ominous phenomenon watched for by the scholars. In a report from Nabû-mušēṣi, the appearance of Jupiter inside the lunar halo together with Scorpius was the subject of the list of quoted omens from the "canonical" series *Enūma Anu Enlil* and from the "extraneous," or noncanonical, series:

> This night, the moon was surrounded by a halo, [and] Jupiter and Scorpius [stood] in [it].

> If the moon is surrounded by a halo, and Sagmegar (Jupiter) stands in it: the king of Akkad will be shut up.

> If the moon is surrounded by a halo, and Nēberu stands in it: fall of cattle and wild animals.

> The star of Marduk at its appearance is (called) Šulpa'e; when it rises 1 "double-hour" it is (called) Sagmegar; when it stands in the middle of the sky, it is (called) Nēberu.

> If the moon is surrounded by a halo, and Scorpius stands in it: entu-priestesses will be made pregnant; men, variant: lions, will rage and block the traffic of the land.

> These (omens) are from the series.

> If the moon is surrounded by a halo, and Šulpa'e stands in it: the king of the Westland will exercise supreme power and bring about the defeat of his enemy's land.
>
> This (omen) is extraneous.[85]

In later astrological context six (not all consecutive) zodiacal signs in which the moon is surrounded by a halo are mentioned in §25 of TU 11. The meaning of the passage is not clear, except that the lunar halo is associated with weather: "Leo, Virgo, Scorpius, Sagittarius, Capricorn and Pisces: together six places, where, if the moon will be surrounded in them by a halo, a . . . wind will rise on the eighth day."[86]

Solar omens were also divided into two principal parts, the first dealing predominantly with the sun at sunrise[87] and the second with eclipses. Attention to the dates of sunrise, cloud formations, parhelia, and the proximity of the sun to other celestial bodies is characteristic of the first part of the section on Šamaš. For example:

> If the sun is red like a torch when it becomes visible on the first of Nisannu, and a white cloud moves about in front of it, variant: stands at its side, and the east wind blows: in Nisannu the east wind will blow and in that month, on the 28th, 29th or 30th an eclipse of the sun will take place and during that eclipse [. . .], variant: in that month the king will die and his son will seize the throne, and the land [. . . , variant:] the land will be happy, the sky will [. . .] its rains, (and) the earth its produce in the proper season.[88]

This is followed by an omen for the sun being yellow at rising on the first of Nisannu. Omens for the sun's rising on the first day of other months are given, including whether it appears in a cloud, or whether one of the four winds blows during the event, or whether various stars are present (GUB) at the rising of the solar disk (*šamšatu*). Parhelia (up to eight) on different dates are ominous, as in: "If a normal disk is present and one disk stands to the right (and) one to the left: If the king treats the city and his people kindly for reconciliation and they become reconciled, the cities will start vying with each other, city walls will be destroyed, the people will be dispersed."[89] The bulk of the first half of the solar omens deals with the sun's rising, although there are some for setting, including setting in proximity to other celestial bodies (for example, the moon, the Wagon constellation, Jupiter, Centaurus, Scorpius).

The second part of the solar omens is devoted to eclipses. These omens are particularly bad for the king. To illustrate, the first three omens of *Enūma Anu Enlil* Tablet 33 read:

If on the first day of Nisannu the sun is eclipsed: The king of Akkad will die.

If on the first day there is an eclipse and its light is red and the day is dark: In its month there will be a lunar eclipse and in that year the king will die.

If on the first day there is an eclipse and its light is cool: The land will diminish; funerary places will be established in the land; there will be plague; there will be wailing, variant: imprisonment without escape.[90]

Omens for solar eclipses focus on the days around conjunction when solar eclipses can occur, but they are also constructed for every other day of the lunar month, in keeping with the importance of schematic completeness to the limit of conceivability, on a continuum from the normal to the abnormal and finally the unobservable.

The solar eclipse omens also include the appearance of the bright planets Venus and Jupiter, invariably a good omen for the king:

If (Šamaš) makes an eclipse and Venus and Jupiter appear: Peace for the king, but famine for the land.[91]

If on the day of the eclipse, Venus and Jupiter are seen with him: the city, king, and his country will be well, lions will become wild and block passage on the road.[92]

If an eclipse occurs and Venus and Jupiter are seen: The reign of the king of the universe will be long lasting; in place of him, in that year, someone will die; a notable person will die.[93]

On days when a solar eclipse cannot occur, however, the presence of Venus and Jupiter signal dire consequences:

If on the sixth, seventh, ninth, tenth, or fifteenth day, there is a (solar) eclipse (and) Venus and Jupiter stand beside him, variant: stars come out beside him: The walls will be destroyed; a cabal and the guard will overthrow the country.[94]

Planetary omens focus on the conjunction of planets with the fixed stars in the Path of the Moon, some synodic phenomena, especially the heliacal risings and settings of Venus, and various optical phenomena (brightness/dimness, colors). The positions of planets relative to the moon, or to the horns of the crescent moon, fixed stars, and each other figure prominently, with indications of the planet being in front of (*ana* SAG), behind (*ana* EGIR), to the right and left, and to the upper/lower right/left. Also prominent are the dates of appearances (themselves of ominous importance relative to lucky/unlucky calendar days), times relative to sunrise and sunset.

As an example of the interest in synodic phenomena, an observation of the last visibility of Venus as an evening star ("Venus se[t] in the west.") is given in a report, followed by three omens from the series: "If Venus in [*Ṭebētu*] from the 1st to the 30th day disappears in the west: the harvest of the land will prosper. If Venus keeps a stable position: the days of the ruler will be long; there will be truth in the land. If Venus moves in the Path of Ea and sta[nds: the god]s will have peace for the Westland."[95] Of the three omens, only the first corresponds directly to Venus's last visibility, but given a last visibility, other elements would be taken into account, as for example what "Path" the planet was in, reflected in the third of the omens cited. The significance of the observed phenomenon for the king and the state was assessed on the basis of the portents selected from the series.

Getting the observation right was of obvious importance, particularly in the case of phenomena that were difficult to see, such as the heliacal phenomenon of Mercury close to the horizon, as in the following report from the scholar Balasî: "Concerning Mercury, about which the king my lord wrote to me: yesterday Issar-šumu-ēreš had an argument with Nabû-ahhē-erība in the palace. Later, at night, they went and all made observations; they saw (it) and were satisfied."[96] The visibilities of the planets, including the sun and moon, were of keen interest to the scribes because they portended good or ill for the king, country, and population. The system of interpretation for a positive or negative outcome was constructed with respect to the periods of visibility and invisibility and based on a sense for what was normal and what anomalous. The reports are abundant evidence to this effect, as, for example, that Venus was seen to rise early,[97] or set early,[98] or that the moon was visible early at the beginning of the month.[99]

Hanson's idea that observation does not exist without an idea, a purpose, or an influence is well supported by the observational program of the Neo-Assyrian scholars. The reported observations were motivated by the idea to interpret heavenly phenomena. Although divination was the motivation and purpose of the diviners' observational program, it was not the only result of it. In addition to delivering on recommendations to the king, the empirical attitude of the scholars established the foundation for further material gains in astronomical knowledge.

THE REGULAR WATCH

Recording the observational particulars of the "regular watch" (*naṣāru ša ginê*) made use of the second two reference systems for observing celestial phenomena, systems (3) and (4), mentioned before. These are (3) the counting (MUL.ŠID.MEŠ/*kakkabū minâti*[100]) or norming, or "Normal Stars" (after Epping's "Normalsterne"[101]) and (4) the zodiacal signs. Normal Stars are found in the Diaries, Normal-Star Almanacs, and Goal-Year texts. They are ecliptical stars found within approximately 10° north or south latitude of the ecliptic.[102] The Normal Star system was used exclusively in observational contexts, while the zodiacal signs (classified as LU-MAŠ/*lumāšu*) seem to have been used for both observation and computation and are attested in Diaries, Almanacs, and ephemerides. The Babylonians were also interested in the longitudinal coordinates of the Normal Stars, as shown in two late Babylonian star catalogs (BM 46083 and 36609+).[103]

The units of measure with respect to the Normal Stars were the cubit (KÙŠ = *ammatu*) and the finger (ŠU.SI = *ubānu*). Positions were given as so-and-so many cubits *e* "above," SIG "below," *ana* IGI "in front of," or *ár* "behind" a Normal Star. In the late period the cubit was reckoned as 24 fingers, or 2° of arc[104] (cf. the Old Babylonian equivalence 1 KÙŠ = 30 ŠU.SI). Whether the Normal Star units in cubits and fingers were an alternate and compatible coordinate system to that of ecliptical longitude, measured in zodiacal signs or "degrees" (UŠ) has been discussed by Steele, who concludes that they were.[105] This, as he points out, is consistent with the fact that the Babylonian zodiac was sidereally fixed.

Use of the signs of the zodiac is attested in the latter part of the fifth century. The signs consisted of twelve equal (or near equal) 30-degree parts, indicated by the term UŠ, for which there is no known Akkadian equivalent. The signs were classified as LU-MAŠ/*lumāšu*, based on an older term for star, or zodiacal constellation. The convention of the zodiac almost certainly relates to the division of the ideal year into twelve thirty-day months, thereby striking a correspondence between calendar months and zodiacal signs: "[the sun] completes in one year (its course through) the twelve signs of the zodiac (literally: completes 12 *lumāšī*), each month it leaves one sign of the zodiac (*ištēn* MÚL *lumāš*) behind."[106] The five planets and the names of the zodiacal signs were summarized in a late Babylonian text as follows:

MÚL.BABBAR Dele-bat GU$_4$.UD GENNA AN	Jupiter, Venus, Mercury, Saturn, Mars
ḪUN MÚL.MÚL MAŠ.MAŠ ALLA A	Aries Taurus Gemini Cancer Leo
ABSIN RÍN GÍR.TAB PA MÁŠ	Virgo Libra Scorpius Sagitarrius Capricorn
GU ZIB.ME	Aquarius Pisces[107]

The Babylonian zodiac was normed with the stars, not at the vernal equinox, which was taken variably as Aries 8° (System B) or 10° (System A).[108] Observational texts giving positions of planets relative to zodiacal signs regularly employ the phrases in the "beginning" (= 0° – 5°) and "end" (= 25° – 30°) of a sign, probably indicating an observational function for the signs in addition to their use in calculation.[109]

Diaries were designated by the rubric *naṣāru ša ginê ša ištu* x MU.y.KAM *adi qīt* MU.y.KAM "regular (celestial) observation which (goes) from the xth month of the yth year to the end of the yth year." They provided positions of the moon and planets relative to the Normal Stars and the zodiacal signs (after the middle of the fifth century BCE), calendar dates of important lunar phenomena (the Lunar Six phenomena, see above), eclipses (observed and computed), equinox and solstice dates (always computed), and Sirius data (always computed). They also included observations of atmospheric phenomena (various clouds, mist, fog, rain) and the direction and quality (gusty, strong) of winds. Terrestrial phenomena, such as a regular account of commodity prices (for one shekel of silver how much barley, dates, sesame, wool, *kasû*, and *saḫlû* could be obtained[110]), the level of the Euphrates, historical events, and occasional bizarre occurrences (such as wild animals seen in the city, or anomalous births, that is, *izbus*) were also the object of "the regular watch." The Diaries constitute the bulk of surviving nontabular astronomical texts.[111] As Hunger and Pingree put it: "That someone in the middle of the eighth century B.C. conceived of such a scientific program and obtained support for it is truly astonishing; that it was designed so well is incredible; and that it was faithfully carried out for at least 700 years is miraculous."[112]

A Diary normally collected observed (as well as computed) data for six (or seven) months of the year, though Diaries for shorter periods (as short as for a single night) are preserved. Those for six months or a year are compilations made from shorter term records. Observation focused on the positions of the moon (and the duration of its visibilities around syzygy) and of the planets on consecutive days of the month, providing, in addition, the locations of the planetary synodic phenomena by means of cubits in relation to the Normal Stars, the number of degrees for the Lunar Six, and the number of fingers of magnitude of lunar eclipses. The following example is from the earliest preserved Diary, from the middle of the seventh century BCE (–651), a line from which was already quoted above:

> [Month I, . . . ,] the moon became visible in a cloud; it was bright? and high. [Beginn]ing of the night, overcast. Night of the 1st, the river level rose. The 1st, the sun was surrounded by a halo. Night of the 2nd, (and) the 2nd, overcast. The south wind blew. The 3rd very overcast. In the afternoon it rained. Night of the 4th, (and) the 4th, it rained. The 6th, middle part of the day, the sun was surrounded by a halo. It was split toward the east. The 7th, the moon was sur-

> rounded by a halo. The river level rose. Night of the 12th, overcast. The 12th, the 13th, the 14th, overcast. The 14th, one god was seen with the other. The river level receded. Mercury's last appearance in the east behind Pisces, and Saturn's last appearance behind Pisces; I did not watch because the days were overcast.[113]

This Diary becomes fragmentary, but toward the end (col. iv 10′) records a battle between "the troops of Babylonia" and the "troops of Assyria," and continues to note cloud banks, rainbows, wind, haze, a lunar and a solar halo, Mars in the region of Aries, and another battle "in the province of Sippar" between Babylonia and Assyria. Although this text displays certain variations as compared to later example, the elements of the Diaries archive seem to be well represented.

From nearly a century later, in the thirty-seventh year of Nebuchadnezzar II, the second extant Diary begins as follows. Note that the positions of the moon and of Saturn are given by reference to stars in the Path of the Moon, the Bull of Heaven and the Swallow respectively, not one of the Normal Stars. For the night of the eighth, however, the moon is positioned with respect to the Normal Star β Virginis (The Single Star in front of the Furrow):

> Year 37 of Nebukadnezzar, King of Babylon. Month I, (the 1st of which was identical with) the 30th (of the preceding month), the moon became visible behind the Bull of Heaven; [sunset to moonset:] . . . Saturn was in front of the Swallow. The 2nd, in the morning, a rainbow stretched in the west. Night of the 3rd, the moon was 2 cubits in front of [. . .] it rained?. Night of the 9th (error for: 8th), beginning of the night, the moon stood 1 cubit in front of β Virginis. The 9th, the sun in the west [was surrounded] by a halo [. . . The 11th] or the 12th, Jupiter's acronychal rising. On the 14th, one god was seen with the other; sunrise to moonset: 4°. The 15th, overcast.[114]

Additional reports in this Diary include that someone was killed "by the command of the king," that a fox entered the city, a wolf killed two dogs in Borsippa, and that there was disease. Additional month sections record the positions of the moon, planets, winds, storms, a summer solstice (month III), information on prices, and the level of the river. Other Diaries mention *izbus* such as in No. -418 during the month of the intercalary *Addaru*, a ewe gave birth to a lamb without a jaw (line rev. 9′). Important military and political events are also noted, as in No. -330, which records the entrance of Alexander the Great into Babylon.

As in the Neo-Assyrian reports, it was important to the authors of the Diaries to note whether a phenomenon had been observed or not, though this was carried out on a more consistent and extensive scale. The Diaries made frequent use of the expressions "when I carried out the observation (for such-and-such) I did not see (x)" (*kî* PAP NU IGI/*kî attaṣar ul āmur*) and "I did not make the observation" (NU PAP/*ul attaṣar*):

> (Night of the 15th, lunar eclipse of 1 finger); when I watched, I did not see it. (-302/301: 19′)
>
> (The 28th, 25° before sunset), solar eclipse; when I watched I did not see it. (-291 B 23)
>
> Mist, when I watched I did not see it. (-256:21′)
>
> I did not make the observation because the days were overcast (-651:8)
>
> The 26th (moonrise to sunrise) 23°; I did not observe the moon. (-567:11)
>
> [Venus was] 1 2/3 cubits above α Virginis; I did not observe (it). (-391:3)[115]

This terminology was established already in the seventh century, as seen in the following text giving observed and computed dates of first and last visibilities of Mars (and one position in year 9).[116] As John Britton noted,[117] it is an exceptionally early compilation of planetary data with regard to the visibility and nonvisibility of synodic phenomena:

> Year 1, not observed, (intercalary Month) VI_2; Year 2, not observed, II 20 last visibility, not [observed] VI 30 first visibility, not obser[ved]; Year 3, not observed, (intercalary Month) XII_2; Year 4, no watch, (Month) III 27 last visibility, not watched for, (Month) VII 25 first visibility, not observed; Year 5, not observed; Year 6, no watch, (Month) VI 15 last visibility, not watched for, (Month) IX first visibility, not watched for, (intercalary) XII_2; Year 7, not observed; Year 8, (Month) V 20 last visibility, (Month) XII 10 first visibility; Year 9, (intercalary Month) VI_2, (Month)VII 4 towards the middle of the Crab; Year 10, (Month) VII 13 last visibility.[118]

Where the text notes that no observation was made (NU PAP and NU ŠEŠ, variant logograms for the verb *naṣāru* "to watch"), the date has been computed. The reverse of the text contains data from the sixth century, and according to Britton's analysis, is more consistent with Diaries, giving the locations of first and last visibility in cubits with respect very probably to Normal Stars. From the fourth century onward, use of the phrase NU PAP "I did not observe" accompanied the dates of equinoxes, solstices, and heliacal phenomena of Sirius derived not from observation but regularly computed in accordance with a predictive scheme.[119]

The two earliest Diaries share with *Enūma Anu Enlil* and the reports the terminology for the opposition of sun and moon seen in the reports, that "one god was seen with the other."[120] This expression drops out of use after the sixth century in favor of the Lunar Four. A number of other phenomena observed in Diaries correspond to ominous phenomena found in *Enūma Anu Enlil*. Among such phenomena known also from omens are nonperiodic lunar phenomena such as "earthshine" (AGA *apir* "he wears a corona") at first visibility, halos, and the appearance of meteors together with eclipses. Regular attention to lunar and solar halos in the Diaries may also be continuous with the frequent reporting of halos by the Neo-Assyrian scholars. One scholar's letter notes: "I will send (the king) a tablet (with omens) from the lunar halo (TÙR *ša* Sin)."[121] Both the reports and the Diaries connected the appearance of halos with the onset of clouds and rain. The Diaries occasionally report the sun's rising or setting "in a box" (*ina pitni*/NA_5), some kind of cloud formation, which also occurs in the solar omens of *Enūma Anu Enlil*.[122]

Among planetary phenomena that had ominous significance, the positions of the planets with respect to the counting stars may be an extension of the attention in planetary omens to conjunctions of planets with fixed stars. *Enūma Anu Enlil* Tablet 50–51, for example, has omens for planets (Jupiter and Mars) "approaching" (TE/*ṭeḫû*) or "reaching" (KUR/*kašādu*) various fixed stars (GÀM and GÍR.TAB).[123] Venus omens include the planet "reaching" and "passing by" stars as well: "If Venus reaches the Pleiades," or "If Venus stands at the Pleiades for two days and passes (them)."[124]

Periodic phenomena were the primary concern of the Diary writers, and certainly eclipses feature prominently among those periodic phenomena watched for by the scribes. Both lunar and solar eclipses were regularly reported in Diaries, as they also were in separate eclipse reports and compilations of consecutive eclipses arranged in eighteen-year cycles (see below). Diaries, eclipse reports, and Goal-Year texts all recorded both observed and computed eclipses (Almanacs and Normal Star Almanacs have only computed eclipses). Due attention was given to the date and the time of onset, which were the two aspects of eclipses that could be predicted by the Saros period of 223 months (dates and times) as well as the mathematical astronomical Systems A and B (dates).[125] The ability to predict eclipses, which began in the seventh century among the writers of the Neo-Assyrian letters and celestial omen reports to the kings, did not preclude the ominous significance of eclipses.

During the Neo-Assyrian period, the lunar eclipse prompted the performance of a number of rituals against its ill portents, most dramatic perhaps being the substitute king ritual in which a nonroyal couple was placed on the throne for the period of danger (determined in accordance with the time of the eclipse, for example, the period for an eclipse of the evening watch was one hundred days) and then killed (sacrificed) at the close of that period. A different ritual against the evil of the lunar eclipse was still known in the Hellenistic period.[126] In view of this, the eclipse ritual extant from the late period is not merely antiquarian, but reflective of practice and of the idea that, despite its being predictable by calculation, the eclipse was still viewed as a dangerous omen.

Numerous cloud conditions and bad weather (thunder and lightning, rains and winds, fog, mist, hail) are mentioned in the Diaries because they prevented the observation of other phenomena of interest. *Enūma Anu Enlil* contains a wealth of weather omens in a section devoted to atmospheric phenomena. Diary -308:16 reports that "(at sunset) lightning flashed continuously in the north, it thundered once." Similar omens are known, such as "If Adad produces thunder and lightning,"[127] or "If Adad thunders and white (flashes of) lightning flare up."[128] These observations may or may not have been because thunder, lightning, and other atmospheric or optical phenomena had been part of the omen series. The principal reason for mentioning clouds and bad weather in Diaries seems to be because observation could not be carried out, but the seasonality of the weather was certainly also of interest alongside other phenomena thought to be periodic.

Illustrating the view that weather and its consequences were periodic, and therefore inherently predictable, is the Late Uruk text TU 11 §§ 23–24. This shows that an attempt to correlate the occurrence of seasonal rains and floods with planetary periods, where the periods for Jupiter (72), Venus (16), and Mercury (13) are known elsewhere from astrological prediction of weather,[129] and those for Mercury (46), Saturn (59), and Mars (79 and 47) are Goal-Year periods:

> §23 In order for you to calculate (lit., "make") rain and flood: 72 of Jupiter; 64, variant: 16, of Venus; 46, variant: 13, of Mercury; 59 of Saturn; 79, variant: 47, of Mars.
>
> §24 In order for you to calculate (lit., "make") rain and flood: you return behind you . . . of 9 (times?) 60, 3600, 3600, 3600, 3600, 3600, 3600, 3600, 3600, 3600, and 10,30 which [broken]- of Saturn you return. 40 years of Venus, 30 of Mercury, you set(?) behind you, and (there will be) regular(?) flood. Secondly, to calculate (lit., "make") rain: 30 years of Saturn, 41 of Jupiter, . . . you add(?) 83 . . .-[broken]- . . . In these months rain will come from the sky. In 72 years a

> comet which had appeared in the Tails, will appear (again) in the Tails, in 36 years it will correspond (lit., "answer"). In 21 years rain will correspond to rain, flood to flood. In 21 years an earthquake will correspond to an earthquake. In 654 years-[broken][130]

Another Uruk text also reflects the thinking that a systematic correspondence existed between weather and the planets:

> If at the beginning of the cold season, (that is) in *Ulūlu* or *Tašrītu*, Mercury or Venus are seen in the east, there will be rain in this year. If at the beginning of the warm season, (that is) in Addaru or Nisannu, Mercury or Venus are seen in the west, there will be flood in this year. If they are seen together in the east, there will be no rain. If they are seen together in the west, no floodwaters. Bright (planets Mercury or Venus): much rain and floods. Dim: little.[131]

The same text employs the astrological aspect of opposition (planets assuming positions in signs 180° apart) in order to predict rains, flooding, and economic prospects:

> At opposition if Jupiter stands in the Lion (Leo) and Saturn (stands) in the Great One (Aquarius): you predict rain and flood for the lands. You predict a rise in business, an abundance of grain.
>
> If they stand with one another in the Great One (Aquarius) or the Lion (Leo): rain and floods. You predict a rise in business.
>
> When Jupiter and Saturn stand steadily in their positions (and) the Moon or the Sun undergoes an eclipse with them: a severe famine will take place and the people will be greatly con[fused?].
>
> At opposition of Jupiter in the Lion (Leo), Mars in the Great One (Acquarius): heavy rain. In the Tails (Pisces) and the Furrow (Virgo): ditto (heavy rain).
>
> (If) Jupiter becomes stationary in the Bristle (Taurus), Mars in the Scorpion (Scorpius): heavy floods. (If) Jupiter stands in the Old Man (Perseus) or the Great Twins (Gemini), or Mars (stands) in Pabilsag (Sagittarius): you predict rain and flood for the lands that month.[132]

An early Seleucid Uruk text similarly correlates market price fluctuations with the periods of planets: "If you want to make a prediction for the market price for barley[broken] notice the movement of the planets. If you observe the first visibilities, the last visibilities, the stationary points, the conjunctions, . . . the faint and bright light of the planets an zodiacal signs and their positive or negative latitude . . . , your prediction for the coming year will be correct."[133]

The combination in Diaries of astronomical data with political and economic events (as well as other topics found in omen apodoses, namely, the mention of disease,[134] of wild animals in the city streets,[135] and of famine and the necessity to sell children[136]) is further testimony to their thematic relationship to the celestial omen literature. The practical use of celestial divination after the Neo-Assyrian period is not attested in later Babylonian, Persian, or Macedonian royal courts; however, scribes continued to preserve and copy the omen series until late in the Hellenistic period. Despite the resonance between celestial omens and the contents of Diaries, no direct evidence confirms that the ongoing observations recorded in these texts were used for divination. Nonetheless, as others have remarked,[137] the Diaries were immersed in, or at least had their inception in, a culture in which ominous celestial phenomena were correlated with mundane events of a political, military, and economic nature.

Late astrological texts from Uruk show that such correlations were being made in innovative ways, no longer simply copying *Enūma Anu Enlil,* but applying new elements based on the practice of observing or calculating the positions of planets in the zodiac and using their periods of return to ecliptical positions to gauge the recurrence of terrestrial events. The evidence of the Diaries and the Late Uruk astrological texts with procedures for calculating celestial, meteorological, and worldly phenomena in correspondence with one another show that the traditional omens were no longer the sole expression of the idea of such correspondences.

Another important collection of observations focused on eclipses.[138] The beginning of the tradition of eclipse reporting coincides with that of the Diaries in the middle of the eighth century. Indeed, many observations of eclipses are preserved in Diaries, as described above, but independent eclipse report texts are extant both as later compilations, arranged in groups of eighteen years, and as contemporary records of individually observed eclipses. Both the Diaries and the eclipse reports include computed eclipses, or eclipse possibilities, as well as observed occurrences. Eclipse reports, both lunar and solar, show a marked continuity from the eighth to the first century BCE. Some eighth-century lunar eclipse reports were limited to the time of occurrence by the night watch and the side of the disk to be darkened, while later reports (sixth to third centuries BCE) could include the time of onset, magnitude, duration, positions relative to a fixed star, positions of planets visible during the eclipse, and lunar data concerning the intervals between sunset/sunrise and moonset/moonrise around opposition of sun and moon. One early collection of lunar eclipse reports,[139] probably the earliest extant observational text from Babylon,[140] dates to –747/–746, as follows:

> At the command of Bēl and Bēltiya may it go well.
> 1,40 accession year of the kingship of [. . .]
> Month XII, 5 month (interval), the 14th, morning watch (an eclipse) occurred; it set eclipsed.
> 2,10. Year 1, month VI, [the n]th(?), (an eclipse) occurred; north. . . . [. . .]
> [. . .] the south wind blew(?). It set eclipsed. Month VI was int[ercalary].
> [Month XI, the 1]4th (an eclipse) occurred. 1,40° to clearing/or, before sunrise, to[tal(?).
> [Year 2, Month] V, the 14th, a total (eclipse) occurred.
> [Month XI,] passed.
> [Year 3, Month V, pass]ed. Month VI was intercalary.
> [. . .] total [. . .]

One of the very latest attested lunar eclipse reports dates to –66. Its contents attest to the continuity in the style of eclipse reporting as well as the increased detail in its contents:

> Year 180 (Seleucid Era), i.e., year 244 king Arsaces and Pir'uštanā, his wife, the lady, Month X 15 1 ME, mist, measured. As the moon rose, two thirds of the disk on the north-east side were eclipsed. 6° night duration of maximal phase, until it began to become bright. In 16° night from south-east to north-west it became bright. 23° total duration. Its eclipse had the "garment of the sky." During its eclipse, north (wind) blew. During its eclipse Venus, Saturn and Sirius stood there, the other planets did not stand there 1 1/2 cubits in front of α Leonis [eclipsed]. At 16° before sunset, on the 15th 9 NA, mist, [measu]red.[141]

In both of these eclipse reports, as well as in parts of Diaries not concerned with eclipses, the prevailing winds were noted.

The four winds appear in connection with lunar eclipses in *Enūma Anu Enlil* Tablets 15, 16, and 20, and such connections were preserved in Hellenistic astrology up to the early fifth century CE in the *Apotelesmatica* (I, 21) of Hephaistio of Thebes.[142] In Tablet 20 the presence of the wind was used to make a connection between the eclipse and a geographical region on earth to be affected by the eclipse.[143] In *Enūma Anu Enlil* winds were also a phenomenon given as a consequence of a solar eclipse, but in such contexts the wind was not identified by its direction:

> If on the first day of Abu there is a solar eclipse: there will be ill winds.[144]

> If on the thirteenth day (of Month IX): The south wind, variant: a destructive wind will rise up.[145]

> If on the 28th day (of Month X), (for) the city, king and his people, peace in the beginning of the year, a biting wind will come up and diminish the crop.[146]
>
> If (there is a solar eclipse) on the 28th day (of Month XI): the city, king, and his people will have peace, (but) in the spring the west wind will come up and diminish the crop;
>
> there will be trouble in the land; variant: a flood of waters will carry away the land.[147]

Other hints of celestial divination in the background are the attention to the planets visible during the eclipse and in the reference to the "garment," or "cloak" of the sky (*nalbaš šamê*). The "cloak" feature is known from the omen series as a term for a cloudy sky, covered over as with a cloak: If the moon at its first appearance is as dark as the "garment of heaven" (*nalbaš šamê*).[148] In the report cited above, it is uncertain what the "garment of heaven" describes about that particular lunar eclipse apart from its suggestion of "darkness."

Solar eclipse reports are attested from the fourth to the first centuries BCE. Eclipses that "passed by," that is, were expected but did not occur, were also noted. A solar eclipse report from the year −9, nearly the end of the cuneiform writing tradition, is preserved:

> [. . .] . . . Arsac[es(?) . . .]
> [. . .] the 28th day, solar eclipse, from [. . .]
> [. . .] it began; 23° of day to the inside of the sun . . . [. . .]
> [. . .] its . . . were clear (?); 2° [maximal phase(?);]
> Venus, Mercury, Mars(?), or: eclipse) . . . the remainder(?) [. . .]
> Sirius, which had set, in its non-[. . .]
> In its eclipse, . . . stood there(?) . . . [. . .]
> people broke pots [. . .]
> They broke. In 23° of day it cleared from north [and west]
> to south and east. 48° onset [maximal phase,]
> and clearing. In its eclipse, the north and west winds blew.
> At 1,30° of day before sunset. The 28th, 17;30 KUR, measured.[149]

The details of this solar eclipse report are fragmentary, but it is clear that similar data as were given in lunar eclipse reports, including the prevailing winds and the planets visible during totality, are still present.

Features of the reporting of lunar and planetary phenomena in Diaries and eclipse reports alike intersected with phenomena that had ominous significance. This is not to say that the observational program represented by the Diaries and eclipse reports was focused solely on ominous phenomena.

Indeed, Hunger and Pingree note that the Diaries exclude much that was of interest in omens.[150] This stands to reason in virtue of the different aims of these corpora. Although the use of celestial divination in regular interaction between scholars and kings is not attested to in later Babylonian sources in the manner of the Neo-Assyrian court (communiqués could have been in Aramaic by then), the fact that scribes preserved and copied the omen series until late in the Hellenistic period suggests that divination maintained intellectual currency. It is also noteworthy that the professional title for scribes who wrote and copied astronomical and astrological texts was generally *ṭupšar Enūma Anu Enlil* "scribe of the omen series *Enūma Anu Enlil*, though the title does not appear in Diaries.

Asger Aaboe, suggesting that Diaries and celestial divination were intimately connected, said, "the Diaries can be viewed as collections of raw material for omens, and they provided in the process the observational basis for constructing the theories behind the ACT texts [Astronomical Cuneiform Texts]; and, finally, the ACT texts predict precisely the core of the celestial phenomena recorded in the Diaries."[151] Aaboe was followed by Noel Swerdlow,[152] but opposed by Pingree and Hunger, who did not find a connection between Diaries and omens.[153] The evidence is ambiguous on this question, although it must be said that the claim that the purpose of the Diaries was for the production of additions to *Enūma Anu Enlil* is not borne out by textual evidence and remains purely conjectural. On the other hand, the treatment of the phenomena as predictable does not preclude the significance of those same phenomena as signs.

The ambiguity of the observational texts with respect to the question of the relevance of divination or astrology can be understood without attempting to see how the observational texts were or were not in service of divination or astrology. Regardless of the relevance of divination, observation was obviously an indispensable source of knowledge and one in which a tremendous amount of effort over the long term was invested. If the character of knowledge in the divinatory compendia was schematic and "constructed," in order to make sense of phenomena from a divinatory perspective, that of the Diaries and the eclipse reports was aimed at taking account of the phenomena either by observation or computation for the date in question.

The Diaries provided a picture of the sky on a certain date. If the desired phenomena were not observed, computed dates were supplied, accompanied by the remark that the observation was not made (NU PAP). As noted before, some phenomena (solstice, equinox, and Sirius first visibility, acronychal rising, and last visibility) were regularly computed. Estimated dates were also provided ad hoc for heliacal phenomena seen after their ideal date of synodic appearance, when the planet was "high," that is, too high in the sky for a first visibility, and sometimes noted as "faint," upon observation. In such cases, the writer noted the discrepancy, giving (by some method—an estimation?) a date for the ideal visible phenomenon, as in the following examples from the fourth and third century:

> [. . . Mercury's first appea]rance [in the west] 8 fingers above Jupiter, 4 fingers back to the west; it was high, (ideal) first appearance on the [nnth].[154]

> The 17th, Venus' first appearance in the west in Taurus; it was small, sunset to setting of Venus: 9°30′; it was high to the sun; (ideal) first appearance on the 15th.[155]

This practice goes back at least to the fifth century, as attested in the earliest horoscope (of 410 BCE): "*Du'ūzu* the 30th, Saturn's first visibility in Cancer, (it was) high and faint; around the 26th, (the ideal) first visibility."[156] Jennifer Gray pointed out that such ideal dates are incorporated into the planetary summaries at the end of the month sections of Diaries after –284, and that from –260 ideal dates are always included if they are recorded in the Diary proper.[157] Therefore, while observation, keeping the watch, and recording what was seen were the primary methods of the observation statements, quantitative and predictive methods, even estimations, were not only not excluded but also were fully integrated as supplementary to the program of "the regular watch."

THE EMPIRICAL ATTITUDE OF CUNEIFORM SCHOLARS

The Babylonian scholars did not debate the values of experience versus reason as sources of knowledge or ask the question as to which sort of description of the world, the rational or the empirical, was more justified or more "real." The opposition between such traditions of knowledge is a critical issue for seventeenth- and eighteenth-century philosophy, but not in our present context. Nonetheless, observation assumed a major role in the culture of cuneiform knowledge. Observation statements grew out of the context of celestial divination, first in the environment of the Neo-Assyrian period scribes, both Assyrian and Babylonian, and then at Babylon for what developed into the Diaries program. Observation was a method that coexisted with those of the hermeneutic and analytic methods of divination and astrology as well as the mathematical methods of prediction. Each of these methods was cultivated independently by the scholars, but functioned interdependently within the framework of the cuneiform corpus of knowledge of heavenly phenomena.

From the point of view of the definition of Empiricism since Francis Bacon and the growth particularly of British Empiricism and the responses to it, the work of the Babylonian diviners and astronomers, the liver inspectors and physiognomy experts, or even the Late Babylonian mathematical astronomers, cannot even remotely be implicated in the history of Empiricism, and we should not expect that it would. That history, according to Bas van Fraassen's account, is the "story of recurrent rebellion against a certain systematizing and theorizing tendency in philosophy: a recurrent rebellion against the metaphysicians."[158] The idea of an "antimetaphysical tradition in philosophy"[159] as a way to understand Empiricism does not help to penetrate the context of thought in Near Eastern antiquity as represented by its observational records, their contents, and their methods. The Diaries, as the most exemplary source for the elevation of observation to high epistemic value, are not clear testimony to the "de-animation of the heavens."[160]

The empirical dimension of cuneiform knowledge might, however, be profitably understood in terms of Van Fraassen's useful phrase, the "empirical stance," which was meant to represent not a theory about science but rather an attitude valuing observation as a source of knowledge, and observing as a procedure for science. Such an attitude can therefore accommodate various and changing positions on knowledge and objects of inquiry throughout history, be they physical phenomena of a uniform natural world or a wholly different conception of phenomena in the separate visible realms of a divinely designed order. The empirical stance, or attitude, developed within cuneiform scribal scholarship, with regard to the observation of ominous phenomena in general and celestial phenomena in particular, produced centuries of sustained observation as well as various methods of prediction reconcilable with divination, astrology, or magic. Doubtless this is why Hellenistic Greek writers gave priority in the astronomical sciences to Babylonia and the "Chaldeans." In his *Bibliotheca Historica,* Diodorus of Sicily said, "a man can scarcely believe" "the number of years which, according to their statements, the order of the Chaldeans has spent on the study of the bodies of the universe . . . for they reckon that, down to Alexander's crossing over into Asia, it has been four hundred and seventy-three thousand years since they began in early times to make their observations of the stars."[161] Roughly a century later, in his *Natural History,* Pliny invoked Epigenes as an authority on the antiquity of Babylonian astronomical observations, saying they went back 720,000 years.[162]

Apocryphal reports aside, the purpose of Babylonian, and Assyro-

Babylonian, observational programs, their intellectual contexts, and the understanding of what was being observed, differed from later efforts to know nature. And yet, as between the two corpora of observation statements, the reports and the Diaries, a greater attention to representing a picture of the heavens on the date in question, giving observed as well as computed phenomena for completion's sake, was introduced as a new purpose in the Diaries, beyond that of tracking and interpreting ominous phenomena.

This chapter aimed to show the way in which observation functioned as the key to much cuneiform scribal scholarship as well as the importance of the material presented for the history of observation itself. Recalling Daston and Lunbeck's suggestion that the history of observation might simply be the history of science "in its vast entirety," it is clear that Babylonian and Assyro-Babylonian observation of the phenomena, and the well-developed empirical attitude of its practitioners, forms a strong kinship bond connecting cuneiform knowledge to science, and therefore belongs to any historical consideration of observation and of science.[163]

CHAPTER EIGHT

Prediction and Explanation in Cuneiform Scholarship

qība tašakkan "You make a prediction"
anniu pišeršu "this is its interpretation"

> *Šumma izbu* is difficult to interpret. The first time that I come before the king, my lord, I shall (personally) show, with this tablet that I am sending to the king, my lord, how the omen is written. Really, [the one] who has [not] had (the meaning) pointed out to him cannot possibly understand it.
>
> SAA 10 60 rev. 1–14

PREDICTING AND EXPLAINING IN THE CUNEIFORM CONTEXT

Predicting and explaining in science are both aimed at phenomena, or constituent parts of phenomena. With respect to the cuneiform world, the question now remains as to what and how phenomena were regarded by the learned scribes as objects of predictive and explanatory knowledge. The selection of phenomena, and how to understand, compute, or observe them, are direct and indirect functions of social-intellectual motives. These aspects of ancient knowledge were situated in institutions of scribal scholarship, whether in the palace or temple, and developed over time. The clearest evidence for the social and intellectual context of scribal scholarly work and purpose, as said before, comes from the seventh century, when scribes were involved with celestial observation and calculation, as well as with watching for and interpreting omens for the benefit of the Assyrian king. But celestial divination predates the seventh century BCE by some one thousand years, and continued to figure in the kinds of texts the scholars produced well into the last centuries BCE.

Before the rise of Assyria, celestial divination, as part of a diverse corpus of divinatory texts, was already geared toward the concerns of the state. These concerns center on the strength of the king and the prosperity of the economy, as is clear from the kinds of events correlated with celestial phenomena in earlier omen texts. The celestial omen texts from second millennium scribal centers seem to converge around a central core of omens, which in turn were absorbed into a version identifiable as Middle Assyrian (ca. eleventh century) and then again in the version attested at Nineveh in the seventh century.

A limited number of astronomical texts also stem from the Middle Assyrian period, such as the Astrolabe, with its scheme for the length of daylight. Prediction, therefore, had at least two forms of development already taking place during the second millennium BCE—the divinatory and the astronomical. And just as divinatory prediction had multiple forms and emphases, so too did the astronomical.

After the fall of Assyria, divination and astronomical activity persisted in the Neo-Babylonian, Persian, and into the Seleucid and Arsacid periods, although much less information is available to fill the gap in our understanding between the production of tablets to do with celestial observation and prediction and their possible purpose vis-à-vis the monarch. This chapter acknowledges that much more can be said as to both the diachronic and synchronic pictures of scholars in their cities and in relation to their kings, temples, and gods, but takes a different tack. It considers what kinds of prediction and explanation are characteristic of the cuneiform scholarly texts that have been relevant to this study, paying particular attention to the phenomena deemed predictable and the use of schemata (in divination) and models (in astronomy) to predict them.

Prediction in the realm of knowledge of the heavens can be said to have been *from* the phenomena (in celestial divination) and *of* the phenomena (in astronomy), but there is an asymmetry about the two kinds of prediction. To account for that asymmetry some might prefer the term *prognostication* for the divinatory class and reserve *prediction* for mathematical astronomy. Strictly speaking, the results of the mathematical methods of Babylonian astronomy are computed values for phenomena on dates in the future or the past, that is to say, predictions and retrodictions were generated from the computational methods used. Despite potential objections to the terminology of *prediction* employed here, as a general term for many kinds of forecasts, prognostications, or projections, or indeed for the process of forecasting, projecting, or calculating, it seems to suit present purposes.

Before turning to the two basic predictive methodologies of cuneiform knowledge, there is the most basic and fundamental question of the nature of phenomena in the cuneiform texts, that is, how the compilers and copyists of the tablets containing observations, omens, and calculations of phenomena understood them. Although the means by which scribes engaged with phenomena differed considerably from the evidence of one text category to another, for example, omens to ephemerides, I propose that there was not as much difference as different approaches and methods might suggest.

When Otto Neugebauer posited the separation between "the men of the Diaries," by which he meant scribes whose interest in the heavens was for celestial omens, and "the authors of the ACT [Astronomical Cuneiform Texts, meaning ephemerides],"[1] it could have been assumed that the former group took celestial signs to be, as he said, "events beyond the power of arithmetical rules,"[2] whereas the astronomers' systematic analyses of periodic phenomena required a new conception as a direct result of a new grasp of the phenomena. As it now seems that the community of scribes who produced the various astronomical texts, as well as astrological texts,[3] was one integrated group, changes in ideas about the phenomena must be acknowledged within this community over time. Just as the astronomical Diary texts reflect a fundamentally new idea about celestial observation,[4] predictive methods too reflect a new idea in the prediction of the phenomena.

To return to the question of the nature of phenomena in the cuneiform texts, let us consider how a scientific phenomenon is viewed in the modern idiom of the philosophy of science. Daniela Bailer-Jones said, "A phenomenon is a fact or event in nature, such as bees dancing, rain falling, or stars radiating light. A phenomenon is not necessarily something as it is observed. . . . To conjecture thus is not to take the bee's movements as something happening entirely at random. It is *treating* what is observed as a phenomenon. . . . This seems to suggest that picking out a phenomenon has something to do with distinguishing the causal processes that make up that phenomenon."[5] This definition underscores the crux of the present discussion, that ever since the introduction of the conception of nature, the idea of what a phenomenon is has radically departed from what seems to be the case in cuneiform texts.

Once again the question of the scaffolding of the cuneiform intellectual world presents itself, both phenomenologically, in terms of the scribes' lived experience of the phenomena, and metaphysically, in terms of how they might have understood the phenomena to be constituted. It might be said, as mentioned before, that the notion of "divine design," as in the expressions "the designs of heaven and earth" (*uṣurāt šamê u erṣeti*) or "the enduring designs of the gods" (*uṣurāt ilī kīnāti*), provided such a scaffolding, but these expressions operated on a different level of meaning and belonged to a different sort of discourse from that which pertains to nature, not having a physical, or material, equivalence. "The enduring designs of the gods" does not convey notions of material essence or physical uniformity, only that all was subject to, or a manifestation of, the divine will that imposed order upon the world. As such the concept of divine design cannot substitute for nature as a category for the material world as a whole in the way "God's creation" was able to later in history. Nor can it substitute in any meaningful way for the notion of an internal teleological property of things. Consequently, apart from and outside the discourse of nature, we cannot assume that phenomena in cuneiform sources were understood to be part of a physical order equivalent to nature.

Bailer-Jones noted the important distinction between something merely seen and something that can be identified as a phenomenon. Understanding implies having a context for the many observations of a phenomenon that necessarily precede the construal of it *as* a phenomenon. In this sense, James Bogen and James Woodward made a compelling distinction between data about a phenomenon and the phenomenon itself.[6] And Bailer-Jones suggested that phenomena have "the potential to be theoretically explained," which can be the case "before any data are collected."[7] Indeed, she said, "it is also sometimes hardly possible to interpret data without having a phenomenon in mind."[8]

Close attention to phenomena is certainly characteristic of cuneiform intellectual culture. However, defining phenomena as facts or events in nature and as objects in a causal network, or that they are potentially explainable within a uniform order such as that of nature, is not consistent with the various forms of cuneiform attention to observed facts or events in the visible realms of their cosmos. In essence, two views prevailed in cuneiform antiquity, one that identified phenomena as signs to be interpreted, the other as objects to be modeled (mathematically or otherwise) and predicted. Both views focused on instances, or occurrences, of a phenomenon, and there is no reason to think that the construal of such a phenomenon was different in each case. In the astronomical context, Mathieu Ossendrijver has adopted the term *event* to refer to such occurrences, and *event frame* as "the fundamental reference frame of Babylonian mathematical astronomy,"[9] to which all coordinate systems and units (such as the mean synodic month and the mean *tithi*) are related. How we understand the formation of the concept of the event and the event frame is an important part of our historiography of cuneiform knowledge of phenomena. That it is continuous and not irreconcilable with the view of phenomena as occurrences of signs is noteworthy.

Much scholarly literature is available that explicates the predictive methods of late Babylonian astronomical tables. The field began less than 150 years ago with Johann Nepomuk Strassmaier and Joseph Epping's collaborative decipherment of astronomical cuneiform texts (ephemerides and Normal Star Almanacs) at the end of the nineteenth century,[10] and Kugler's exposition of the lunar theory at the start of the twentieth.[11] Otto Neugebauer's ACT (*Astronomical Cuneiform Texts*, 1955) and HAMA (*A History of Ancient Mathematical Astronomy*, 1975), as well as his collaborative papers with A. J. Sachs, and Peter Huber's study of nonmathematical astronomical texts and eclipse reports throughout the 1950s and 1960s established another platform for study of Babylonian astronomy as an integrated part of the history of the "exact" sciences. Asger Aaboe's focus on lunar and planetary tables and his interest in the relation of observation to theory, working on the assumption of the primacy of zodiacal longitude, was an important part of the literature from the 1960s to the early 2000s. Noel Swerdlow's *Babylonian Theory of the Planets* (1998) was a comprehensive exposition of the relation of observation and theory on the assumption of the primacy of calendar dates. Mathieu Ossendrijver's *Babylonian Mathematical Astronomy: Procedure Texts* (2012) offers the newest full treatment of the Babylonian mathematical astronomy. The papers and collaborations of John P. Britton, Lis Brack-Bernsen, John M. Steele, and Mathieu Ossendrijver, have provided further analysis of Babylonian astronomical prediction, from Goal-Year methods to the complexities of the Systems A and B lunar theory. In terms of uncovering the mathematical structures of the cuneiform astronomical texts, these works represent the state of the art in the field from the late 1980s to the present.

The unifying project of all these works was reconstruction of the characteristic Babylonian quantitative approach to astronomical phenomena. This includes the nonmathematical method of the Goal-Year texts and the ways that the various computational systems (principally known as A and B, but including other variants) modeled the phenomena with zigzag and step functions for computed predictions of synodic phenomena and daily positions of the planets and moon. Recent work has determined relations among the nonmathematical astronomical texts—the dependence of Goal-Year texts on the Diaries and the dependence of Almanacs and Normal Star Almanacs on Goal-Year texts.[12] Ongoing effort is underway to understand how mathematical and nonmathematical astronomical texts relate to each other, as well as how the astronomical texts in general relate to astrological texts such as horoscopes, or any of the texts that employ astrological schemes (triplicities, terms, hypsomata, dodecatemoria, or the calendar text scheme). The distinction between astronomical and astrological thus made is in accordance with the aims of particular text genres, not a matter of demarcating between science and nonscience, or even astronomy and astrology, which cuneiform texts did not do.

This chapter does not set out to present a digest of the major aspects or results of the abovementioned expository literature on Babylonian predictive astronomy, nor to detail various quantitative models from the tablets themselves. Rather, it sets out on a philosophical foot to put the question of the nature of prediction to this material and to extend discussion of the matter of prediction to the broader scope of cuneiform knowledge. This of course means that divination comes once again into view. Discussion is thus opened up to two broad categories of prediction developed within cuneiform scribal knowledge.

Two sets of evidence relevant to prediction in cuneiform texts present a diverse array of text types. The first is the vast repertoire of omen texts. The second consists of the many texts containing computed astronomical phenomena, including Diaries, Goal-Year texts, Almanacs, Normal Star Almanacs, ephemerides, and horoscopes, which employ various kinds of quantitative methods, including and especially those that used a variety of predictive models. Together these sources represent the entire chronological span of cuneiform scholarship, the omen texts beginning in the Old Babylonian period and the last datable cuneiform texts being astronomical Almanacs of 75 and 79/80 CE.[13] They can also be considered as representative of the full range of areas, objects, and methods developed in cuneiform culture for prediction.

As the present focus is on the nature or style of prediction itself, entrée to this question can be no better made than through the observations and analyses of the subject by Daniel Kahneman and Amos Tversky. The categorical division just made between kinds of prediction in the two cuneiform source groups (divinatory and astronomical) relates in a certain way to Kahneman and Tversky's discussion of the psychology of prediction in their classic paper of 1973.[14] They distinguished between intuitive, or category, prediction on one hand and mathematical, or statistical, prediction on the other. They explored the difference in the rules that determine each, showing a tendency for a judgmental heuristic of what they called representativeness (as mentioned in chapter 5) to dominate intuitive predicting, even when other factors are more determinative of the outcome: "People predict by representativeness, that is, they select or order outcomes by the degree to which the outcomes represent the essential features of the evidence . . .

[and] the ordering of outcomes by perceived likelihood coincides with their ordering by representativeness and that intuitive predictions are essentially unaffected by considerations of prior probability and expected predictive accuracy."[15] Such prediction was described as intuitive because it stems from a certain kind of cognitive process, defined in Kahneman's later work as belonging to the first of two systems.[16] "System 1," so-called, produces the heuristics of judgment involved in intuitive prediction because it is guided by associative memory.[17] System 1 "runs automatically"[18] and operates to construct meaning and coherence from associations and associative memory. Kahneman traced the associative mechanism to Hume's *An Enquiry concerning Human Understanding*, where section 3.2 describes three principles of the association of ideas—namely, "resemblance, contiguity in time or place, and cause or effect."[19] Clearly, resemblance played the key role in Kahneman and Tversky's representativeness.

Representativeness as a heuristic is relevant to divinatory texts for the role analogy played as the foundation of many omen statements, as discussed in chapter 5. This argues for omen divination as a form of intuitive prediction, which operated with categories of things that can be related to one another, frequently to find likeness (representativeness). Although Kahneman suggests that "the work of associative thinking is silent, hidden from our conscious selves,"[20] this does not imply that the process of forecasting or prediction that results from intuitive thought is made without conscious judgment. Indeed, in a later paper with Tversky,[21] they showed that confidence in predictions correlates with the degree of representativeness between the outcome and the input, calling this "the illusion of validity." More omens than can be decoded by reference to an analogical relationship between the sign and the outcome demonstrate even more clearly the role played by intuition (which includes philological considerations of relationships between signs, logograms, and words) in the relations seen between categories of ominous signs and the social, political, or economic events correlated with them.

Representation and analogy played a different role in astronomical prediction, in the sense that mathematical models with zigzag and step functions were applied in the table texts, and in the procedures describing them, to all synodic events relevant for the moon, sun, and planets. When we speak of Babylonian astronomical models, it should be clear that we are speaking wholly in modern, but useful, terms for the purpose of historical analysis. Such models determined the recurrence of synodic events by means of underlying periods and period relations. That the periodic nature of diverse celestial phenomena could be similarly theorized is an example of the use of analogy in the application of predictive models. How the models themselves were meant to represent the behavior of a phenomenon in some observational or theoretical way presents another analogical dimension. Apart from the question of their representativeness, the various models produced in both Early and Late Babylonian astronomy, functioned as heuristic devices for analyzing the periodic nature of planetary, lunar, and calendrical phenomena, thus belonging categorically to Kahneman and Tversky's mathematical, or mathematical model-based, class of prediction. Certainly, however, the different systems applied models to represent the phenomena differently. System A modeled phenomena directly, in terms of their position in the zodiac, whereas System B did so indirectly, meaning it did not model a phenomenon (or synodic event) per se, but via the event number in the table.

In applying Kahneman and Tversky's classifications to the cuneiform context, it is not difficult to see that the class of numerical or model-based prediction will refer to the kind of prediction represented by astronomical texts of a mathematical character. This classification can be applied to any of the astronomical texts that computed phenomena on the basis of mathematical models constructed so as to generate a value y_{n+1} from its preceding value y_n in the scheme. Regardless of a model's simplicity or complexity, or its degree of fit with observed phenomena, models are found from the entire chronological range of Babylonian astronomy, from the scheme for the visibility of the moon in *Enūma Anu Enlil* Tablet 14 to the Late Babylonian ephemerides. In this way, the models of planetary phases and of the complex aspects of the lunar theory, while much refined and subtler mathematically as compared to earlier astronomical texts, are not so different in the sense that they represent the same kind of cognitive tool, the results of what Nancy Nersessian called model-based reasoning. This cognitive process combined, again in her terms, simulative, analogical, and imagistic thought.[22]

What both Early and Late Babylonian astronomical models also have in common is the objective to derive the next phenomenon, the "event," in Ossendrijver's terminology. These events differ from one astronomical text group (such as Astrolabes, Goal-Year texts, or ephemerides) to another, and represent such phenomena as the length of day, synodic appearances of the moon (new or full moons) and planets (first appearances, stations, acronychal risings and last appearances), or Lunar Six phenomena. No matter how schematic or ideal, the aim of astronomical models was inherently predictive, as well as being continuous with the aims of celestial divination, to know when a phenomenon will appear again. Of course the phenomena and the quantities generated in mathematical astronomical tables are not strictly speaking identifiable as divinatory signs, nor do they correspond to those of the omen series.

While in mathematical terms the late ephemerides introduced an unprecedented theoretical treatment of the phenomena of interest, it is also true that a certain common ground between early and late astronomy can be defined in terms of the quantitative expression of periods and the interest in finding relations between periods, that is, integer equivalences between quantitative parameters, the periods and their relations derived in the early material being quite schematic in relation to those derived later (e.g., 12 schematic months = 1 year, versus 235 mean synodic months = 19 years). Period relations become a central element underlying the models constructed in late mathematical astronomy. Despite this commonality, the later use of functions is considerably more refined, not only introducing the step function (System A) but also modifying the zigzag function, for example, in the lunar theory (e.g., Column Φ) to better respond to observational data.

For the predictions from signs in divinatory texts, the applicability of Kahneman and Tversky's class of intuitive prediction is perhaps less clear. The dispute as to whether and in what way Assyro-Babylonian omens were predictive will be reviewed further in this chapter, section "The Question of Epistemic Modality in Divinatory Texts." Kahneman and Tversky's interest, however, was in how people think when engaged in forecasting. As they said, "Any significant activity of forecasting involves a large component of judgment, intuition, and educated guesswork. Indeed, the opinions of experts are the source of many technological, political, and social forecasts. Opinions and intuitions play an important part even where forecasts are obtained by a mathematical model or a simulation."[23] In fact, prediction from signs is not generally understood as prediction in the modern philosophy of science, but was in Hellenistic philosophy.[24] It seems to me to have been a legitimate area of predictive knowledge in cuneiform intellectual culture as well. The inferential character of Babylonian omens is only a part of how the system was used for prediction, the other dimension being interpretive and involving much that can be called intuitive in relation to the cultural meanings attached to elements and categories of signs. The adaptation of the term *intuitive prediction* thus seems justified to me, based on the influence of intuition and judgment through the use of analogy and association in the practice of making inferences from ominous signs.

Divinatory prediction in the cuneiform context concerns the expectation of regular recurrence of certain events given some other event. When ominous phenomena are in question, we might say $\mathcal{Q}$ is the consequent of P, where P and $\mathcal{Q}$ are phenomena of different classifications (e.g., celestial and terrestrial). Astronomical prediction also concerns the expectation of regular recurrence, but of phenomena of the same classification, for example, eclipses, or the synodic phenomena of a planet. In principle, each class of prediction, the divinatory (or intuitive) and the astronomical (or mathematical model-based), implies expectations about recurrence. The two classes of prediction differ in the kind of phenomena being predicted and in the nature of the patterns or periodicities observed or conceived in each domain. Each presupposed its own visible or conceptual order relevant to what was being predicted.

The element of expectation is perhaps the chief criterion determining the predictive nature of each system. Expectation is, moreover, a determining factor in the definition of anomaly, which, ever since Thomas Kuhn's *The Structure of Scientific Revolutions*, has been theorized as an important driver of scientific change.[25] As considered in chapter 4, the concept of anomaly and the way it was defined by established epistemic norms, either in the realm of divinatory ideals or in the form of numerical mean values employed in astronomy, was a feature of the knowledge systems under consideration.

The cuneiform scribes did not normally engage with the question of the nature of prediction in divination, but a statement from the *Diviner's Manual* of the seventh century clarifies the conditions under which a sign will indicate a future event that is expected to occur:

> When you look up a sign (in these omen collections) be it one in the sky or one on earth and if that sign's evil is confirmed(?) then it has indeed occurred with regard to you in reference to an enemy or to a disease or to a famine. Check (then) the date of that sign and should no sign have occurred to counteract (that) sign, should no annulment have taken place, one cannot make (it) pass by, its evil (consequences) cannot be removed (and) it will happen.[26]

Unlike the later Hellenistic philosophers, as documented in Cicero's *De Natura Deorum*, *De Divinatione*, and *De Fato*, the Assyro-Babylonian diviners did not debate how or why ominous signs indicated future events, but seem to have taken signs to be a divine language (discussed in chapter 6, section "Determinism and the Gods"). The mechanism of divine decree, if it can be so described, seems to underpin the system, rendering divinatory prediction neither a matter of physics nor of iron-clad fate. The passage in the *Diviner's Manual* presumes the inevitability of the consequent of a sign if nothing is done to prevent it, and the bulk of the tablet is devoted to the methods for such prevention, saying "these are their *namburbis*" (*annû namburbīšunu*), that is, "these are the methods to dispel bad omens." The *namburbis* here are neither ritual in nature nor involve incantations to the gods, but proceed by establishing the months and days in the calendar according to the length of the year, risings and settings (of planets), heliacal risings of stars, and intercalations by observing the Pleiades in relation to the moon.[27]

The methods given in the *Diviner's Manual* required either observation or computation (of lunar visibility in the event of cloudy weather), and the procedures to check dates and intercalations were based upon calendar schemes in *MUL.APIN* and *Enūma Anu Enlil*, fixed in accordance with the ideal twelve-month 360-day year (*Diviner's Manual*, line 57). On this basis, the text explains methods for correcting calendar dates with an eye to "undoing" (*pašāru*) untoward portents. Here explanation has a particular function in accordance with the predictive aim of divination.

Complementary to the theme of prediction is explanation, well established as a core value of science. The complex history of philosophical accounts of how science explains nature goes back to Hellenistic antiquity. But the forms and aims of Assyro-Babylonian explanation were not directed to an understanding of the natural world, but rather to the meanings of words in texts, and also to the predictive methods (or, in our terms, models) found in texts. There is no reason why these explanations would depend upon notions of uniform or universal laws, of mechanical causality, or of phenomena defined in terms of being facts or events in nature. Explanation in cuneiform texts was not determined or contextualized by the discourse of nature. It had, instead, a hermeneutic character, in terms of interpreting texts and also of interpreting the world of ominous phenomena.

A look at modes of explanation, that is, with respect to divinatory versus astronomical prediction, shows a direct relationship between prediction and explanation in cuneiform scribal practice. This relationship was formed, however, of rather different objectives as compared with the tradition of natural science. A once major claim from the history of the philosophy of science was that prediction and explanation had a kind of inverse relation, such that, in Hanson's words, "explaining P is simply predicting P after P has actually occurred," and "predicting P is just explaining it before it has occurred."[28] The epigraph to Hanson's paper quotes Carl Hempel, who, by the early 1940s, had thought about the mutual implication of prediction and explanation in the context not of science, but of history. The epigraph reads: "An explanation . . . is not complete unless it might as well have functioned as a prediction; if the final event can be derived from the initial conditions and universal hypotheses stated in the explanation, then it might as well have been predicted, before it actually happened, on the basis of a knowledge of the initial conditions and general laws."[29] Hanson's own view opposed the ideal picture Hempel conveyed. Hanson argued that in the history of science before Newton, no theory both explained and predicted in accordance with Hempel's ideal, and even then, it was not to be long-lived.[30] The expository texts in cuneiform do not bear relation to prediction in that way either, as will be more fully described below. What is obviously lacking is the notion of general causal law, either to describe the occurrences of natural or of historical phenomena.

Not very long after Hempel's essay on the "function of general laws in history," his paper with Paul Oppenheim introduced the influential covering law model, or deductive-nomological model of explanation in science.[31] As Heather Douglas pointed out, Hempel and Oppenheim's covering law model, with its emphasis on the logic of explanation, satisfied the contemporary antimetaphysical bent of midcentury philosophers of science.[32] It dealt neatly with causality, truth, and logical demonstration, leaving aside talk of metaphysical grounds for explanation. Subsequently, Hempel and Oppenheim's keystone paper engendered an entire generation's concerns with explanation in science, and still represents the classic epistemic view. Wesley Salmon called the paper "the fountainhead from which the vast bulk of subsequent philosophical work on scientific explanation has flowed—directly or indirectly" and "the division between the prehistory and the history of modern discussions of scientific explanation."[33]

Douglas noted that the logical-empiricist attention to explanation as an epistemic problem acted to downplay prediction as an independent subject. She suggested that explanation and prediction could each shed light on the other, that they "are best understood in light of each other and thus that they should not be viewed as competing goals but rather as two goals wherein the achievement of one should facilitate the achievement of the other."[34] Although Douglas's stakes were in defining criteria for modern scientific explanation, and her conclusion was that such explanations were measurable in terms of the degree of success in generating testable predictions—not something relevant to the cuneiform material—her claim that it is useful to consider prediction and explanation in terms of each other will nevertheless be of interest for examining predictive and expository/explanatory knowledge in cuneiform (below in the section "Cuneiform Explanation as Interpretation of Texts"). In quite a direct way, prediction, in divination and in astronomy, necessitated explanation, and explanation, in the form in which it was given by the scribes, was predicated upon and also facilitated the achievement of prediction.

THE QUESTION OF EPISTEMIC MODALITY IN DIVINATORY TEXTS

One approach to the question of the nature of omen statements as predictions is through the nuances of relationship between the antecedent and consequent clauses in cuneiform divinatory texts. The Akkadian term for the consequent in celestial divination is EŠ.BAR/*purussû* "(divine) decision," a term that stems from the juridical language used to describe the relation between the divine and the world that underpinned the practice of divination, as discussed in chapter 6. Given the shared casuistic formulation between omens and laws, this perspective on the formulation of omen statements suggests that the conceptual environment of these texts was perhaps not predictive, but one of case judgment. But omens must be read in the light of divinatory practice, not the practice or function of case law per se. Therefore, it seems to me not a matter of prediction *versus* judgment, but of how to see both meanings active on different levels of engagement.[35]

The apparent simplicity of the If *P*, then *Q* formulation makes for many possible interpretations of its meaning and function. No unified theory has ever succeeded in this regard.[36] But the specific use of conditional statements as omens in the cuneiform repertoire suggests that on grammatical, logical, and semantic levels, the consequent follows the antecedent in an omen statement. The practical application of omen series is seen in the letters and reports sent by the Neo-Assyrian scholars either to warn or assuage the king in the face of ominous signs, such as in the examples given in previous chapters. In view of those sources, if the omen statements in the series were not predictions, the practice of divination was nothing if not prognostic. The *Diviner's Manual*, quoted before, said, "When you know the sign and they ask you to save the city, the king, and his population from the hands of the enemy, from pestilence, and famine, what will you say? They will say to you, how will you make (the evil consequences of omens) pass by?"[37] Because the observation of signs was for the purpose of knowing what would happen, as plainly stated in the *Diviner's Manual*, and as reflected in the many reports and letters the scholars sent, it seems justifiable to see a predictive aim in the practice of divination.

From the standpoint of what it meant for the scribes to know what would happen on the basis of a sign, epistemic modality has been introduced as the way to read the conditional force of omen statements. Markham Geller has rightly pointed out that the Akkadian verb does not distinguish between the future "will happen" and the modal "might/may/could/should/can happen."[38] "Let us suppose," he writes, "that the basic argument can also mean: 'if abc, then xyz *may* occur.'" David Brown agreed that omens "should *not* be translated: If A is observed, then B will take place, but instead . . . as: If A is observed, then B *might* take place (if the appropriate counter measures are not undertaken)."[39] The choice of the English modal serves the purpose of clarifying the translator's understanding of whether divination deals with predictions of certainty (*Q* will happen) or with possibilities and contingency (*Q* might happen). Because the verb is not marked, it is, however, difficult if not impossible to decide between the two based solely on the expression of the omen statement itself. Here is where the larger context of divinatory practice must be taken into account.

I have a slightly different view on this issue, but from an essential point of agreement concerning the contingent nature of prediction by interpretation of signs. Analysis of the meaning of an ominous sign *P*, as "*Q* is thought/believed to be the case unless some action nullifies it," does not illustrate epistemic modality. Epistemic modality may appear to be supported by the use of ritual means to avert bad omens, because the possibility that a portent would not occur was directly related to the success of measures taken to avoid it. Many sources make it plain that such measures (either ritual *namburbi*s or those described in the *Diviner's Manual*) were an integral part of scribal learning for those who had to observe, report, and make recommendations on the basis of omens. But the necessity of performing a *namburbi* indicates confidence in the outcome of an omen, which will—not might—occur unless steps are taken against it. In other words, there is a subtle yet distinctly different semantic force between "the thing will happen unless" and "the thing may or may not happen."

It is not necessarily the case that a lack of marking in the Akkadian as to the modal sense of the verb in the consequent is the same as a lack of confidence in the outcome of the antecedent on the part of the ancient scholar. The interpretation of epistemic modality in omens is, as Geller said, to circumvent the problem of causality in the conditional clause, as well as, I think, to avoid the related implication that omen divination was an expression of hard determinism, or fatalism, wherein a sign indicated a necessary or unavoidable consequence. I could not agree more, that neither causality nor fatalism explains the relationship between antecedent and consequent in omen statements. But in view of the notion of divine verdicts and the fact that divination was practiced in concert with other forms of communication with the gods, some of which were designed to appeal to and appease an angry god, thus undoing the evil of signs, it seems clear, as suggested before (chapter 6, section "Divinatory Signs and Causality"), that the idea of divine causality is at the heart of cuneiform divination's rationale.

Epistemic modality, on the other hand, indicates doubt or a judgment about likelihood, and Geller has suggested that probability replaced causal certainty in the way the antecedent and consequent relate to each other.[40] It seems to me, however, that the interpretations of signs were not subject to doubt or to degrees of possibility (or probability). At stake is the idea of certain knowledge, which, I argue, is precisely what the scribes (and kings) expected from the divine message delivered through divination, particularly, though not exclusively, via extispicy. The criterion of certainty and the value attached to it was expressed as the ultimate and unquestioned divine approval in the form of the "firm yes" (*annu kēnu*).[41] The "firm yes" is extant most abundantly from sources from the Neo-Assyrian period, though hints of it from before the seventh century come from the use of the expression in earlier royal annals, and as well in the existence of Middle Assyrian omen texts dated roughly to the twelfth century. The divine "yes" is already attested in Old Babylonian Mari, where a letter says, "my (extispicy) omens were good and the god answered me 'yes.'"[42] As evidence of the continued value of divination as a source of certain knowledge, a sixth-century inscription of the Babylonian king Nabonidus proclaims, "Shamash and Adad answered me repeatedly with a firm 'yes' by writing favorable signs in my extispicy divination."[43]

All the aforementioned sources refer to extispicy, but after the reign of Esarhaddon, celestial signs were also taken to reveal certain knowledge from the gods as to the future actions of the king. As reported in his annals, Esarhaddon's plan to restore the Babylonian Marduk Temple Esagil was met by positive celestial signs sent by the moon and sun: "Every month, the gods Sin and Shamash together (that is, the moon and the sun) at their appearance, answered me with a firm 'yes' concerning the renewing of the gods, the completion of the shrines of cult centers, the lasting stability of my reign, (and) the securing of the throne of my priestly office."[44] The positive response from the gods gave the green light to Esarhaddon's building plans: "I trusted in their firm 'yes' and I mustered all of my craftsmen and the people of Karduniash (that is, Babylonia) to its full extent. I had them wield hoes and I imposed baskets (on them, meaning corvee labor) . . . I had Esagil, the palace of the gods, and its shrines, Babylon, the privileged city, Imgur-Enlil, its outer wall, built anew from their foundations to their parapets."[45]

In light of the "firm yes," and the "favorable/unfavorable" schematic of the *Diviner's Manual*,[46] it seems that probabilities, inherently statistical in nature, were not a concern of the scribes as much as a binary idea of the portended consequence either happening or not happening. Probability involves some kind of calculus on the degree of certainty, or the likelihood of an outcome. Indeed, this is what Kahneman and Tversky suggested was *not* the typical mechanism of prediction people in circumstances of uncertainty used. They said, "in making predictions and judgments under uncertainty, people do not appear to follow the calculus of chance or the statistical theory of prediction. Instead, they rely on a limited number of heuristics which sometimes yield reasonable judgments and sometimes lead to severe and systematic errors."[47]

Geller identified a calculus of probability in the determination of the "period" (*adannu* "a period of predetermined time"[48]) of validity, or danger, of the predicted consequent.[49] Heeßel analyzed the process with respect to extispicy, and in particular with respect to a procedure for calculating the "stipulated term" (*adannu*) in a commented text on liver omens.[50] The *adannu* is also known in lunar eclipse omens, in reference to the length of time various eclipses remain valid as signs (a term lasting from one hundred to three hundred days depending upon the time of the eclipse's occurrence), as well as in reference to the six- (and five)-month interval between successive eclipses.[51] The feature of the "stipulated term," the period of time during which the portent's danger had validity, furthers the argument for the predictive nature of cuneiform divination, as Geller and Heeßel have discussed. However, the concept of the *adannu*-term does not indicate the operation of probabilistic reasoning, nor of degrees of validity, or probability, of a sign, so much as a temporal window for its validity, again, in a somewhat binary fashion: valid/invalid. That quantitative procedures were introduced to divination, both to liver omens and celestial omens, to determine this term by means of a coefficient (*uddazallû*),[52] is highly suggestive for the treatment of the celestial phenomena as predictive in accordance with their own *adannu*s, or periods.

What counts as a prediction and how predictions—or forecasts, or judgments concerning future situations, or mathematical predictions based on periodic recurrence—are thought to be possible depends upon our understanding of the world within which something is conceived of as predictable. Whether things are predetermined, fixed in a causal network, or affected by future contingency depends on specific ways of seeing the world. The intuitive prediction of cuneiform divinatory texts does indeed point to a contingent world of divine influence and causality, but not a contingent world of chance. This nuance in the underlying rationale of divinatory prediction can be variously understood or interpreted. Whereas Geller and Brown offered epistemic modality (*Q* may or may not happen) as indicative and explanatory of that underlying contingency, I have taken a different tack in assuming confidence in the outcome of a sign (*Q* will happen, unless), where magical means appealed to the deity in some way to undo the sign and its meaning. At root, however, both approaches aim to explicate Assyro-Babylonian divinatory prediction as a matter of divine causality together with the legitimate use of magic to improve future prospects.

Prediction by mathematical models has a different aspect and operates on the basis of different rules, and yet it is possible to bring out some relationships between the model-based predicting of astronomical texts and the intuitive predicting of divination as two different applications of schematics and analogical thinking. The difference that Kahneman and Tversky defined as one between the statistical norms of a mathematical model versus the intuitive rules of representativeness is meant to characterize the asymmetry between two kinds of predictive methods, not that thinking intuitively and thinking to produce mathematically determined predictions are mutally exclusive processes. How we interpret the significance of this asymmetry in the context of the cuneiform texts, however, is a significant historical question, with consequences for the historiography of cuneiform knowledge.

In his monograph *Mesopotamian Planetary Astronomy-Astrology*,[53] Brown focused on the dissimilarity between the two predictive corpora, omens and astronomy, where divinatory prognostication used idealized schemes of a phenomenon's behavior that were not designed for actual, or accurate, prediction, and, therefore, in his view, only superficially resembled astronomical prediction. Astronomical prediction, on the other hand, emerged within the context of celestial divination during the eighth and seventh centuries, but for the purpose of a realization, not an idealization, of the behavior of cyclical phenomena that would accord with observation. In the monograph and elsewhere, Brown drew the historical conclusion that mathematical astronomical prediction indicated a fundamental changeover in worldview from that of divinatory prediction.[54] He argued that astronomical prediction had a decisive and transformative impact on thought in the cuneiform world and offered the notion of disenchantment as explanatory of that change. Accordingly, progressive change in astronomical methodology developed in tandem with a "change in religiosity amongst intellectuals," meaning a diminution in regard for divine influence in the realm of the phenomena.[55] He said,

> The predictability of phenomena, formerly believed to have been meaningfully altered by gods wishing to inform humanity of their decisions, served to distance the gods from Mesopotamian-based mankind, by establishing the non-arbitrariness of some parts of nature. Is this discovery one of the steps in the gods' gradual withdrawal from all aspects of nature to a position of the first-mover, and the creator of the laws of physics, so far as European thinkers go? Is this not a step in the so-called "disenchantment" of nature? Was it not this idea that spread from Mesopotamia to the pre-Socratic Greeks—that predictability was possible, for the celestial mechanism could be understood in terms of *how* (by what means) it runs, irrespective of *why* (to what end) it runs?[56]

Two issues in the way Brown shaped his historiography of Assyro-Babylonian prediction are of interest here. One is the implication that under the influence of the gods the appearances of phenomena are viewed as arbitrary, not predictable. From this point of view, the object of numerical schemes was not to predict, but merely to gauge whether a celestial sign was propitious or unpropitious in accordance with an ideal value, such as thirty days as the length of the month. The ideal numerical schemes, such as that of the length of the day in *MUL.APIN* or the Astrolabes (discussed in chapter 4), were, in Brown's view, only meant to represent the ideal situation, and therefore were not meant to be predictive. The corollary then is that the discovery of the predictable nature of phenomena signaled a process of disenchantment, whereby the phenomena were no longer produced by the gods, but had become the results of their own inviolable and eternal laws.[57] I would concur with Lehoux, who noted that the expected shift from arbitrary or willful appearances of the gods to the phenomena being rigorously and dependably predicted has no historical evidence to support it.[58] The mostly early modern idea of the disenchantment of nature is not only anachronistic for ancient Mesopotamia, but, as Lorraine Daston put it, was part of the "mythic origins" of the scientific revolution as well.

Sometime between 1550 and 1650, she said, "So runs the myth, nature lost its soul. No longer animated or active, nature was reduced to brute, passive, stupid matter. The Scientific Revolution transformed creative nature into a machine, blindly obedient to cause and effect. . . . The very enslavement of nature to a metaphysics of regularity, necessity, and uniformity recommended it as an arbitrator: blind, impartial, inexorable. . . . This is of course a stick-figure caricature of the nuanced, eloquent, and vigorously argued narrative of disenchantment."[59] Only rarely in the history of science before modernity, say in Epicurean cosmology, was the natural world construed without the influence of the divine. As William Shea pointed out in his review of John Brooke and Geoffrey Cantor, *Reconstructing Nature: The Engagement of Science and Religion*, it was in the nineteenth century that God began to have less relevance for nature, and science and religion came to represent competing rather than complementary worldviews.[60] In view of the persistent importance of God and theology in the history of science until relatively recent times, it seems anachronistic to see the emergence of such a notion as disenchantment in the ancient Near East when no stakes in resolving the relationship between the divine and nature are voiced in the cuneiform sources.

To claim with Brown such a progressive historical shift between enchanted divination and disenchanted astronomy would put the Babylonian and Assyrian scholars in a line of thinkers from Lucretius and Democritus to Bertrand Russell and Otto Neurath, "Apollonians" in their attitudes to science and their guardianship of rationality.[61] Gerald Holton described the "new Apollonians" as looking to undertake "the antimetaphysical mission of liberating mankind from the enchantment and terror of superstition."[62] Assyro-Babylonian scholars are wholly out of place in such a lineup, because cuneiform predictive texts, even classified into the different predictive modes or strategies of the intuitive and model-based types, cannot be correlated with the dichotomous attitude that would construe intuitive prediction as irrational/unscientific and model-based or computational prediction as rational/scientific.

MODELS, ANALOGIES, ANALOGUES

Models, whether mechanical, kinematic, evolutionary, computational, heuristic, analogue, or other kinds, have long prevailed as a tool of science. Understanding the nature and function of models has raised critical issues in the philosophy of science: how models represent, what models are in an ontological sense, how we know things through models, the relation of models to theory, how models function as explanations, and the bearing of models on the realism/instrumentalism, or realism/antirealism, debate. While the Babylonians may not have had a counterpart to our concept of model, historical analysis of the way that celestial phenomena were represented in schemes attested on cuneiform tablets can, I submit, be profitably addressed in terms of this concept. Exploring in particular the use of models and analogy in cuneiform texts from the entire range of its tradition of scholarship is another important dimension in the study of the relation of cuneiform knowledge to science.

Evidence for the earliest models in astronomy comes from texts such as the Astrolabes, *Enūma Anu Enlil*, and *MUL.APIN*. Foundational for modeling celestial and calendrical phenomena was the recognition of periods, however schematic or ideal, and eventually of period relations. This foundation is traceable to celestial divination, which employed the notion of anomalous occurrence in terms of a phenomenon appearing at an abnormal time, expressing the notion variously as *ina la minâtišu* "not according to its count," *ina la simanišu* "not at its time," or *ina la adannišu* "not at its appointed time." Attention to the periodic nature of phenomena culminated in the recognition of two kinds of relations between periods. One identified an integral number of cycles made by one heavenly body (such as the sun) with an integral number of cycles made by another (such as the moon), as in the calendrical cycle fundamental to the Late Babylonian calendar, the nineteen-year cycle, where nineteen years = 235 synodic months and sun and moon return to initial positions, or very nearly. The other correlated integral numbers of phenomena with integral numbers of some time unit, such as eleven synodic phenomena of Jupiter = twelve years, or fifty-seven synodic phenomena of Saturn = fifty-nine years, or the well-known Saros cycle where thirty-eight eclipse possibilities = 223 synodic months.

Prediction of astronomical phenomena throughout the cuneiform tradition, whether by models or empirical methods, was characteristically quantitative. Before the development of parameters such as the year or the synodic month, the Astrolabes, *Enūma Anu Enlil*, and *MUL.APIN* employed ideal time units—namely, the 360-day year or the thirty-day month. These units expressed the periods used in early models of phenomena such as the yearly variation in the length of daylight and the monthly variation in the visibility of the moon at night. *Enūma Anu Enlil* Tablet 14, perhaps based upon a mathematical scheme produced in the second millennium (Old Babylonian period), is an example of the kind of modeling characteristic of Babylonian astronomy in its earliest stage.

Based on a linear scheme for the length of daylight/night in the ideal year known from the Astrolabes (discussed in chapter 4), *Enūma Anu Enlil* Tablet 14 consists of a scheme for the changing duration of lunar visibility at night during the equinoctial months, as well as a section where difference coefficients are given for calculating its visibility in other months (always 1/15th of the length of night for the month in question). It modeled the nightly change in lunar visibility by a zigzag function, taking into account two components affecting the duration of visibility, the first being the daily change in elongation (distance of the moon from the sun from the observer's perspective), and the second being the change in the length of night. The text separated the two factors into different sections and combined them in creating the coefficient for the moon's duration of visibility at night, calculated as 1/15th night. The computation of lunar visibility throughout the year is modeled with a linear zigzag function of daily difference 0;2,40°, a maximum value of 16 and minimum of 8. From the value for the constant difference (0;40°) between the difference coefficients for the moon's visibility in each of the twelve months, the daily difference for the lunar visibility can in principle be interpolated for any date. Evidence that the scribes made such interpolations is lacking, but the potential for such interpolation is embedded in the scheme.

The model in *Enūma Anu Enlil* Tablet 14 similarly reflects the state of knowledge of its time. It did not account for the additional effects of lunar latitude or velocity upon visibility, refinements to come later, and used ideal parameters for the day, the month, and the year. It also differs from the later models of lunar phenomena in its lack of a positional coordinate system as was developed later in the form of the zodiacal circle of twelve equal signs ("longitude" or position reckoned along the ecliptic from Aries, and "latitude" or distance perpendicular to the ecliptic reckoned as "height" or "depth"). To recognize, as the scheme in *Enūma Anu Enlil* Tablet 14 does, the problem of the moon's visibility at night in terms of the effects of the date in the schematic month (effectively "lunar elongation") and the variation in the length of night throughout the year, and to resolve that complex problem with combinable functions, is already evidence of a special kind of reasoning now recognized as "model-based."[63]

The study of models in science and the reasoning that produces them has only recently come into its own. In 1999 Nancy Nersessian said that "in traditional philosophical accounts, what I am calling model-based reasoning practices are considered ancillary, inessential aids to thinking . . . fringe topics in the philosophy of science."[64] In 2006, however, she introduced her subject by saying that "modeling is now widely recognized as a signature feature of the sciences—contemporary and past. A growing body of research in history and philosophy of science establishes that, in contrast to the standard image of scientific reasoning as hypothetico-deductive or logic based in nature, much reasoning is model based, that is, through and with models."[65] As discussed above (chapter 5) in relation to demonstrative logic being too limiting a criterion for rationality, this statement shows that equating reasoning with logic is equally limiting for an account of problem solving in science both philosophically and historically. Elsewhere Nersessian noted that "many philosophers now agree that the basic units for scientists in working with theories are most often not axiomatic systems or propositional networks, but models."[66]

This line of inquiry seems particularly useful in relation to cuneiform astronomical texts in mitigating a positivistic narrative account for Babylonian astronomical theory formation that judges whether a text is more or less scientific based on the accuracy of its results. Taking Bailer-Jones's definition of a scientific model as "an interpretive description of a phenomenon that facilitates access to that phenomenon,"[67] a range of astronomical models can be discussed within the history of cuneiform knowledge, where emphasis is not put on progressive accuracy.

There is doubtless a developmental dimension to the history of Babylonian astronomy. The fact of a full realization of the analysis of lunar and planetary theory in the ephemerides of the Seleucid period is a primary index of development in knowledge. The progressive dimension, however, need not be singled out as the chief feature, or even a characteristic value, of cuneiform knowledge of the heavens. Both the late astronomical and astrological traditions were inclusive of methods and models carried over from the past. Where we would perhaps view the older methods as incompatible with the newer, clearly the ancient scribes did not. The use of older schemes, such as those known in *MUL.APIN*, in late texts is well attested.[68] The coexistence of a plurality of models further mitigates a progressive reconstruction of astronomical knowledge in cuneiform texts.

The use of models in the cuneiform tradition is tied to that of astronomical theorizing. Babylonian astronomical models depend upon the conceptual framework of theory, including the parameters, the units in which parameters are expressible, as well as available coordinate systems. Models can retain or reject older forms, reflecting variously tradition or innovation. The period between the Neo-Babylonian Empire and the arrival of Alexander the Great (i.e., between ca. 626 and 330 BCE) was long viewed as the major developmental phase in Babylonian astronomical theorizing, producing the greatest innovative push in terms of what and how phenomena could be modeled. From the point of view of the progressive narrative, astronomical texts from this period were dubbed an intermediate stage of Babylonian astronomy, after A. J. Sachs's designation for texts belonging to "stages later than MUL.APIN and earlier than ACT," although he noted that "the boundaries in both directions are not sharp."[69] More recent investigations of pre-Seleucid astronomical texts suggest a modified reconstruction, pushing the boundary for predictive theorizing back into the seventh century, contemporaneous with the height of activities among the celestial diviners at the Neo-Assyrian royal court.

If attention to observation of the characteristic intervals of visibility of the moon at syzygies (as represented by attention to the Lunar Six) is characteristic of intermediate astronomy, it is already evident as early as the mid-seventh century in the reign of Nabopolassar of Babylonia and Assurbanipal of Assyria, thereby extending the boundaries of the "intermediate" stage of cuneiform astronomy. By 643 BCE Lunar Sixes were being collected, suggesting intensive investigation of the moon's periodicity. Study of these texts in particular have led Peter Huber and John Steele to conclude that toward the end of the seventh century, methods for prediction of the Lunar Sixes as well as eclipses were taking shape.[70]

As further evidenced in Goal-Year texts, period relations for the moon and five planets were used as a method of predicting the next occurrence of a phenomenon from an observation of that phenomenon recorded one Goal-Year period prior to the Goal-Year.[71] Goal-Year texts were in regular use by the Seleucid period, presenting the raw data for predicting the dates and zodiacal signs for synodic phenomena of Jupiter, Venus, Mercury, Saturn, and Mars (in that order), and the moon, giving data for the Lunar Sixes as well as eclipses. According to the rubric given these texts by the scribes, they included data for "the first day, (synodic) appearances, passings and eclipses, which were established for (goal-)year x" (UD.1.KAM IGI.DU$_8$.A.MEŠ DIB.MEŠ *u* AN.KU$_{10}$.MEŠ *ša* DIŠ MU x *kunnu*).[72] Goal-Year data were then introduced into Almanacs and Normal Star Almanacs.[73]

Goal-Year texts, Almanacs, and Normal Star Almanacs are classified as nonmathematical astronomical texts to distinguish them from the ephemerides of Systems A, B, and other variant systems. The distinction can be defined in several ways, one of which is that mathematical texts are tabular and nonmathematical are not. Nonmathematical texts have an observational emphasis, inasmuch as the Diaries are the principle corpus and the data in the Goal-Year texts derive from them. This only underscores how both observation and prediction are integral to nonmathematical texts, not only in terms of Goal-Year methods, but also where certain phenomena in the Diaries can be supplied by computation (for example, the dates of solstices and equinoxes).

Not all models are predictive, but the quantitative Babylonian astronomical models were characteristically so. Models were part of a complex of mathematical tools and concepts used in the Babylonian approach to predicting phenomena. Ossendrijver has clarified this complex as a three-tiered structure, from elementary operations and numbers at the most basic level, to algorithms and functions, to the computational systems (of which Systems A and B are the best known), with corresponding concepts at each structural level.[74] Models, or modeling, reside in the second level with algorithms and functions, and the two basic computational systems adopted different functions for modeling solar, lunar, and planetary phenomena.

The Uruk Scheme is a good illustration of a deterministic model for calculating the cardinal phenomena of the year in accordance with the nineteen-year cycle without taking account of astronomical concepts or observational data. Britton pointed out with respect to the Uruk Scheme, that "from a perspective of accuracy and mathematical sophistication the Uruk Scheme was a distinct step backwards from the System A solar model. Nevertheless, from roughly −350 onwards, virtually all recorded solstices and equinoxes and nearly all recorded Sirius phenomena are consistent with and evidently computed from this scheme."[75] The Uruk Scheme represents the cardinal phenomena, as well as the synodic visibilities of Sirius (first appearance, opposition, and last appearance) as a function of the nineteen-year cycle. The phenomena are determined by the parameter value (12; 22, 6, 20 synodic months) and are thus fully predictable. Later the nineteen-year cycle would play a role in modeling lunar variation due to lunar anomaly (such as of lunar longitude, latitude, intervals between syzygies, or length of the month) in eliminating zodiacal anomaly over 235 months, thereby isolating the effects of lunar anomaly.

An excellent example of the coexistence of older and newer methods for astronomical prediction is the Uruk text (TU 11), written by the scribe, Anu-uballiṭ, son of Nidinti-Ani, a *kalû* priest of Anu and Antu from Uruk.[76] This text has various procedures or explanations of both astronomical and celestial divinatory nature divided into twenty-nine sections, and attests to the use of a plurality of observational and computational methods, including Goal-Year methods alongside others of a different but still observational nature, such as using the height of the first lunar crescent above the horizon at first visibility as an indication of month length. Still another section uses the daylight scheme from *Enūma Anu Enlil* and *MUL.APIN* to calculate if the following month will be full or hollow by using the coefficient 1/15 of length of night for the daily retardation of the moon.[77] The perpetuation of the older scheme for length of night well into the Seleucid period is also seen in the texts that deal with the risings times of the zodiac, using the *MUL.APIN* scheme in their calculations as well.[78] Referring to TU 11, Brack-Bernsen and Hunger said: "as in other tablets from the intermediate period, we find methods which belong to the most advanced lunar theory associated with a primitive attempt to deal with visibility problems or solar positions. It is the occurrence of column Φ in early texts, giving the momentaneous position of the moon in its anomalistic cycle, next to primitive schemes, which puzzled Neugebauer, and still puzzles us, today."[79]

Despite a willingness to employ older traditional methods of prediction, mathematical modeling of the phenomena proceeded apace within the community of Babylonian astronomers. The latest developed astronomical texts used functions (zigzag or step) to account for the difference between a position (or date) y_n and the next in sequence y_{n+1}, reckoned as longitude or time (in *tithi*s). In System A this difference was taken to vary in accordance with a conceptual model of the ecliptic subdivided into "zones" of longitudinal progress.[80] The simplest version consists of two such zones of progress, fast and slow, where the successive positions of a synodic phenomenon are represented by a step function. In this way, System A modeled the synodic arc—the characteristic change in longitude between successive synodic events (Δλ)—directly with a step function of longitude, that is, $\Delta\lambda = f(\lambda)$. The period in accordance with this model is found in terms of the number of synodic events (Π) to the number of revolutions of the ecliptic (Z), in Neugebauer's formulation expressed as $P = \Pi/Z$, where Π and Z are relatively prime and Z is nearly always regular.[81] The model for the solar synodic arc, for example, was constructed of the parameter 12;22,8 months for the length of the year, and the sun's progress at a rate of 30° per month on one arc of the ecliptic (the length of the fast arc = 3,14°), 28;7,30° per month on the slow arc (= 2,46°).

The component parameters of System A's way of modeling the phenomena—namely, the rate of longitudinal progress and the size of the ecliptical arcs in which the successive synodic events takes place, gave System A a greater degree of flexibility to model various synodic phenomena as compared with System B. In the System A lunar theory, step functions were used in columns B (for the longitude of the sun each month corresponding to the longitude of the moon at new moon and as +180° for full moons) and J (derived from B and depending on solar anomaly). In tying the computation of consecutive phenomena to positions on the ecliptic (synodic arc Δλ), or indeed to the dates of their appearances (synodic time Δτ), System A's model was tied more directly to concrete phenomena as compared with System B.

Nonetheless, the object of System A's model does not seem to have been to describe the actual motion of a celestial body, but to use "fast" and "slow" arcs of the ecliptic as a way to deal with the problem of anomaly.

The goal of the Babylonian astronomical models was prediction, but how this goal was achieved was not uniform from one "system" to the other. These differences raise questions about aspects of modeling that have been a concern of analytical philosophy as well as the philosophy of science. As Mauricio Suárez explained: "The interest from analytic philosophy is related to the notion of reference, and the metaphysics of relations; the interest from philosophy of science is related to an attempt to understand modelling practices. These two distinct forms of inquiry into the nature of representation may be distinguished as the 'analytical inquiry' and the 'practical inquiry.'"[82]

Practical inquiry into the Babylonian astronomical models will observe that the zigzag function in early Babylonian astronomy and in System B, and the step function in System A, modeled astronomical phenomena for the purpose of prediction from a mathematical scheme. How such prediction was used, however, is not entirely clear. Brown suggested that early astronomical texts created schemes for determining when a phenomenon deviated from conceived norms. Whatever their goal, the early schemes were mathematically descriptive of the phenomena they modeled. Late astronomical texts may have enabled the calculation of lunar and planetary positions for the construction of horoscopes, but to date this hypothesis has not been convincingly demonstrated. The philosophy of science's more practical interest in modeling practices is therefore somewhat frustrated by our lack of information on the scribes' objectives for producing predictive texts and tables. The ancient predictive goals, however, did admit a plurality of models, and that pluralism seems to be a characteristic feature of the approach to astronomical prediction in cuneiform culture.

As for the metaphysics of the relation between scientific models and the phenomena being modeled, the question remains as to whether or how the Babylonian astronomical models were thought somehow to be representative of the phenomena and, by extension, the world. Any answer to this question is pure speculation for lack of commentary on the nature of the models from the scribes directly. As said at the outset, the notion of a model is used here heuristically, without attributing that notion to the ancient scribes. However, it must be admitted that a model need not be a literal description or representation of the world and yet still have a connection to the world. In other words, the nonuniform motion of celestial bodies could be modeled with a step function determining longitudinal displacement, or change in longitude, of the body at consecutive synodic moments without purporting to say that the sun or a planet moved at these various rates or "velocities" around the ecliptic. Indeed, the tables were not focused on motion as such, but on synodic events. Where the planet was in between synodic events was either not of interest or could be derived by interpolation. Daily motion tables are attested, though they are far fewer in number.

Even more removed from a description of "physical reality" than System A was System B, which analyzed the differences not between actual consecutive occurrences of the event in question, but between event numbers in the table. As Aaboe put it, "In System B, $\Delta\lambda$ is not tied closely to the ecliptic; rather, the independent variable in the computation of $\Delta\lambda$ is the number n of a synodic phenomenon in some sequence of synodic phenomena of the same kind for a given planet or, if one wishes, the line-number in the ephemeris. $\Delta\lambda = y(n)$ is then computed as a zig-zag function of this line number n."[83] The differences between consecutive values of the function that were tied to event numbers were thus considered as varying continuously between extrema and modeled by means of a zigzag function where the period was constructed from the difference (d) between entries (y_n) in the table and the amplitude (Δ) of the function, so that the period P is equal to $2\Delta/d$.

The zigzag function is the earliest attested mathematical tool for modeling periodic phenomena, as seen in the Astrolabe's model for variation in the length of daylight, the model for lunar visibility durations in Tablet 14 of *Enūma Anu Enlil*, and a Neo-Babylonian model for the seasonal hours.[84] These early zigzag functions are based on simple integer periods, such as the 360-day year, or the thirty-day month. This approach to visible changes in the phenomena that occurred with time was certainly in the background of modeling with zigzag functions in later astronomy. The difference between the early and the later comes down to differences in the periods used to structure the models and to distinct differences in what the referents of the values were in the line entries of zigzag functions.

Comment on the Babylonian tradition, even raising the question as to its place in the history of scientific modeling, is colored by our own tradition of the use of models in physics to say something about reality, a tradition that goes back to James Clerk Maxwell's methods of "physical analogy." Our scientific values follow from a position that the best models are true to the structures of nature, but our philosophical values range in views about truth and reference. Models are not the real world but have a relation to it, as Ronald Giere suggested,[85] in the manner of a map. And like maps, whose representational objectives vary, models are neither true nor false, but correspond variously to "the real world." Indeed, neither a model nor a map can represent or correspond fully, as Borges showed in his story "On Exactitude in Science," about the useless but rigorous map of the Empire "that was of the same scale as the Empire and that coincided with it point for point."[86]

The coexistence of Systems A and B models further argues against the idea that a physical representation was of any comparative value among Babylonian astronomical models. Given the interest of Babylonian astronomy to account for synodic (or daily) positions of the celestial bodies, it does, however, seem justified to view their models as representations, or analogues, that bear in a mathematical descriptive manner on the phenomena. We should note here too that Babylonian astronomical theory had a powerful empirical underpinning (as discussed in chapter 7), which no doubt influenced the way in which the phenomena modeled in the tables were understood. Further argument for the representational or analogic nature of Babylonian astronomical models is that period relations, often referred to as the backbone of Babylonian mathematical astronomy, anchored the models based upon them to the world without representing the world in a direct physical sense. As in Giere's analogy to maps, models may focus quite partially on some domain of the world's multitude of phenomena. There will be various motivations for representing those phenomena, but phenomena of the world are nonetheless what models represent. From this perspective, Babylonian models can be said to be tied to reality without attempting to describe reality. They can be viewed as representations, or analogues, in terms of Nancy Cartwright's earlier view of models as simulacra, "fictional" with respect to reality. Both kinds of model would be "representative," but only as simulacra without correspondence to the world.[87]

Bailer-Jones said this about representation: "The term . . . *representation* introduces a relationship between a symbolic construct, a model, and a phenomenon or object of the world, which could, in epistemological terms, be considered a mixture between truth and falsity. The problem is that models do not *mirror* or *replicate* the world in every single detail. . . . So the relationship that the concept of representation is expected to capture is that symbolic constructs—models—stand in some relevant relationship to the world, without being *replications* of it."[88] Answering her own question, "why do models not tend to be copies of reality?" Bailer-Jones pointed to the "tensions or inconsistencies between models of the same object or phenomenon [that] arise when the models rely on different sets of principles that are at odds with each other."[89] This statement seems apropos of the difference between the models of Systems A and B, which describe the same phenomena but on different principles and using different algorithms and functions. Differing in the manner of calculation of synodic arc and time, or even of daily progress along the ecliptic, Babylonian models all shared a common objective to predict and a disinterest in cosmology or physical explanation.

What Bailer-Jones also stressed about models, however inconsistent or partial their representation of phenomena might be, was their epistemological function. That is, in her words, "models are important forms of knowledge (or of developing knowledge). . . . In other words, models can carry crucial and illuminating information about empirical phenomena despite potentially being inaccurate, inconsistent, and/or incomplete."[90]

The (modern) epistemic value placed on models in science is most interesting in view of the many-layered historiographical debate concerning the status of models mathematical astronomy produced in antiquity, the Middle Ages, and the sixteenth century, as well as the different attitudes that developed regarding the division between mathematical astronomical knowledge and physical cosmology. That division, since Pierre Duhem's claim that ancient astronomy's aim was solely to "save the phenomena,"[91] has in modern times been refracted through the lens of the epistemological dichotomy of instrumentalism and realism.

In an early study, Geoffrey Lloyd reviewed Duhem's claims about Greek astronomy, focusing on the evidence of Ptolemy's works, especially the *Almagest* Bk 1.3 and 7 and Bk 2 of the *Planetary Hypotheses*, in terms of the concerns of instrumentalism versus realism.[92] Bernard Goldstein also traced the complex history of the use of the term "phenomena" in ancient astronomy and what "saving" them may have meant in the time of Ptolemy (second century CE).[93] That particular history encompassed not only a wide range of phenomena treated in Greek astronomical works but also differences in the treatment of kinematic models reconstructed from pre-Ptolemaic times to Ptolemy's *Almagest*. Goldstein concluded that "it is not so easy to distinguish physical from mathematical phenomena," and only in retrospect to decide "what belongs to astronomy and what does not."[94] In his view,

> the authors in late Antiquity who came after Ptolemy, including Proclus and Simplicius, took what Ptolemy had done as completely natural without appreciating his reasoning. Following them, Duhem found it rather straightforward to make the distinction between mathematical astronomy and physical astron-

> omy. As Duhem put it: On the one side there was astronomy—geometers like Eudoxus and Calippus formed mathematical theories by means of which the celestial motions could be described and predicted, while observers estimated to what degree the predictions resulting from calculation conformed to the natural phenomena. On the other side there was physics proper, or so to speak in modern terms, celestial cosmology—thinkers like Plato and Aristotle meditated on the nature of the stars and the cause of their movements.[95]

Subsequently, Duhem's astronomers and physicists have been understood in terms of the intrumentalist or realist stance respectively, although Duhem himself did not use that terminology. Gad Freudenthal has given an account of the origins of the terminology and how it entered the philosophy of science through Reichenbach, Morgenbesser, and Popper.[96] Accordingly, "'instrumentalism' is the thesis that 'scientific theories are useful and that scientists are justified in using them even if the entities they countenance are fictional' (Morgenbesser 1969, 201). Thus it is opposed to the position called 'realism,' which holds that scientific theories consist of statements reflecting (more or less precisely) reality, and positing entities that exist."[97] He noted that the problem of the status of astronomical knowledge (as "instrumental") was the consequence of conflicting ideals of science attributed to "the interaction of two fairly independent theoretical traditions: Greek logic and natural philosophy, especially in their Aristotelian version, on the one hand; and mathematical astronomy of Babylonian-Greek ancestry on the other."[98]

Freudenthal's characterization of ancient mathematical astronomy as of "Babylonian-Greek ancestry," is of material importance to this discussion, as the impact of the transmission of Babylonian astronomy during the early Hellenistic period (second century BCE) was that it resulted in changes to the treatment of celestial phenomena within Greek astronomy. Babylonian predictive models, of both Systems A and B type, were taken aboard in Greco-Roman astrological methods of prediction.[99] Geminus, in the first century BCE, was clearly aware of a Babylonian model of daylight length and of System B's model for the daily anomaly of the moon in longitude.[100] Babylonian eclipse observations were used by Ptolemy in the *Almagest*.

Alan Bowen has analyzed the impact of the reception of Babylonian predictive astronomy and horoscopic astrology in the Hellenistic scientific world in terms of its influence on astronomical thinking accompanied by a new demarcation between astronomy and physical theory.[101] He said: "Specifying how this reception actually took place is difficult, partly because our sources are incomplete and partly because there really was no single, particular (as opposed to general) cause. Rather, what one finds on examining specific documents is a tangle of different considerations in Babylonian horoscopic astrology and its Hellenistic interpretation working together in various, related contexts to produce a new definition of astronomy."[102] Such an intertwining of considerations can be seen in the incorporation of Babylonian System B's zigzag model of nonuniform progress of the moon in longitude by Geminus, who regarded celestial motion as uniform and constant, in what James Evans and J. L. Berggren called "the happy coexistence of philosophical principle and arithmetic convenience that was characteristic of Greek astronomy in his day."[103]

Duhem's representation of Greek astronomy's aim to "save" the phenomena as an instrumentalist one has subsequently been revised by historians of astronomy. Lloyd noted that the distinction between astronomy and physics in Ptolemy, Geminus, and Proclus was not one of separation to the point of being conceived of as divorced from or excluding one another.[104] Nor did he agree with Duhem's characterization of the ancient Greek astronomers as instrumentalists. Goldstein too discussed the fact that Geminus and Ptolemy were each concerned with the physical reality of the phenomena.[105] As he said, "according to Ptolemy the phenomena are 'real' and not illusions, for they are the criteria by which the models are judged, not the other way around."[106]

Geminus is an interesting vehicle for showing that the division between supposed realism in physics and instrumentalism in mathematical astronomy is a misrepresentation of the critical difference between these forms of knowledge. A passage testifying to Geminus's "realism" and classification of astronomical knowledge in the Aristotelian mode comes down to us in Simplicius's commentary to Aristotle's *Physics*, more than five hundred years after Geminus.[107] Simplicius says that Alexander of Aphrodisias quotes a text of Geminus, concerning the *Meteorologica* of Poseidonios. The work gives us the following passage:

> It is for physical theory to inquire into the substance of the heavens and of the heavenly bodies, into their power and quality, and into their coming into existence and destruction. Through these [investigations], it can certainly offer demonstrations concerning size, shape, and ordering. Astronomy, on the other hand, does not attempt to speak about anything of that sort. Instead it demonstrates the order of the celestial [bodies] after declaring that the heavens really are a cosmos, and speaks about the shapes, sizes, and distances of the Earth, the Sun, and the Moon, about the eclipses and conjunctions of heavenly bodies, and about the quality and quantity in their movements.[108]

As the passage continues, it focuses further on how demonstrations from the astronomer (*astrologos*) and the physical theorist (*phusikos*) differ:

> Now astronomers and physical theorists will in many cases propose demonstrating essentially the same [thesis] (e.g., that the Sun is large, that the Earth is spherical), yet they will not follow the same procedures. For, whereas [physical theorists] will make each of their demonstrations on the basis of substance, or power . . . astronomers [will do so] on the basis of the [properties] incidental to shapes or to sizes, or on the basis of the quantity of motion and of the time interval appropriate to it.
>
> Also physical theorists in many cases will deal with the cause by focusing on the causative power; whereas astronomers, since they make their demonstrations of external incidental properties, are not adequate observers of the cause in explaining that the Earth or the heavenly bodies are spherical, for example. Sometimes they do not even aim to comprehend the cause, as when they discourse on an eclipse.[109]

As the passage indicates, the stakes of the divide are not reality, but demonstration and explanation. Bowen pointed to the distinction between the two kinds of explanation, one causal (physics) and one noncausal (astronomy), and put it that the physical theorist can explain "why something is as it is," whereas the astronomer explains only "that the thing is as it is."[110]

This reflects the beginnings of a complex history of views on the relation between astronomical models (hypotheses in the form of circles and orbs) and the world, that continued into the sixteenth century.[111] As a way to understand the many examples quoted from sixteenth-century astronomers that reflect on the status of astronomical models, and bear on the context in which Copernicus was accepted or rejected, Peter Barker and Bernard Goldstein proposed that what was central to the terms of the discussion was "the status of astronomical demonstrations according to the time honored Aristotelian categories of *quia* and *propter quid*."[112] *Quia* (lit. "that," from Aristotle's "knowing that" or *τὸ ὅτι*) demonstration was in essence demonstration from effects without stating causes, and was thus nonexplanatory. *Propter quid* (lit. "on account of which," from Aristotle's "knowing why" or *τὸ διότι*) demonstration was from the a priori, as cause to effect, and was thus explanatory. This refocuses the terms of the debate from "reality" to whether and how astronomy offered causal, or explanatory, knowledge.[113] The division between astronomy and natural philosophy thus came down, in Jamil Ragep's words, not to "the actual set of doctrines but rather on the way to prove them."[114]

Changing views on the status of astronomical knowledge with respect to whether it provided a causal account (explanation) of nature relates to the question raised before about the meaning of "saving the phenomena." The cuneiform astronomical texts were brought into the history of these views by Greek astronomers and natural philosophers aware of their methods, but the missing factor in the cuneiform tradition of knowledge of the heavens was the very thing we regard as indispensible—namely, physical nature. Theon of Smyrna said it in his *Exposition on Mathematical Things Useful for the Reading of Plato*:

> The Babylonians, Chaldaeans, and Egyptians—eagerly sought out some starting points and hypotheses with which the phenomena fit [and] by means of which there is judging of what has been detected before and foretelling the future, some by introducing certain arithmetical techniques (*methodous*) (as the Chaldaeans) and others [by introducing] geometrical ones (as the Egyptians). Since they all produced techniques that are incomplete, that is, without *phusiologia*, it was necessary to investigate this [matter] physically (*phusikos*) as well at the same time.[115]

In Greek philosophical terms, the Babylonian astronomical models did not supply a representation of the world satisfying *Greek* epistemic criteria for explanation or causality. But in Babylonian terms, if we take the ephemerides as mathematical models of synodic events for prediction, we can see in another light how they served to represent the phenomena. What is "missing," in a sense, as Theon indicated, was "knowledge of nature" (*phusiologia*), though there was certainly a connection between the world and the phenomena by means of their periods. As implied before, the way Babylonian models represented was not with respect to the motion of celestial bodies. Neither Systems A or B models were cosmological, nor purported to describe the physical nature of celestial motion in any literal sense, much less a causal sense. Each of the late Babylonian models for nonuniform travel in the ecliptic effected a mathematical representation, without intending to represent "physical reality" or to account for that nonuniform motion in a causal way. In terms of the epistemic values of the cuneiform tradition, the tables with their models of periodic phenomena were not lacking in anything, indeed were considered by their writers and copyists to be a form of knowledge of the highest order, classified as "the wisdom of Anu-ship" (*nēmeq anūtu*).[116] The Late Babylonian ṭext (TU 11), mentioned above, and giving, among other things, rules for calculating month lengths and intervals of lunar visibility around full moon, described its content as a "tablet of the secret of heaven, secret knowledge (*pirištu*) of the great gods."[117]

CUNEIFORM EXPLANATION AS INTERPRETATION OF TEXTS

The ancient roots of explanation are generally sought in the exploration of causality in Greek philosophy, most fully developed by Aristotle (*Physics*, *Metaphysics*, and *Posterior Analytics*). R. J. Hankinson said the Greeks may not have invented causal explanation, but that the questions to which explanation was directed were unified in being causal in nature, as well as ontological, as they were "questions that call for an elucidation of the structure of the world."[118] Goldstein and Bowen discussed a passage from Pliny's *Natural History* in which an account is given of the Roman general Sulpicius Gallus (second century BCE), who was the first Roman to give an explanation of eclipses as natural phenomena.[119] Thales, too, is mentioned in the same context as being the first Greek to explain solar eclipses. They note that the verb used, usually translated as "to predict" (*praedicere*), has in this context the force of explanation by giving causes. The passage connects Hipparchus to the same program of knowing nature: "After them [Sulpicius Gallus and Thales], Hipparchus proclaimed the daily progress (*cursum*) of each star for 600 years, [Hipparchus] who understood the months and days of the nations, the longest daytimes and geographical locations of places, and the appearances of the peoples, and who, as time has shown unequivocally, was partner in the plans of nature (*consiliorum naturae particeps*)."[120]

Causal explanation of nature, in Hankinson's view, distinguished scientific from other forms of explanation: "Science is concerned not with the explanation of some individual event, or some particular ephemeral occurrence, but rather with producing a quite general account of why *things of this sort* happen in the way they do *in this type of circumstance* . . . if our scientific explanation of a particular earthquake is to be scientifically adequate, it must in principle be capable of explaining why any earthquake occurs."[121] Explanation that emphasizes causation and the reality of causal connections has an ontological thrust that distinguishes it from the formal-logical view of Hempel-Oppenheim. According to Jan Faye,

> the ontological view considers a scientific explanation to be something that involves causal mechanisms or other factual structures. The idea is that facts and events explain things. In particular, causes explain their effects. A cause tells us why its effect occurs. A scientific explanation is an objective account of how the real work is connected. The cognitive representation of the facts of the matter does not contribute to the meaning of explanation. An explanation if both true

> and relevant if, and only if, it discloses the causal structure behind the given phenomenon. Furthermore, an everyday account counts as an explanation if it is reducible to science talk about causal processes.[122]

Although distinct in their conceptions of what makes a statement explanatory, both the ontological-causal and the logical-formal notion of explanation are at least in part predicated on the existence of the conception of the laws of nature. Not only Aristotle's focus on nature but also the historical weight of the idea of the laws of nature, belong to a shared conceptual background, common to thinking about explanation, and about explanation and its relation to science in the various Western modes.

In his 1992 article "Hellenophilia versus the History of Science," on the other hand, David Pingree took his perhaps most strident historicist stance with regard to the sciences of other cultures and delivered his clearest indictment of Western historians who failed to appreciate the scientific nature of non-Western bodies of knowledge, in particular what constituted their form of explanation. Inter alia, he said: "Babylonian divination is a *systematic explanation* of phenomena based on the theory that certain of them are signs sent by the gods to warn those expert in their interpretation of future events; there is no causal connection, but only one of prediction so that appropriate countermeasures may be undertaken . . . In themselves, texts describing the rules of the interpretation of omens Enūma Anu Enlil for celestial omens and Šumma ālu for terrestrial ones—are scientific; they provide *systematic explanations* of phenomena."[123] Pingree did not specify or define what he meant by explanation, but singled out explanation and prediction as criteria important in omen texts, precisely in order to place these texts within the privileged purview of the history of science.

Pingree's view that the predictions (apodoses) of omen statements were explanations is clearly at odds with ontological-causal, logical-formal, or any other more recent formulations of explanation (constructive empiricist, cognitive, and others). Omen apodoses provide neither answers to a why question nor reflect upon world structures, nor do they offer mechanical-causal explanation. They do not presuppose a conception of uniformity in the cosmos or of universal laws of nature. The "systematic explanation of phenomena" that Pingree pointed to in Babylonian divination viewed the relationship of antecedents to consequents of omens neither as cause to effect nor as an account for why ominous phenomena appeared. In the place of causal explanation of either of these varieties, cuneiform scholarship developed other modes of explanation appropriate to the various contexts and norms of its knowledge, and this had principally to do with texts.

As divination was a scribal scholarly endeavor it stands to reason that its development would involve philological techniques consistent with other scribal scholarly practice. Ominous signs and cuneiform signs were related in the sense that both were to be read and interpreted. Explanation, therefore, took on a number of functions depending on different text types. Variously, explanatory texts could focus on elucidation (of words by means of synonyms), exposition (of a phenomenon by means of description), or instruction (by means of the procedural steps involved in making, calculating, or performing something), all arguably subsumable under the rubric interpretation.

Explanation was a vital part of the cuneiform project of knowing the world of signs, the correspondences between things, and the meaningful relationships between words and the world. In the form of commentaries and procedures, explanation found its way into other areas of scribal knowledge apart from divination, for example, into mathematics, medicine, and astronomy. Indeed, Eckart Frahm characterized the intellectual tradition of the cuneiform scribes as one "that conceptualized the world as a text in need of interpretation."[124] This observation serves particularly well for describing the rationale of divination, where the reciprocity of word and world seems reflected in the metaphors of the heavenly writing and the liver as the tablet of the gods. The constellations and planets were the manifest form of the gods' writing upon the surface of the sky, while in a kind of mirror image, the gods also wrote marks, colors, and features on the surface of the liver. That these divinatory media were thought of as parallel is conveyed in the incipit to Tablet 16 of the *bārûtu* series: "If the liver is a mirror of heaven."[125] The expository text titled *iNAMgišḫurankia*,[126] with material concerning celestial divination, metrology, the calendar, and number symbolism, explicitly states that it is concerned with "whatever pertains to the design of heaven and earth/underworld, corresponding pairs of (things in) heaven and earth/underworld, (and) things of the *Abzu*."[127] In this and related commentary texts, explanation and exposition of texts is raised to a high learned philological exercise.[128]

Explanation was carried out in a similar way with respect to the intuitive framework of divinatory forecasting as well as in response to the model-based prediction of astronomical texts. Despite the difference in modes of prediction, explanation in one context focused on words found in texts, in the other on the calculation rules for generating (table-)texts. In lieu of ontological-causal or logical-formal explanation, which would not have answered the needs of cuneiform scribal scholarship, a hermeneutic orientation to explanation developed to elucidate and explicate the meanings of texts. For both intuitive and model-based prediction, therefore, explanation was directed toward words, referring to words in the omen series, and procedures, referring to the algorithms of ephemerides, respectively. An intertextual dimension was thus at the core of explanatory method in the context of prediction.

The most explicit form for explanation in the diviner scholar's practice was to designate an interpretation of a sign as "its interpretation" (*pišeršu*), as discussed in chapter 6, section "Determinism and the Gods" above. Frahm regarded omen divination as "an exegetical act."[129] In much the same way as Pingree saw the nature of omen apodoses, Frahm too said,

> It is worthwhile to note in this context that a key term from the Babylonian "theory" of divination resurfaced in the commentary tradition of later civilizations. The term in question, *pišru*, seems to designate omen apodoses, which were interpretations of sorts of some natural phenomenon. *Pišru* is derived from *pašāru*, which in Mesopotamian dream interpretation refers to the reporting of one's dream to another person, the interpreting of an enigmatic dream by that person, or the dispelling or removing of the evil consequences of such a dream by magical means. *Pišru* is never used to label Mesopotamian text commentaries, but the earliest exegetical treatise available to us from Jewish tradition, the commentaries from Qumran, are called peser, and the Arabic word tafsir, which designates commentaries on the Qur'an, is another cognate of the same root.[130]

Numerous reports from the Neo-Assyrian scholars refer to the explanation of omens with the term *pišru*. The following illustrate:

> Tonight Saturn approached the moon. Saturn is the star of the sun, (and) the relevant interpretation (*pi-še-er-šú*) is as follows: it is good for the king. The sun is the star of the king.[131]

> If the earth quakes in Nisan: His land will [defect] from the ruler . . . now [its] interpretation (BÚR-[*šú*]) is this: because it quaked in Adar and Nisan after [another], therefore (the omens) 'it kept quaking, att[ack of the enemy]' applies.[132]

> Mars reac[hed] Cancer and entered it. I kept wat[ch]; it did not become stationary, it did not stop; it tou[ched] the lower part (of Cancer) and goes on. (Its) going out (of Cancer) remains to be s[een]. When it will have gone out I shall [send] its interpretation (*pi-šìr-šú*) to the king my lord.[133]

These passages make clear that in the context of divinatory reporting, explanation meant interpretation. The sense of interpretation in the first two of the three reports just quoted (the third is ambiguous) represents one order removed from Pingree's and Frahm's statements that the omens themselves produce "explanations" in the form of the apodoses. The third example could well mean that the scholar would tell the king what the series indicated was the outcome of Mars' "going out (of Cancer)." But each reflects the central concern of divination to interpret signs, either directly in the form of the antecedent-consequent relation, or on a metalevel where the diviner provided a further analysis or clarification of the meaning of an ominous sign.

The metalevel explanations in divinatory reports take the form of philological commentary on the words used in the omen text, as in the following, which first quotes an omen and then explains the verb (*ṭerû* "pierce"[134]) of the protasis: "If the moon's right horn at its appearance pierces (*ṭirāt*) the sky: there will be stable prices in the land; a revolt will be staged in the Westland. 'Its right horn pierces the sky,' as it says, means it will slip into the sky and will not be seen; DIRI—pronounced dir—is 'to slip,' said of a horn."[135] The author, Nabû-ahhē-erība, has peppered his report with glosses, another explanatory technique, whereby the logographic writings of words are syllabically written below the line in a smaller hand to distinguish glosses from the main text. In the case of this particular report, the logogram AGA for "crown" (*agû*) is glossed *a-gu-u* (line 1) and the "right horn" which "pierces the sky" is a phrase for which the words written with logograms, "its right horn" (SI ZAG-*šú*) are glossed syllabically *qar-nu i-mit-ti-šú*, and "sky," written AN-*ú* is glossed *šá-mu-u* (line 5). The sign DIRI is also explained by a gloss to its pronunciation as "dir" (rev. line 5). The principal explanation in the passage, however, is the elucidation of the verb *ṭerû* on analogical-phonological grounds with the verb *ḫalāpu* "to slip in or through." The association to *ḫalāpu* is based, not on the Akkadian pronunciation of *ḫalāpu*, but on the sound of its logographic spelling, whereby /ṭer/ can be associated with /dir/. A late medical commentary by Enlil-kāṣir, son of a *kalû*, concerning incantations for difficulty in childbirth, addresses itself to this same word *ṭerû* differently, saying, "with her horn she roots up the ground (*ina qarnišu qaqqar ṭerāt*), *dārû* means 'permanent,' also *ṭerāt* means 'she shelters,' because *dārû* means *ḫaṣānu*."[136] This interpretation rests on the phonic similarity between /ṭer/ and /dar/. This commentary to an incantation for difficulty in childbirth titled "The Big Cow of Sin," is related to and follows the well-known composition "Cow of Sin" (*littu ša Sin*) in one of its manuscripts.[137] The related incantation contains the line "(I am pregnant) . . . with my horns I root up the soil."[138] The horns of the moon in the omen, and of Geme-Sin the cow in the incantation, have a further internal relation as the moon-god was frequently personified as a bull, his horns, that is, his crescent shape, becoming his principal symbol. The association between the omen for the moon's horn in the omen report and that of the incantation commented on in the medical commentary text hangs on the use of the verb *ṭerû,* which required some thought on the part of the scholars as to what its meaning or meanings could be in context. What is noteworthy is that each explanation, in the report and in the commentary, did so on the basis of phonological similarity. The phonological resemblances /ṭer/, /dar/, and /dir/ gave rise to Akkadian synonyms that were worked in to the analysis of the meaning of "horns" (of the moon, or of Geme-Sin the cow) constructed with *ṭerû* "to pierce."

The explanation of words in this fashion, by phonological analogy and etymography, is quintessential cuneiform scribal philology, which stemmed ultimately from the translation methods of the compilers of lexical lists.[139] A particularly relevant word list for the present context is that designated as *ṣâtu,* a term that Frahm noted was "one of the most complex and difficult terms used in Mesopotamian scholarship."[140] One of its usages denoted a particular kind of lexical list (or perhaps a specific list), which *The Assyrian Dictionary* translated as "explanatory word list (commentary based on traditional interpretations, lit. excerpted words)." Sometimes it clearly referred to different categories of lists, but in later first-millennium contexts, such as that which concerns divinatory texts, it referred to commentaries.

The understanding of the term *ṣâtu* as a commentary is of importance in the present context because the nature of the *ṣâtu* commentary, derived from its root verb (*w*)*aṣû* "to go out," was to excerpt, or take out from a main text, or, alternatively, to reveal or explain.[141] The above discussed examples illustrate this form of explanation. The explanatory function of the *ṣâtu* is noted in the colophon of a number of exemplars of the medical plant list *Uruanna*: "First section (of the list) Uruanna = *maštakal*. Plants which are 'explained' (BÚR = *pašrū*) in *ṣâtu-* and synonym (*lišānu*)-lists."[142] Indeed, *ṣâtu*s are known for the omen series *Enūma Anu Enlil, Šumma ālu, Šumma izbu,* and SA.GIG "symptoms," clearly a kind of commented text with special emphasis on the intuitive predictive texts that constituted divinatory knowledge.[143]

Turning to explanation in model-based prediction, astronomical texts have different predictive and explanatory properties compared to divinatory material. Given that they are both parts of the intellectual output of scribes, however, it is perhaps not surprising that they share the element of their common intertextual nature. Exemplary of explanatory texts on the astronomical side are the collections of procedures that explain rules for computing table texts, thus establishing a kind of intertextual reference to the tables themselves. Other astronomical procedure texts that do not relate directly to ephemerides, such as the Seleucid Uruk TU 11, or the instructions for constructing a sundial, do not display the same interdependence with other texts, at least that we know of.

Also common to most procedural texts, as Ossendrijver pointed out, is the second-person address, as though in or from a dialogue, perhaps the vestige of an earlier rhetorical form.[144] The second-person rhetorical form has already been noted in the context of omens (discussed in chapter 7), such as: "Observe his last visibility [on the 28th, variant: 29th of *Kislīmu*, and you will predict an eclipse. The day of last visibility will show you the eclipse."[145] Using the same device, the following procedure for a 4-zone Jupiter System A′ model gives instructions for how to calculate longitudes with a step function, using the characteristic zone (or arc subdivision) boundaries for the planet, designated by zodiacal degrees:

> From 9 Cancer until 9 Scorpius you add 30. (The amount) by which it exceeds 9 Scorpius you [multiply] by 1;7, [30].
>
> From 9 Scorpius until 2 Capricorn you add 33;⸢45⸣. (The amount) by which it exceeds 2 Capricorn you multiply by 1;⸢4⸣.
>
> From 2 Capricorn until 17 Taurus you add 36. (The amount) «by which» it exceeds 17 Taurus you multiply by 0;56,15.
>
> From 17 Taurus until 9 ⸢Cancer⸣ [you add] 33;⸢45⸣. (The amount) ⸢by which⸣ it exceeds 9 Cancer you multiply by 0;53,20.[146]

The instructions to multiply a coefficient at the zone crossings is to account for the change in rate of progress of the synodic event in the new zone. Instead of immediately shifting to the new rate of progress, a modification of the previous rate must be taken into account at the border of two zones. This calculation procedure is carried out in the mathematical generation of values in the tables, but the procedures provide a verbal description of how this is accomplished.

In addition to the second-person address, procedures referring to the subdivision of the synodic arc and describing the interval, or distance (ZI = *nisḫu*) a planet goes from one synodic phenomenon to the next within the total synodic cycle, called a "push," expressed in distance (*birītu*) or days (*ūmū*), simply adopt a third-person descriptive style. Thus from the same text (No. 32) the four zones of the model are clarified (note that half brackets indicating partially preserved signs have been omitted here): "From 9 Cancer until 9 Scorpius the small one. From 9 Scorpius until 2 Capricorn the middle one. From 2 Capricorn until 17 Taurus the large one. From 17 Taurus until 9 Cancer the middle one."[147]

An early Goal-Year procedure text (BM 45728), perhaps the earliest such explanatory astronomical text according to Britton, who dates it to the second half of the seventh century, gives sidereal periods for planetary synodic phenomena and corrections for the dates.[148] The tablet, written by one Labaši,[149] enumerates the periods for the appearances (IGI.DU$_8$.A = *tāmartu*) of the moon, Venus, Mercury, Mars, Saturn, and Sirius (why not Jupiter?) in a procedural manner, although without the intertextual dimension of later procedures. A selected passage reads:

> [Appearance of] Venus. 8 years [you go back] behind you . . . 4 days you subtract. You observe (it).
>
> [Appearance of] Mercury. Your 6 years you go back behind you . . . to it you add . . . 10 you add to the (date of) appearance. You observe (it).
>
> [Appearance] of Mars 47 years you go back [behind you]; 12 days in addition [. . .] 10 you add to the (date of) appearance and you observe (it).
>
> [App]earance of Saturn, 59 years you go back [behi]nd you. To the day (lit. "day by day"), (it = Saturn) appears (again).[150]

In the segment of the text quoted here, the period for observing Venus is given as 8 years − 4 days, Mercury 6 years + 10 days, Mars 47 years + 12 days, and Saturn 59 years "to the day," where the addition or subtraction of days is a feature of similar texts that correct for dates of observations, and the periods for Venus, Mercury, and Mars are those known in Goal-Year texts.[151]

Later astronomical procedure texts explain the computational methods of the ephemeris tables directly, although Ossendrijver pointed out that some are not complete as to their coverage of all the steps required for generating a table.[152] The verbal idiom of the late procedures is partly consistent with that of the divinatory tradition. For example, the phrase used to tell someone to take something into consideration is, literally "hold x in your hand" (*ina qātika tukâl*), or perhaps "bear in mind," already seen in *Enūma Anu Enlil*. The various contexts in which the phrase is employed can refer to a wind (or direction), times, positions, or a Goal-Year. The following shows its usage in *Enūma Anu Enlil* Tablet 20: "You observe his (the god's) eclipse and bear in mind the north (IM.SI.SÁ *ina* ŠU-*ka tu-kal*)."[153] In the late astronomical procedures, times and positions are to be "held in one's hand (or hands)," as in Procedure Text 46 (= ACT 812) rev. ii 1–2: "The day when Venus appears in the west (EF) or sets in the east (ML): you hold the times and positions for the *igigubbû*-coefficients [. . .] in your hands."[154] In other words, the procedure explains how the synodic arc is to be subdivided, and "you" bear these things in mind.

Another feature shared with the divinatory corpus was the collection of explanatory texts into series. Some explanatory texts (*mukallimtu*s or *ṣâtu*s) constituted multitablet series of their own, and a few astronomical procedure texts appear similarly to belong to series.[155] A set of procedures referring to Saturn, giving a number of procedures explaining Systems A, B, and B″ for Saturn, has a colophon with the catchline for the next tablet (concerning Mars) in the series, with an incipit "the displacement (or progress in longitude) of Mars."[156] Similar to the interpretive material prepared and collected for intuitive predictive texts (omens), the mathematical tables too had an interpretive corpus prepared and collected as procedures. The following are the beginning lines of the Saturn procedure (following the invocation to Anu and Antu that "it go well," and again, half brackets have been omitted):

> [For Saturn. From 10 Leo] until 30 Aquarius the small one; [from 30 Aquarius] until 10 Leo the large one.
>
> [In (the region of) the small one: with the Sun] its displacement is 0;5 per "day."
>
> [After the appearance (First Appearance)] for 30 "days" its displacement is 0;5 per "day."
>
> [For 3 month] it moves 0;3,20 per "day," then it is stationary (First Station).
>
> [For 52;30 "days"] it turns back 0;4,13,40 per "day," and (then) it rises to daylight
>
> (Acronychal Rising).
>
> [For 60 "days"] it turns back 0;3,20 per "day," then the second station (Second Station).
>
> [For 3 months] it moves 0;3,35,30 per "day."
>
> [For 30 "days" bef]ore its setting (Last Appearance) it moves 0;5 per "days" then it sets.
>
> [For] 1° it turns back 7;33,7,30.[157]

These lines have a direct reference to columns in an ephemeris and how they were structured in "zones" as well as providing a verbal description for subdividing the synodic arc within the "small" zone.

Explanation for model-based predictive texts, therefore, involved on one hand the specific exposition of the parameters, coefficients, and computational steps, the algorithms for generating astronomical tables, and on the other the production and preservation of such explanatory texts as part of a written repertoire. The standards for what qualifies science as explanatory are local and historical. We can no more apply to cuneiform texts the standards of Aristotelian explanation than we can of Hempel and Oppenheim. Explanation, just as prediction, had particular contexts and aims in its local cuneiform environment. These contexts and aims changed over the course of the millennia in the ancient Near East, but remained essentially a scribal concern.

The large body of predictive knowledge, the intuitive as well as the model-based, necessitated explanation and was treated systematically for explanation, which meant explicating words, interpreting signs, and describing verbally the mathematical functions, algorithms, and parameters required for producing tables of astronomical phenomena. For the cuneiform scribes, explanation of predictive texts was not judged in terms of demonstrative criteria, nor did it concerned itself with causes or of the way predictive models might describe nature's physical structures or laws. The goals of explanation as such, as attested in commentaries, omen reports, scholars' correspondence, and astronomical procedures, were not only about phenomena but also about texts, with a hermeneutic objective.

An exploration of the context of explanation must recognize the importance of schemes (both divinatory and astronomical), models, analogies (both divinatory and astronomical), and analogues. As divination and astronomy both had predictive aims, explanation was embedded in the entire predictive undertaking. Each branch of cuneiform predictive knowledge was tied to programs of observation and a tradition for recording and dating those observations. From our point of view, observation, prediction, and explanation belong within the purview of science. Although the methods of investigation into the world of perception and experience, and ideas of what was usual, unusual, regular, irregular, normative, and anomalous were determined by the particular phenomena—many but not all ominous phenomena—that interested the Assyrian and Babylonian scholars over time, in terms of the observational, predictive, and explanatory dimensions of cuneiform knowledge, particularly in its persistent attempts to grasp an order of things and to resolve what is anomalous into a system, it is hardly possible not to see the features of its kin in the later history of science.

Conclusion

Unifying the great periods in the history of science is the idea that each period is marked by a special relationship between science as inquiry and nature as object of inquiry. The sociologist George Homans put science and nature into a metaphorical dialogue: "What makes a science are its aims, not its results. If it aims at establishing more or less general relationships between properties of nature, when the test of the truth of a relationship lies finally in the data themselves, and the data are not wholly manufactured—when nature, however stretched out on the rack, still has a chance to say 'No!'—then the subject is a science."[1] The allusion here is to Francis Bacon, putting nature to the question, a euphemism for being tortured on the rack.[2] Asking questions of nature ("putting nature to the question") or reading the Book of Nature are old metaphors for science that are still readily understood. "Man and nature" are at the center of the story of science, and the connection between science and nature is taken for granted. From the perspective of an Assyriologist trying to come to terms with the meaning of cuneiform knowledge in relation to what we call science, science and nature are less a foundation for historical analysis and more an assumption needing qualification for cuneiform texts.

In the following formulation of the essence of science in history, Alistair Crombie appealed to a different metaphor, but still found the conception of nature to be the very backbone and foundation, the sine qua non of science:

> The study of its history shows that science in each period is characterized by a definite and particular concept of nature, which is intimately connected with

> the method and purpose of asking and answering questions about natural phenomena. The concept of nature affects the concept of man and thus that of human good and knowledge. The dominant concept of nature and its associated method and conception of the purpose of science, though seldom without their critics, characterize the main movements. These movements form the backbone and ribs of the history of science. . . . In the history of science in Europe three main concepts of nature may be distinguished: nature as consisting of things whose essence is an active principle of operation (substantial form, secondary causes), nature as mechanical, where only proximate causes are the object of science, and nature as a process, which once more raises the question of final causes.[3] The first of these ruled men's minds from the beginnings of Greek science until the seventeenth century, the second from Galileo and Newton to the end of the nineteenth century, since when it has shared its place with the third. During the first period the primary purpose of science was understanding and utility was regarded as subordinate. Since the seventeenth century the primary purpose of science has been understanding such as will give power over nature and improvement of the practical arts. The concept of nature does not determine the concept of man but may profoundly affect it. During the first period man was regarded as part of nature until Christianity, sharply distinguishing spirit and matter, separated him from it. The mechanical concept of nature emphasized the dualism between mind and matter and in the hands of materialists reduced man to a machine. The evolutionary view once again asserted the unity of nature in an emergent process of which man is a part.[4]

In this summary statement Crombie established science and nature as orthogonal coordinates, within which humankind is to be placed within a science-nature grid. The schematic character of the statement as a whole would not be widely accepted today, but the science-nature grid remains firm. In Crombie's first designated period, in which nature was an active principle, the divine in the form of gods or God was part of nature. The conception of nature as machine, Crombie's second stage in the schema, belongs to a particularly modern sensibility, the transition perhaps epitomized in Schiller's poem "The Gods of Greece," where first nature is imbued with divinity only to end up as "dull" and "godless."[5] The machine then gives way to the idea of process, wherein human beings' most fundamental understanding of their place in the world is conceived in terms of their relation to nature.

Well off the bottom edge of Crombie's timeline of the history of science sits the two-thousand-year-long intellectual culture of the Assyrian and Babylonian scribes of the ancient Near East. A certain sympathy with cuneiform material might be found in Crombie's earliest conception of an animated nature, if not for the fact that any reference to the gods and nature, whether distinct or one and the same, was excluded from cuneiform texts, because neither the discourse of science nor of nature were a part of its idiom. Crombie's metaphor of the backbone and ribs of the history of science was meant to show that without the backbone representing the dominant concept of nature, the ribs would not support the associated methods and conception of the purpose of science. But this is in fact a most imperfect representation of the historical "body" of science.[6] It is an image inadequate to other historical evidence, such as that from the ancient Near East, China, India, and the Islamic world. It is a representation of the history of science as a singular integral body with a singular, universal, transhistorical, unified nature, reconstructed from the Western tradition. Such a depiction has already undergone reappraisal and revision in favor of the local, the historical, and the diverse. But a reappraisal of the relationship of science to nature requires another kind of consideration from the perspective of the ancient Near East.

To measure the intellectual achievement of the cuneiform world solely in terms of what came later runs the risk of diminishing the integrity as well as the complexity of its understanding of things. While much of the material of interest to this study long predates the pre-Socratics and Plato, nearly as much is contemporary with Aristotle, the Stoics Chrysippus and Diogenes, the astronomers Apollonius of Perga, Hypsicles, and Hipparchus. The Babylonian scholars of the period when Greek *physikoi* began to explore matter and motion remained unaffected by the interests of Western philosophy in knowing nature or in using nature heuristically. For two thousand years, Babylonian knowledge of the heavens was not structured by a classification of the moon and the planets as phenomena of nature, nor were their cyclical appearances understood in terms of physical laws. Models of astronomical prediction were neither dependent upon a geometrical geocentric cosmos nor constructed to account for planetary motion as such.

Instead of the motion of heavenly bodies the Babylonians were concerned with periodicity. Theirs was a thoroughly quantitative approach and did not depend upon a physical framework. At the same time, astronomical observation, prediction, and explanation were the tools of scholars who found it relevant to make correspondences through many forms of relation and analogy between the above (heaven, AN = *šamû*) and the below (earth/netherworld, KI = *erṣetu*), not only for divination and astrology but also for medicine. Celestial divination, astrology, and astral medicine were integral to the observational and predictive goals of astronomy as it developed in the traditions of *ṭupšarrūtu*.

It is true that late Babylonian mathematical astronomy provided the foundation for all later forms of Western astronomical science. It is equally the case that the Babylonian idea of the relevance of the stars for human life was the impetus for later astrology, still an important part of celestial knowledge as science even in the early modern world. As the foregoing discussion has emphasized, however, the Assyro-Babylonian astral sciences were but part of a greater totality of knowing the world within the learned traditions of their scholars. As a whole those scholarly disciplines were attuned to the regularity and irregularity of phenomena, to what was normative and what anomalous. The preoccupations of cuneiform scholarship, missing the necessity to organize knowledge within a unique and eternal natural world, was nonetheless heavily invested in observation and prediction, mathematical methodologies, divination from signs, and medical diagnostics.

The interest in understanding and describing perceptual and intellectual phenomena, that is, things seen and things imagined, runs through and across the disciplinary boundaries of the scribal scholarly corpus. An understanding of the totality of *ṭupšarrūtu*, as Crombie suggested for Western science, would be the basis for a full exposition of what might rightfully be called the cuneiform scientific imagination. It has not been the goal here to undertake such a comprehensive account of the totality of the scribal scholarly corpus, and thus the following remarks must be considered partial and colored by the materials that have been in focus—divination, astronomy, and astrology.

It is worth recalling that in an address to the American Academy of Sciences in 1996, Gerald Holton spoke about the "art of the scientific imagination," for him a set of forces at work in the making of science. The forces of the imagination were complementary to other components making science what it is, such as the language of mathematics, the skepticism of public science, the rationality of forming and testing hypotheses. Among the many tools of the scientific imagination, he singled out three: the visual, the thematic, and the analogical. With respect to the cuneiform world I have focused on the analogical in previous chapters, but themata in cuneiform scribal scholarship would also be a fruitful area for further research.

The thematic imagination derives from Holton's finely developed notion of themata as a definable part of the work of science.[7] In Holton's analysis, themata took their place beside eight other intersecting forces and influences on science in the mind and in the society producing science—namely, scientific content ("facts, data, laws, theories, techniques, lore"),[8] historical trajectory, context of discovery, the relation of the public and the private, the personal psychobiography of the scientist, influence of sociological setting, influence from the cultural environment, and the logical analysis of the work itself. Themata were proposed as key to approaching many questions not answerable by other components of analysis, questions to do with choices made about the direction of scientific investigation, the underlying continuity of science despite its obvious historical changes, or why certain models of explanation persist despite evidence and understanding to the contrary.

Holton ferreted out well-defined themata in the history of science from Greek antiquity to modernity, some of which come in the form of antithetical doublets, such as the plenum and the void, evolution and devolution, constancy and change, complexity and simplicity, mathematics versus mechanistic models of explanation, and others. The themata of Western science since the Greeks make sense within a framework of science as the investigation into nature and the attempt to explain the phenomena of nature as parts of an integrated whole, though some seem to have the capacity to apply beyond the borderlines of the conception of nature (constancy and change, complexity and simplicity, for example).

It seems telling that the themes that emerged and persisted in the history of Western science do not resonate in cuneiform texts. Not surprisingly, "the most persuasive characteristic of modern science from its beginnings has been simply the generally accepted thema of . . . the belief that nature is in principle fully knowable. Kepler found support for this belief in equating the mind of God and the mind of man on those subjects which can be understood in the exact sciences. In more recent times, Heisenberg has said: 'Exact science also goes forward in the belief that it will be possible in every new realm of experience to understand nature.'"[9] This, again, points to what must be a fundamental discontinuity in the overarching aims of cuneiform knowledge vis-à-vis Western science, that is, that cuneiform knowledge was not concerned with the investigation and theorization of nature, but established structures and meanings of phenomena in a different way.

Instead of the order of nature, a differently structured system provided the framework for developing the themata of the cuneiform texts dealing with heavenly phenomena, including the observed, conceived, and imagined ominous phenomena, such as risings, settings, colors, brightness, directions and winds, binary or otherwise schematic structures of interpretation of their meaning, the predictability of the visibility of certain signs (especially eclipses), schematization, and models for prediction in divination/astrology and in astronomical computation. All of these themata were an active part in the imaginative structures of cuneiform traditions of knowledge of the heavens. Research in other areas of the repertoire of *ṭupšarrūtu* will undoubtedly find the themata that characterize the content and goals of those corpora, for example, of divination, medicine, and magic.

What seems particularly useful about Holton's idea of themata transposed to the cuneiform world is that, as he said, they are not to be confused with metaphysics, paradigms, or worldview, although they are nonetheless embedded in paradigms and worldviews. Through periods of "normal science" and revolution alike, antithetical themata (as in the examples cited above) can persist, and can be attached to individuals (in his terms, "scientists," which strictly speaking did not exist in Near Eastern antiquity) more than to social and cultural context. Translating this relationship to the cuneiform evidence must allow for a different identification of the contents of texts with certain scribes. Authorship is impossible to discern from within the many generations composing the community of copyists and "owners" of divinatory and astronomical tablets. As copying and authorship are two different things, the influence of themata on individual scribes is at present difficult to analyze, although progress in the study of the names and affiliations of the late Babylonian scribes might provide the tools to begin analysis of this kind.[10] A consideration of themata might, however, also be useful if it is tied to texts rather than individuals. As a matter of texts, the thematic imagination surely plays its own determinative part in addition to the influences of metaphysics, paradigm, or worldview.

Themata were only one of the regions on Holton's map of the scientific imagination. Another was analogy, which can be readily identified in cuneiform scholarly texts. Determining which phenomena were ominous and developing models for prediction of celestial phenomena, some of which were related to ominous signs, each depended upon the imaginative strategies of analogy, whether to relate phenomena from different domains, as celestial to terrestrial, or to use mathematical analogues for description and prediction of lunar and planetary appearances. Outside of the astronomical/astrological tradition other kinds of scribal knowledge, for example, those of lexical lists, mathematics, or of a variety of technical instructions, are subjects for separate study and provide other contexts in which to consider the exercising of association and analogy within the bounds of *ṭupšarrūtu*.

The scribal imagination shaped cuneiform knowledge on different levels. Most basic was that of the potential of the cuneiform script for hermeneutic expansion by association, analogy, and polyvalence of various kinds. Another level can be identified in the conception of the phenomena as signs from the gods, and still another in the phenomena as objects of rigorous mathematical schematization and the construction of models for the purpose of prediction. Coexisting with models for predicting the phenomena was the persistent idea of the influence of the gods through their cosmic "designs," a metaphor for the establishment of norms and order in the world.

The cuneiform epistemic repertoire changed significantly from its early manifestation in the omens of *Izbu*, *Enūma Anu Enlil*, and the extispicy series, in the cataloging of stars and planets in *MUL.APIN* and the ideal scheme of heliacal risings in the Astrolabes, the magical and medical corpora, to the observational and predictive astronomy of the period post-600 BCE. Major changes emerged in a number of pivotal periods—for the first millennium we immediately think of the seventh and the third centuries, that is, the periods of the Neo-Assyrian and the Seleucid Empires respectively, when massive political and social change no doubt had an impact on scholars' lives and livelihoods. In addition to developments in mathematical astronomy, the later period introduced birth astrology and astral medicine, which would be influential to other cultures well into the premodern era.

Despite development from older traditions, none of the new developments in intellectual interests entirely supplanted older tradition, suggesting that progress was not an epistemic value for the scribes. The positivistic story of scientific progress is not bolstered by the cuneiform evidence, as each new development coexisted with older material. This is seen in the perpetuation of omens and in the kinds of schematic associations known from omen divination into birth astrology. Scribal attitudes about scholarly knowledge as "secret knowledge of the great gods" (*pirišti ilāni rabûti*) applied to astronomy and to expository texts concerning rituals and theological subjects alike.

The relationship of the gods to the world, and the world as suffused with the divine, might be as appropriate a unifying feature for Assyro-Babylonian *ṭupšarrūtu* as understanding God's creation through nature was a unifying feature for medieval and Renaissance Christian and Islamic sciences. Indeed, the importance of the divine in the activities associated with knowledge by the *ummânū* seems inescapable, just as the insistence upon God's role and relationship to the physical world was integral to medieval and early modern science. However, despite the presence of the gods in many of the texts produced by the *ummânū*, cuneiform knowledge was an embodiment of ideas not adequately analyzable from the perspective of the philosophy of "religion and science." In particular, as Peter Harrison stated, "philosophy of religion deals with the existence of God, exploring such questions as whether the scientific study of nature provides evidence for God's existence and whether scientific investigation relies on implicitly theistic assumptions about the uniformity of nature or the reliability of our cognitions. Also relevant in this context is the issue of how God interacts causally with the world (which takes in ideas about divine action, providence and miracles)."[11] From this perspective cuneiform knowledge lies outside the main concerns of the philosophy of religion and science, because it is situated in history before either nature *or* God became terms of discourse.

In view of the early and the later developments of astronomy, to characterize the main drive for cuneiform knowledge as a desire to know the gods is reductive and incomplete. In the same way that hellenistic Greek astronomy coexisted with ideas about the intelligence and souls of the stars, ideas about the gods and the heavenly bodies remained a viable conception in the cuneiform world. The sources considered in this study might be seen as unified by a certain axis of knowledge, not between human and god, but between the human knower and the ordered phenomenal world. Of course this relationship is arguably the defining feature of what we consider scientific as opposed to other kinds of knowledge. Around this axis many kinds of knowledge in cuneiform scholarship—divination, magic, and predictive astronomy alike—were structured, and the investigation of regularity and irregularity, norms and anomalies was organized within the particular epistemic, ontological, and metaphysical bounds of that axis.

In the introduction to this book, the question was raised: What kind of science is it that does not have nature as its conscious object of inquiry? From many perspectives established in the philosophy of science the answer would be "no kind." The problem comes from the naturalism not only of today's sciences, but also many of today's histories of science. However, kinship between the cuneiform world and the later West concerning inquiry about the phenomenal world is recognizable on several fundamental levels. First, despite the different framework, or frameworks, within which regularities, anomalies, uniformities, and nonuniformities were discernible and conceivable in the cuneiform world, a sense of an order of things is evident. Of course "order" is always in relation to a set of things, or the presumed set of "all things." The cuneiform scribes in comparison, say, to Western natural philosophers from any period, ordered the world in a different way and ordered phenomena within that world to suit their purposes, whether it was for divination or predictive astronomy. Second, in a much related way to later science's goal to make the order of nature intelligible, the Assyrian and Babylonian scribes viewed their world as inherently intelligible, readable, and interpretable. This view was subject to changes in detail or emphasis over the course of its two-millennia-long history, but the persistence of the importance of reading signs, and the development of methods for predicting phenomena with ominous significance—and, in the course of which, calculating phenomena that were not part of the divinatory enterprise—strongly suggests this was a consistent component of *ṭupšarrūtu*. The devotion of much ancient scholarly effort to various kinds of divination through omens is explicit demonstration of the intellectual commitment to an intelligible world. The world may not have been defined as, or identified with, physical nature, but it was ordered, regular, and intelligible. And third, efforts to predict phenomena known to be periodic resulted in the development of quantitative methods and models to achieve predictive goals. In a similar way to later science, as represented in the Western tradition, cuneiform knowledge systems satisfied the desire to understand the phenomena of subjective experience, perception, and imagination, and they did so by means of schematic models, both qualitative (divinatory interpretive schemes) and quantitative (mathematical descriptions of predictable astronomical phenomena).

None of the categories defined in relation to nature, laid out by Lorraine Daston as the supernatural, preternatural, artificial, and unnatural, serve in any but an anachronistic way to describe what the Assyrian and Babylonian scribes observed and predicted in their conceptual and experiential world. Daston has defined the changes in conceptions of all of these categories between the early modern and the modern period, noting that "the contrast between early modern and modern versions of the non-natural is striking. In the current metaphysical vernacular, the artificial has been swallowed up by the natural, the supernatural has shrunk to a philosophical possibility, the unnatural rings archaic, and the preternatural no longer exists at all."[12] That these categories can come into being and pass away should underscore the fact of the prehistory of them all.

As the foregoing chapters have shown, lines of continuity as well as discontinuity from cuneiform texts to later traditions of astrology, astral medicine, natural philosophy, and the exact sciences force us to consider where to position the learned cuneiform world in the history of science. These continuities are balanced by discontinuities in the epistemic values as well as the practical goals of the cuneiform scribes, each of which reflect attitudes about knowledge within a conceptual framework that separates the cuneiform tradition from the Hellenistic Greek or any other tradition of science. The basic discontinuity, I would argue, is in the fact that no equivalent of the idea of nature as the essential makeup of things, or as a universal material realm of being operating in accordance with its own eternal laws, or of everything that does not belong to the sphere of human culture, can be offered by way of a translation in cuneiform sources. The "designs" of the gods do not offer an epistemological or an ontological equivalent.

Therefore, in claiming that there was no framework of nature to structure the investigation, description, or prediction of phenomena in cuneiform texts, the seemingly reasonable question of what other framework may have stood in for nature proves unreasonable. There is no equivalence and, as has been proposed here, the texts that investigate, describe, and predict phenomena do so in a way that does not require what we would call a unified physical framework or the idea of an essential makeup of things. The universe was considered in its parts, various "heavens," earth, underworld, and the subterranean *Abzu*, as domains of gods or of places for visible phenomena. Phenomena were considered in terms of what was regular and irregular, and periods for the celestial phenomena were constructed around a sense of the ideal as well as of in consideration of observational experience.

Despite this substantial conceptual discontinuity, the argument for kinship between cuneiform knowledge and science stands in terms of its engagement with phenomena by means of observational, predictive, and rational methods. Because cuneiform sources do not engage explicitly with the nature of their investigations into phenomena nor of the structure or style of their own knowledge, for the historian "science" in the cuneiform world is very much "an object *made* at least as much found,"[13] if not more made than found. Even if we do not assign the anachronistic word *science* to the bodies of knowledge and intellectual practices of the scribes, they themselves had words—*ṭupšarrūtu* "scribal scholarship" and *nēmequ* "wisdom"—that served to encompass those textual corpora and scholarly practices devoted to knowing the world of their experience. Coming to terms with what might legitimately be called the scientific imagination of the cuneiform world is dependent upon coming to terms with the full range and depth of *ṭupšarrūtu* and to what end it was directed within its particular local spaces.

Our explorations and reconstructions thus to some extent find, but more so, *make*, a history of science in the cuneiform world. What we find, as well as what we make, of these findings is certainly enough, in my view, to establish an intellectual kinship between cuneiform knowledge and its related practices on one hand and what we call science on the other. The point is that a history of science inclusive of cuneiform texts is not dependent on whether the ancient *literati* of the Near East thought about nature, or whether they asked questions of nature in an effort to produce knowledge. That they did not express such thoughts or questions, and yet developed the models (schematic, quantitative, predictive), reasoning styles (empirical, deductive, analogical), and methods (quantitative, analytic, hermeneutic) by means of which to find structures of order in their experiential and conceptual world, as well as to define norms and anomalies, regularities and irregularities worthy of investigation, seems to me to argue for the necessity of granting a place for cuneiform knowledge in our history of science. In doing so we will allow that our history of science can and should be inclusive of yet more variations on the scientific imagination.

Acknowledgments

I am happy to express my thanks to the Institute for the Study of the Ancient World of New York University for the Visiting Research Scholarship in 2013–14 that afforded me the luxury of time for the research and writing of this book. In addition, it is a pleasure to thank Brown University's Department of Egyptology and Assyriology, as well as the Joukowsky Institute for Archaeology, for the invitation to participate in the Babylon at Brown series in the spring of 2014, where I had the opportunity to field its basic premise and profit greatly from discussion with faculty and graduate students.

I owe a particular debt of gratitude to Professor Carolyn Merchant of the University of California, Berkeley, whose enthusiastic support of the project was crucial to its development. Our collaborative research seminar titled "Nature/No Nature: Rethinking the Past, Present, and Future of Nature in the Contemporary Humanities" at the Townsend Center for the Humanities at the University of California, Berkeley, in the spring of 2012, proved enormously beneficial to the conceptualization of this project. I thank the Townsend Center for its support of our seminar and the participants in the seminar for their inspiring contributions.

It is a pleasure to acknowledge the invaluable comments, corrections, and suggestions made by my colleagues who took the time to read and discuss portions of the manuscript. Special thanks are due Alan C. Bowen, Geoffrey Lloyd, Mark Geller, Dorian Greenbaum, Nils Heeßel, Cale Johnson, Clemency Montelle, Eleanor Robson, Aaron Tugendhaft, Niek Veldhuis, Beate-Pongratz-Leisten, Rita Watson, John Steele, Noel Swerdlow, Eduardo Escobar, and Jay Crisostomo. I apologize for all the errors of omission and commission that remain. The comments and suggestions of the two anonymous readers for the University of Chicago Press were also enormously constructive, and I thank them too for their generous and insightful reviews of the manuscript.

Portions of several chapters have drawn on a number of previous publications or articles still in press. Chapter 2 is a revised and expanded version of "A Critique of the Cognitive-Historical Thesis of *The Intellectual Adventure of Ancient Man*," in *The Adventure of the Human Intellect: Self, Society, the Divine in Ancient World Cultures*, ed. Kurt A. Raaflaub (Malden, MA, and Oxford: Wiley, 2016), 16–28; chapter 3 contains material from "Categories, Kinds, and Determinatives" in *Die Sprache des Bewusstseins und das Bewusstsein von Sprache im Alten Orient*, ed. J. Cale Johnson (Berlin: Berliner Beiträge zum Vorderer Orient, in press); chapter 5 is a revised and expanded version of "The Babylonians and the Rational: Contexts of Rational Reasoning in Cuneiform Scribal Scholarship," in *In the Wake of the Compendia: How Technical Handbooks and Encyclopedia Reshape Early Mesopotamian Empiricism*, ed. J. Cale Johnson, Science, Technology and Medicine in Ancient Cultures (Berlin: De Gruyter, 2015), 209–46; chapter 6 contains a revised, reworked, and expanded version of "Divine Causality and Cuneiform Divination," in *A Common Cultural Heritage: Studies in Mesopotamia and the Biblical World in Honor of Barry L. Eichler*, ed. G. Frame, Erle Leichty, Jeffery Tigay, and Steve Tinney (Bethesda, MD: CDL Press, 2011), 189–203, also included in my *In the Path of the Moon: Babylonian Celestial Divination and Its Legacy, Studies in Ancient Magic and Divination* (Leiden: Brill, 2010), 411–24. Chapter 6 also contains revised portions of my article "Where Were the Laws of Nature before There Was Nature?" in *Laws of Heaven—Laws of Nature: The Legal Interpretation of Cosmic Phenomena in the Ancient World*," ed. Konrad Schmid and Christoph Uehlinger, Orbis Biblicus et Orientalis 276 (Fribourg: Academic Press, and Göttingen: Vandenhoeck and Ruprecht, 2016), 21–39. Chapter 7 reworks some material from "Observing and Describing the World through Divination and Astronomy," in *The Oxford Handbook of Cuneiform Culture*, ed. Karen Radner and Eleanor Robson (Oxford: Oxford University Press, 2011), 618–36.

Thanks are also due to Grace Helu-Lara for help in preparing the manuscript. And, for his work on images for the cover art and map, and for enjoyable discussion of the relevance of the philosophy of science to cuneiform texts, I thank Eduardo Escobar. Finally, I am very happy to express my gratitude to editor Karen Merikangas Darling of the University of Chicago Press for her willingness to consider the manuscript long before it took shape and for her steady support.

Abbreviations

Abbreviations used for cuneiform text citations and for Assyriological bibliography follow the abbreviations list of *The Assyrian Dictionary*, vol. 20, *U* and *W* (Chicago: Oriental Institute, 2010), vii–xxix, with the single following exception:

ADRT: *Astronomical Diaries and Related Texts from Babylonia* (A. J. Sachs and Hermann Hunger, *Astronomical Diaries and Related Texts from Babylon*, vol. 1, *Diaries from 652 BC to 262 BC*, Vienna: Österreichische Akademie der Wissenschaften, 1988; A. J. Sachs and Hermann Hunger, *Astronomical Diaries and Related Texts from Babylon*, vol. 2, *Diaries from 261 BC to 165 BC*, Vienna: Österreichische Akademie der Wissenschaften,1989; A. J. Sachs and Hermann Hunger, *Astronomical Diaries and Related Texts from Babylon*, vol. 3, *Diaries from 164 BC to 61 BC*, Vienna: Österreichische Akademie der Wissenschaften, 1996; Hermann Hunger and A. J. Sachs, *Astronomical Diaries and Related Texts from Babylonia*, vol. 5, *Lunar and Planetary Texts*, Vienna: Österreichischen Akademie der Wissenschaften, 2001; Hermann Hunger, *Astronomical Diaries and Related Texts from Babylon*, vol. 6, *Goal Year Texts*, Vienna: Österreichische Akademie der Wissenschaften, 2006).

Note on dates: dates are given as BCE and CE except in the context of texts containing astronomical observations (chapter 7) where the convention of the negative year numbers—for example, -679 (= 680 BCE)—is employed.

Notes

INTRODUCTION

1. Lloyd, *Methods and Problems*, 418.
2. Grant, *A History of Natural Philosophy*, 1.
3. Sahlins, preface to *What Kinship Is*, ix.
4. Glacken, preface to *Traces*, xiv.
5. Ibid., and for de Buffon, see Buffon, *Oeuvres choisies*, 48.
6. Collingwood, *Idea of Nature*, 1.
7. Williams, *Key Words*, 87.
8. Ibid., 219.
9. Ibid., 224.
10. Galton coined the phrase in *English Men of Science*, 12. As quoted by Ian Hacking in *London Review of Books*, 17, Galton said, "Nature is all that a man brings with himself into the world; nurture is every influence that affects him after his birth. The distinction is clear: the one produces the infant such as it actually is, including its latent faculties of growth and mind: the other affords the environment amid which the growth takes place, by which natural tendencies may be strengthened or thwarted or wholly new ones implanted."
11. Turner, "Crisis of Late Structuralism," 14.
12. Lloyd, *Cognitive Variations*, 131–32.
13. Sahlins, preface to *What Kinship Is*, ix.
14. Rochberg, *In the Path of the Moon*, 1–18.
15. Livesey and Rouse, "Nimrod the Astronomer," 203–66.
16. Evans and Berggren, *Geminos's Introduction*, 228–30.
17. Veldhuis, *History of the Cuneiform Lexical Tradition*. See also Van de Mieroop, *Philosophy before the Greeks*.

CHAPTER ONE

1. Collingwood, *Idea of Nature*, 175.
2. Descola, *Beyond Nature and Culture*, 63.
3. Landsberger, Kilmer, and Gordon, *Fauna of Ancient Mesopotamia*.
4. Veldhuis, *Religion, Literature, and Scholarship*, 82, my emphasis.

5. Berggren, "Mathematics and Religion in Ancient Greece and Medieval Islam," 16–18.

6. Merlan, "Old Academy," 33. See also Broadie, *Nature and Divinity in Plato's Timaeus.*

7. De Zorzi, *La Serie Teratomantica Šumma Izbu.*

8. Aaboe et al., *Saros Cycle Dates and Related Babylonian Astronomical Texts.*

9. For discussion of miracles in the biblical text, see Grant, *Miracle and Natural Law in Graeco-Roman and Early Christian Thought*, 153–81.

10. Foster, *Before the Muses*, 457.

11. Ibid., 483.

12. Daston and Park, *Wonders and the Order of Nature*, 13.

13. Ibid., 14.

14. *ina libbi immeri tašaṭṭar šīri* BMS 6:110.

15. *ṭuppi ilāni* JCS 21 [1967] 132: 8 and 14, cf. *ṭuppi ša ilī* YOS 11, 23:16; see Starr, *Rituals of the Diviner*, 30. See also Lambert, "Qualifications of the Diviners," 147, making reference to line 16 of the Enmeduranki text. In all of these passages, the "tablet of the gods" is clarified in an appositive as the *takaltu(m)*, mostly translated as "liver." Note, however, that the CAD s.v. *takaltu* meaning 1a takes it to mean a pouch or bag (for carrying the stylus?).

16. AfO 17 89: 3, and Rochberg-Halton, *Aspects of Babylonian Celestial Divination*, 270–71.

17. Restoration not secured from parallel texts.

18. Starr, *Queries to the Sun-God*, 84, Text Nr. 74: 1–6.

19. Cicero, *De Div.* I.1.

20. Cicero, *De Div.* I.6.

21. Ibid.

22. Ibid.

23. Grant, *Miracle and Natural Law*, 4, 6–7.

24. Ibid., 8–9.

25. Hacking, "Almost Zero," 29.

26. Williams, *Problems in Materialism and Culture*, 69.

27. Ibid.

28. Ibid.

29. Funkenstein, "Disenchantment of Knowledge," 19.

30. Latour, *We Have Never Been Modern*, 7.

31. To date the most thorough exposition is Descola, *Beyond Nature and Culture.*

32. For the variation in the Greek (and Latin) terms for celestial science, see Bowen, "Three Introductions to Celestial Science in the First Century BC," 299 and note 1.

33. Bottéro, *Mesopotamia: Writing, Reasoning, and the Gods*, 11.

34. Pingree, *From Astral Omens to Astrology.*

35. Oppenheim, *Ancient Mesopotamia*, 1–2, 27, 30.

36. Parpola, introduction to *Letters from Assyrian Scholars*, xviii.

37. Selin, *Encyclopedia of the History of Science, Technology, and Medicine in Non-Western Cultures.*

38. Robson, "Uses of Mathematics in Ancient Iraq (6000–600 BC)," 93–114.

39. Ibid., 95, referring to Cooke, *History of Mathematics.*

40. Michalowski, "Presence at the Creation," 381–96.

41. Discussed in, for example, Reiner, *Your Thwarts in Pieces* or Groneberg: *Syntax, Morphologie und Stil.*

42. Wittgenstein, *On Certainty*, 29e, No. 211: "Now it gives our way of looking at things, and our researches, their form. Perhaps it was once disputed. But perhaps, for unthinkable ages, it has belonged to the *scaffolding* of our thoughts." Emphasis in the original.

43. Chang, *Is Water H_2O?* 253.

44. Hesse, *Revolutions and Reconstructions in the Philosophy of Science*, 29; see also Gregersen and Køppe, "Against Epistemological Relativism," 453.

45. Longino, *Fate of Knowledge*, 37.

46. Ibid., 7.

47. Ibid., 1.

48. Chang, "Putting Science Back in History of Science."

49. Shapere, "External and Internal Factors in the Development of Science," 1–9.

50. Ibid., 3.

51. Ibid.

52. Ibid., 6.

53. Lehoux, *What Did the Romans Know?* 208.

54. Kuukkanen, "I Am Knowledge: Get Me Out of Here!," 590.

55. See Rochberg, "History of Science and Ancient Mesopotamia," 47–53.

56. For a detailed discussion of the major cuneiform library types and collections, both Assyrian and Babylonian, see the pioneering study of Pedersén, *Archives and Libraries in the Ancient Near East, 1500–300 B.C.* and, more recently, Frahm, *Babylonian and Assyrian Text Commentaries*, especially chapter 8.

57. Robson, "Production and Dissemination of Scholarly Knowledge," 557–576; Robson, "Reading the Libraries of Ancient Assyria and Babylonia," 38–56, as well as Robson and Stevens, "Scholarly Tablet Collections in First Millennium Assyria and Babylonia." For the Babylonian tablets at Nineveh, see Fincke, "Babylonian Texts of Nineveh," 111–49.

58. http://www.ane.arch.cam.ac.uk/research/gkab.html.

59. Tinney, "Tablets of Schools and Scholars," 577–96, with bibliography. For the history of lexical texts see Veldhuis, *History of the Cuneiform Lexical Tradition.*

60. There is, *pace* Oshima, *Babylonian Theodicy*, xli, a school exercise tablet with a passage from the lunar eclipse omens of *Enūma Anu Enlil* found in the courtyard of a private house in Neo-Babylonian Uruk. See Mauer, "Ein Schülerexzerpt aus Enūma Anu Enlil," 239–42. The excerpt appears to be of *Enūma Anu Enlil* Tablet 18, the section for *Tašrītu* (not *Šabaṭu*, as in Mauer, 240, translation of line 1). The surviving section of Tablet 18 is not well preserved, so identification with the excerpt is based only on the schema of its omens, i.e., the month and days 14, 15, 16, and 20 (21 is not preserved).

61. Rorty, *Philosophy and the Mirror of Nature*, 70–71.

62. Nagel, "What Is It Like to Be a Bat?" 435–50.

63. Putnam, *Reason, Truth, and History*, 5–6.

64. Zwart, *Understanding Nature: Case Studies in Comparative Epistemology*, 4.

65. Ibid.

66. Ibid., his emphasis.

67. Gellner, *Words and Things*, 22.

68. Pearce, "Materials of Writing and Materiality of Knowledge," 167–79.

69. Borger, "Geheimwissen," 189.

70. Geller and Geus, "Esoteric Knowledge in Antiquity—Some Thoughts," 3.

71. Beaulieu, "New Light on Secret Knowledge in Late Babylonian Culture," 107–8. See the full study in Lenzi, *Secrecy and the Gods.*

72. Frahm, *Babylonian and Assyrian Text Commentaries*, 344, and note 1,639 with further bibliography.

73. Heeßel, "Hermeneutics of Mesopotamian Extispicy," 18 and note 9.

74. Ibid., 18 and note 9, citing VAT 99 34 vi 61–63.

75. Lambert, "Qualifications of Babylonian Diviners," 149 (transliteration) and 152 (translation).

76. Ibid., and 152, lines 22–29.

77. Ibid., 149–50, lines 30–67.

78. TU 44 I 1–6, see Linssen, *Cults of Uruk and Babylon*, 252 (transliteration) and 255 (translation).

79. For the "secrecy labels," see Lenzi, *Secrecy and the Gods*, 170–86. A connection between

kingship and *ummânūtu* is also discussed, in ibid., 185–86, and further details can be found in ibid., chapters 2 and 3. For this latter relationship, the foundational work is Pongratz-Leisten, *Herrschaftswissen in Mesopotamien*.

80. Neugebauer, *Astronomical Cuneiform Texts*, and Ossendrijver, *Babylonian Mathematical Astronomy: Procedure Texts*.

81. Rochberg, "Conceiving the History of Science Forward," 515–31.

CHAPTER TWO

1. Fiske, "From Ritual to the Archaic in Modernism," 189.

2. Müller, *Chips from a German Workshop*, 2: 10–11.

3. Neugebauer, *Astronomical Cuneiform Texts*, followed by his *The Exact Sciences in Antiquity*, and *A History of Ancient Mathematical Astronomy*, in which Babylonian astronomy constituted the entirety of Book II. Neugebauer's works were fundamental to everything that came after, including what is now an extensive bibliography on the mathematical astronomy in the works of Asger Aaboe, Noel M. Swerdlow, John Britton, John Steele, Lis Brack-Bernsen, and Mathieu Ossendrijver.

4. Frankfort et al., *Before Philosophy*, 19.

5. Ibid., 237.

6. Ibid.

7. Ibid.

8. Sahlins, *Culture in Practice*, 564.

9. Ibid., 562.

10. Ibid., 564.

11. Ibid.

12. Frankfort et al., *Before Philosophy*, 237.

13. Oppenheim, "Man and Nature in Mesopotamian Civilization," 634–66.

14. See *The Assyrian Dictionary* (henceforth, in the notes, CAD) s.v. *šiknat napišti* and also sub *napištu* 2b.

15. Bottéro, *Mesopotamia: Writing, Reasoning, and the Gods*, 97.

16. Oppenheim, "Man and Nature," 635.

17. Ibid., with a footnote to his *Ancient Mesopotamia*, 248, which expresses a more functionalist approach to cuneiform lexicography, without the *Ordnungswille* thesis.

18. Selz, *The Empirical Dimension of Ancient Near Eastern Studies / Die empirische Dimension altorientalischer Forschung*.

19. Ibid., 51.

20. Landsberger, "Die Eigenbegrifflichkeit der babylonischen Welt," here quoted from the English translation, Landsberger, *Conceptual Autonomy of the Babylonian World*, 5, italics in the original.

21. Despite the terms' continued use, for the best argument against the notion that emic and etic are the equivalent of insider and outsider, subjective and objective, or actor and observer categories, see Boyarin, "Rethinking Jewish Christianity," 9–10, note 9. First setting out the terms is Harris, "History and Significance of the Emic/Etic Distinction," 329–50.

22. Kelley, *Human Measure*, 2.

23. Ibid.

24. Ibid., 3.

25. Strathern, "No Nature No Culture," 174–222.

26. Ibid., 178, citing Goody, *Domestication of the Savage Mind*, 64.

27. See George, *Babylonian Gilgamesh Epic*, 544–45, Tablet I: 102–3.

28. Snyder, *Practice of the Wild*, 5.

29. Ibid., 9.

30. George, *Babylonian Gilgamesh Epic*, 484–85.

31. Ibid., 485.

32. Stathern, "No Nature No Culture," 176.

33. Kirk, *Nature of Greek Myths*, 286.

34. Liverani, *Myth and Politics in Ancient Near Eastern Historiography*.

35. Quoted from the Penguin paperback edition, Frankfort et al., *Before Philosophy*, 11.

36. Wengrow, "Intellectual Adventure of Henri Frankfort," 608, note 100.

37. Davidson, *African Genius*, 18. I thank Adam David Miller for this reference.

38. Ibid.

39. Zhmud, "On the Concept of 'Mythical Thinking,'" 158.

40. Frankfort et al., *Before Philosophy*, 36.

41. Adams, *Philosophical Roots of Anthropology*, chapter 2, "Progressivism: The Tap Root."

42. Needham, *Belief, Language, and Experience*, 159.

43. Geertz, *Interpretation of Cultures*, 43–44.

44. Shweder, "Anthropology's Romantic Rebellion against the Enlightenment," 28.

45. Ibid.

46. Ibid.

47. Ibid.

48. The law of participation was key to Lévy-Bruhl's theory of the "primitive" mind, wherein the classical law of noncontradiction did not apply. That is, in "primitive" thinking, "*p* is *q*" and "*p* is not *q*" are not necessarily mutually exclusive, and the same phenomenon can then exist at once as belonging to different and contradictory orders of being (human and animal, physical and metaphysical).

49. Lévy-Bruhl, *How Natives Think*, 78, emphasis in the original.

50. Ibid., emphasis in the original.

51. Ibid., 79.

52. Cassirer employed this term in a quote from Prescott, *Poetry and Myth*, 10: "The mythmaker's mind is the prototype; and the mind of the poet . . . is still essentially mythopoeic." See Cassirer, *An Essay on Man*, 75.

53. Although Martin Buber's *I and Thou* was not mentioned in *The Intellectual Adventure*, the use of the terms "It" and "Thou" as a way to describe the relation of the human being to objects in the world outside herself (*Before Philosophy*, 12) was certainly drawn from this work.

54. Cassirer, *Philosophie der symbolischen Formen*, 2, and Lévy-Bruhl, *How Natives Think* (the edition of 1926) are mentioned in the list of suggested readings following the introductory chapter. Buber, *Ich und Du* is not. Cassirer is mentioned (*Before Philosophy*, 30) in the context of the perception of time, personalized as analogous to the biological rhythms of human life, rather than as an abstract natural process, but not in relation to the concept of mythopoeic thought.

55. For the passage from the posthumously published *Notebooks*, see below, note 63. Littleton, in his introduction to the Princeton publication of *How Natives Think* in 1985 said, "part of the problem was Lévy-Bruhl's terminology, which remained for the most part the same as it had been in *How Natives Think*. Expressions such as 'prelogicality' (which he eventually abandoned), 'undeveloped peoples,' 'primitive mentality,' and even 'the natives,' no matter how carefully defined, were bound to infuriate a generation of anthropologists that had struggled to free their discipline from the shackles of ethnocentrism, and who, like Boas, were firmly committed to the proposition that all human beings everywhere are endowed with the same potential for cultural attainment. Indeed, this remains a problem for the modern reader, as the author of this introduction, who is firmly committed to what Geertz (1984) has recently called 'anti anti-relativism,' can readily attest." See Lévy-Bruhl, preface, *How Natives Think*, xx. Also Geertz, "Distinguished Lecturer: Anti Anti-Relativism," 263–78.

56. Lévy-Bruhl, *How Natives Think*, chapter 2, "Law of Participation," also chapters 6, 7, and 8.

57. Cassirer, *An Essay on Man*, 78.

58. Cassirer defined nature here with Kant (*Prolegomena to Every Future Metaphysics*, sec. 14) as "the existence of things as far as it is determined by general laws." See Cassirer, *An Essay on Man*, 76.

59. Ibid.

60. Ibid.

61. Ibid., 77.

62. Ibid., 79–80.

63. The key passage, from Notebook VI, August 29, 1938, is, "If I glance over all I have written on the subject of participation between 1910 and 1938, the development of my ideas seems clear to me. I started by positing a primitive mentality different from ours, if not in its structure at least in its function, and I found myself in difficulties in explaining the relationships with the other mentality, not only among us but also among 'primitive peoples' . . . let us entirely give up explaining participation by something peculiar to the human mind, either constitutional (in its structure or function) or acquired (mental customs). In other words, let us expressly rectify what I believed correct in 1910: there is not a primitive mentality distinguishable from the other by *two* characteristics which are peculiar to it (mystical and pre-logical). There is a mystical mentality which is more marked and more easily observable among 'primitive 'peoples' than in our own societies, but it is present in every human mind." See Lévy-Bruhl, *Les Carnets de Lévy-Bruhl*, published posthumously, 129–32, quoted here from the English, in *Notebooks on Primitive Mentality*, trans. Peter Rivière, 100–101 (emphasis in the original).

64. Buber, *I and Thou*, 68.

65. Frankfort et al., *Before Philosophy*, 13.

66. Ibid., 12.

67. Crawley, *Mystic Rose*.

68. Buber, *I and Thou*, 54–59.

69. Dodds, *Greeks and the Irrational*, 184–85.

70. Frankfort et al., *Before Philosophy*, 35.

71. Ibid., 8.

72. Hume, *An Enquiry concerning Human Understanding*, Section 12, Part 3.

73. Williams, *Problems in Materialism and Culture*, 68.

74. Ibid., 68–69.

75. Ibid., 69–73.

76. Ibid., 73.

77. Ibid., 75.

78. Latour, *We Have Never Been Modern*, 7 ("it is a bit more and a bit less than a culture"), 96 (he defines, in accordance with established anthropological practice, other engagements with nature as representation, whereas otherwise, in science, for example, "Nature, for its part, remains unique, eternal, and universal."), and 105–9 ("One society—and it is always the Western one—defines the general framework of Nature with respect to which the others are situated. . . . The solution appears along with the dissolution of the artifact of cultures. All natures-cultures are similar in that they simultaneously construct humans, divinities and nonhumans. None of them inhabits a world of signs or symbols arbitrarily imposed on an external Nature known to us alone").

79. Delany, *Law and Nature*, 54.

80. Ibid., 77, emphasis in the original.

81. See Lloyd, *Cognitive Variations*, chapter 7, "Nature versus Culture Reassessed," 131–50, and Descola, *Beyond Nature and Culture*.

82. Strathern, "No Nature No Culture," 177.

83. Castro, "Cosmological Deixis and Amerindian Perspectivism," 469.

84. Ibid., 470.

85. Robson, "Numeracy," 990.

86. Williams, *Problems in Materialism and Culture*, 73.

87. Latour, *We Have Never Been Modern*, 97.

88. Ibid., 96.

89. Selin, *Encyclopedia of the History of Science, Technology, and Medicine in Non-Western Cultures*. Terms such as "Western" and "non-Western" reify civilizational constructs in a way that makes such ambiguities as are found in the cultural identity of ancient Mesopotamia difficult to account for. Sheldon Pollock put it best when he said, "if we are prepared to look historically, civilizations reveal themselves to be processes and not things. And as processes they ultimately have no boundaries; people are constantly receiving and passing on cultural goods . . . all cultures participate in what are ultimately global networks of begging, borrowing, and stealing, imitating and emulating." See Pollock, *Language of the Gods in the World of Men*, 538–39.

90. Latour, *We Have Never Been Modern*, 97.

91. The critique of these categories is fully explicated in Descola, *Beyond Nature and Culture*.

CHAPTER THREE

1. CAD s.v. *irrū* "intestines," 181 and CAD s.v. *mudû* adj. meaning 1b "knowledgeable," 165.

2. A concise presentation of the essentials may be found in Lenzi, "Akkadian Scholarship: Kassite to Late Babylonian Periods."

3. CAD s.v. **idūtu*.

4. CH §9: 35, see CAD s.v. *mudûtu* usage a.

5. CAD s.v. *mudû*.

6. ABL 1277 rev. 9, see Parpola LAS I, 274–75; Parpola SAA 10 No. 30, for a slightly different translation, and CAD s.v. *ṭupšarrūtu*. Note that I maintain the transcription *ṭupšarrūtu*, but cf. Streck, review of CAD vol. 19 (Ṭ).

7. CAD s.v. *lē'û* usage c.

8. Streck Asb. 254: 15. This may indeed have been a literary exaggeration, if Alan Lenzi and Victor Hurowitz are correct in saying that the meaning of *apkallu* is limited to the sages of mythology. See Lenzi, "Uruk List of Kings and Sages and Late Mesopotamian Scholarship," 148, note 46. The CAD takes the passage more literally, under *apkallu* meaning 3 usage b.

9. Lambert BWL 86: 256–57, translating *lamādu* as "understand," rather than "learn." Note the association of the inside of the divine with the inside of heaven, recognized in commentary texts by the correlation of *libbu* and *qerbu*, see BWL 76 comm. to line 82 *qé-reb // lìb-bi*.

10. *ēkama ilmada alakti ilī apâti*, *Ludlul bēl nēmeqi* II 38, translation modified from Annus and Lenzi, *Ludlul Bēl Nēmeqi*, 20 and 35. It should be noted that there is a double entendre in the use of the word *alaktu* "way," which has a meaning "course (of the stars)" used in oracular contexts, as first pointed out in Abusch, "*Alaktu* and *Halakhah*," 18–23. *Ludlul* II lines 33–37 are also apposite, particularly line 37, which refers to the divine decision using the word *milku* "decision (of a deity)": *milik sa anzanunzê iḫakkim mannu* "who knows the decision of (the gods of) the netherworld (lit., the abyss)?" *milku* is elsewhere used in parallel with *purussû* "(divine) decision," or "omen portent." See CAD s.v. *milku*, meaning 1 usage c. The inscrutability of the divine is attributed to Anu's *milku* (*Anum šamê ša la ilammadu milkišu ma*[*mma*] "Anu in heaven whose counsel nobody understands," BMS 1: 9, see Foster, *Before the Muses*, 760 (III.54b) and the heart (or mind, also expressed with the word *libbu*) of Marduk (*libbu rūqu ša ilammadu ilū gimrašunu* "whose heart is so deep that none of the gods can comprehend it," *Enūma Eliš* VII.118, see Foster, *Before the Muses*, 482).

11. CAD s.v. *aḫāzu* meaning 4.

12. See CAD s.v. *iḫzu*. Note also the rare word *eršūtu* "wisdom" in CAD s.v. *eršūtu* A. Cf. Pingree, "Logic of Non-Western Science," 46 on the *sāstras* "teachings."

13. Finkel, "Adad-apla-iddina, Esagil-kīn-apli, and the Series SA.GIG," 148–49, preserved in two sources (a Nimrud tablet, A 63–64 and a late Babylonian tablet B 27').

14. Beaulieu, "Setting of Babylonian Wisdom Literature," 3–19.

15. CAD s.v. *nēmequ* usage b, CT 2738:43, and Hunger, *Babylonische und Assyrische Kolophone*, 2, Text Nr. 319: 6.

16. See CAD s.v.

17. As far as the celestial omen corpus is concerned, an Old Babylonian literary catalog from Ur includes the incipits (Sumerian/Akkadian) u_4 an-né / *i-nu anum ù* *ᵈen-líl*, see Kramer, "New Literary Catalogue from Ur," 172, lines 49–50. Kramer identified these lines as the bilingual opening to the composition *Enūma Anu Enlil.* He noted a parallel in an Old Babylonian literary catalog from Nippur as well, for which, see Kramer, "Oldest Literary Catalogue," 18, line 58 é-an-ni. Old Babylonian exemplars of the series, however, do not have the same content as those known from later periods, but that does not rule out the fact that the scribes viewed the celestial omens already as a composition. See also Hallo, "On the Antiquity of Sumerian Literature," 176.

18. Scholarship, perhaps supported by the poorly known and poorly attested royal palaces of the Kassite period, continued, and it seems that the omen texts emerging from this period had taken on a special character as organized, standardized, and serialized into multiple tablet collections. Lenzi, "Akkadian Scholarship," note 153 notes the doubtful nature of a scholarly library in the palace of Tiglath-Pileser I.

19. On the question of whether *bārûtu* is in fact the title of the extispicy series, see Koch-Westenholz, *Babylonian Liver Omens*, 25–27.

20. See de Zorzi, "Bird Divination in Mesopotamia," 85–135.

21. The Akkadian *ḫarṭibi* is a rendering of the Egyptian *ḥry-tp* "head/chief," a short form of *ḫry-ḥb.t ḥry-tp* "chief lector priest." I thank Alexandra von Lieven, Rita Lucarelli, and Micah Ross for help with the Egyptian term.

22. Fales and Postgate, *Imperial Administrative Records, Part I: Palace and Temple Administration*, Text 1.

23. Oshima, introduction to *Babylonian Theodicy*, xxxvii–xlvi.

24. Lenzi, "Uruk List of Kings and Sages," 143–44 with notes 23 and 25.

25. CAD s.v. *apkallu*, meaning 1, and Reiner, "Etiological Myth of the Seven Sages," 1–11.

26. K.2486+ 1–22, with ellipses, see Lambert, "Enmeduranki and Related Matters," 132, and Lambert, "Qualifications of Babylonian Diviners," 149 (transliteration) and 152 (translation).

27. George, *Babylonian Gilgamesh Epic*, 708–9, line 86.

28. FgrHist 680 F 1 (4), see Burstein, *Babyloniaca of Berossus*, 13–14. Dalley, *Myths from Mesopotamia*, 182, notes that the Akkadian equivalent of Oannes, Uan, is a pun on *ummânu* "expert."

29. Lenzi, "Uruk List of Kings and Sages."

30. Špelda, "Search for Antediluvian Astronomy," 339.

31. Ibid., 341.

32. Hunger, *Babylonische und Assyrische Kolophone*, 42, Text No. 98, and Neugebauer, *Astronomical Cuneiform Texts*, 18 colophon U.

33. Lenzi, *Secrecy and the Gods*, 170, note 159 argues for the translation "restriction" because of the usage of *ikkibu* in the meaning "limit," in the sense both of being reserved and inaccessible, implying as well as that limit cannot be crossed, hence a transgression.

34. Löhnert, "Scribes and Singers of Emesal Lamentations," 421–47, with previous literature.

35. Jolly, "Magic and Science," 1247. Assyriologists mostly seek to differentiate the cuneiform evidence from that of anthropological study of contemporary cultures; see Wiggermann and Van Binsbergen, "Magic in History, a Theoretical Perspective, and Its Application to Ancient Mesopotamia," 1–34, esp. 6–9; Guinan, "A Severed Head Laughed," 7–40; Abusch and Schwemer, *Corpus of Mesopotamian Anti-Witchcraft Rituals*, 3, where they note that the material from Evans-Pritchard's study of Central African witchcraft beliefs is not comparable. See also Cryer, *Divination in Ancient Israel and Its Near Eastern Environment*, and Dolansky, *Now You See It, Now You Don't*, 27–36.

36. In addition to the relation between science and magic in Near Eastern antiquity is its relation to, or lack of boundaries with, religion. This may also be deemed a problem of terminology, stemming from a lack of a classificatory equivalent to religion in the ancient Near East. This lack of distinction is even more apposite when, for example, the study of a class of physical objects, such as we would consider the celestial bodies to be, is at the same time, from the ancient standpoint, a study of divinities, showing perhaps from an unexpected source how science cannot be reduced simply to sets of propositions. The relation between the categories of science and religion and their impact on the study of ancient Mesopotamia, however, lies outside the scope of the present discussion.

37. *Šurpu* in Reiner, *Šurpu*; Udug-hul in Geller, *Forerunners to Udug-Hul*, Geller, *Evil Demons*, Geller and Vacin, *Healing Magic and Evil Demons*; *Muššu'u* in Böck, "When You Perform the Ritual of 'Rubbing,'" 1–16.

38. Finkel, "On Late Babylonian Medical Training," 153.

39. Heeßel, "Hands of the Gods," 120–30; Heeßel, *Babylonisch-assyrische Diagnostik*.

40. Finkel, "On Late Babylonian Medical Training," 195.

41. Heeßel, *Babylonisch-assyrische Diagnostik*; Heeßel, "Bibliographie zur altorientalischen Medizin," 34–40; Geller, *Ancient Babylonian Medicine: Theory and Practice*.

42. Heeßel, "Astrological Medicine in Babyonia," 1–16; Geller, "Look to the Stars," and Geller, *Melothesia in Babylonia*.

43. See Falkenstein, "'Wahrsagung' in der sumerische Überlieferung," 45–68; Renger, "Untersuchungen zum Priestertum der altbabylonischen Zeit," 203, note 940; Rutz, *Bodies of Knowledge in Ancient Mesopotamia*, 18–19; Richardson, "On Seeing and Believing," 227, where the meaning of *máš-šu-gíd-gíd* is etymologized as "the one who reaches the hand (in)to the goat," and see further the references in notes 6–9; Maul, *Die Wahrsagekunst im Alten Orient*, 24–25 and note 15. For discussion of the meaning of Sumerian *máš* in the title *máš-šu-gíd-gíd* as a reference not to a goat but to the young of the caprine family, which includes all bovids related to both sheep and goats, see Maul, ibid., 29 and note 2.

44. Falkenstein, "'Wahrsagung' in der sumerische Überlieferung," 47, also Finkelstein, "Mesopotamian Historiography," 464, note 12.

45. As cited in CAD s.v. *bâru* discussion section, p. 125. Later, in Old Babylonian, these professions are better attested, as outlined in detail by Renger, "Untersuchungen," and even occur in omen protases: "If he sees a diviner (*bārû*) / an exorcist (*āšipu*) / a physician (*asû*)."

46. Westenholz, "Clergy of Nippur," 299.

47. Gudea *Cyl.*A xii 16–17; xiii 16–17; xx 5 refers to the performance of extispicy; dreams (máš-gi$_6$ "night vision") are found in i 17–18; i 27 and note the use of the word giskim "sign," viii 19; ix 9, and xii 11, see Edzard, *Gudea and His Dynasty*; Koch-Westenholz, *Mesopotamian Astrology*, 32–33.

48. Richardson, "On Seeing and Believing," 225–66, especially 227–33, §§ 2.1 and 2.2.

49. See OB Diri Nippur 316, in Abusch, "*Alaktu* and *Halakhah*," 18–23; Brown, "Astral Divination in the Context of Mesopotamian Divination, Medicine, Religion, Magic, Society, and Scholarship," 77, note 22.

50. Pongratz-Leisten, "King at the Crossroads between Divination and Cosmology," 35–36; Marchetti, "Divination at Ebla during the Old Syrian Period: The Archaeological Evidence," 279–95; Richardson, "On Seeing and Believing," 232–33; Maul, *Die Wahrsagekunst*, 186 and note 16.

51. Koch-Westenholz, *Babylonian Liver Omens*, 13–14 and note 21.

52. Rutz, *Bodies of Knowledge*, 221–63.

53. Goetze, YOS 10 31 ii 38–41.

54. Goetze, YOS 10 31 ii 24–30.

55. Goetze, "Reports on Acts of Extispicy from Old Babylonian and Kassite Times," 94–95. See also the discussion of the ritual context of extispicy in Maul, *Die Wahrsagekunst*, 32–54.

56. Virolleaud, ACh Sin 1:1–8.

57. See above, note 17, and the remarks in Hunger and Pingree, *Astral Sciences in Mesopotamia*, 7, where they question the necessity of identifying the title with the celestial omen series.

58. Virolleaud, ACh Sin 1 : 1–8.

59. Cf. for slightly different translations, Foster, *Before the Muses*, 495; Verderame, *La Tavole I–VI della serie astrologica Enuma Anu Enlil*, Section I.4.1 §0.a and §0.b, 9 and 13; Koch-Westenholz, *Mesopotamian Astrology*, 76–77.

60. Oppenheim, "A Babylonian Diviner's Manual," 200 and 204, lines 38–42.

61. The thirteenth century B.C.E. celestial omens from Emar are found in Arnaud, *Recherches au pays d'Astata Emar* VI, Nos. 650–65. The Harādum text is published in Joannès, "Un precurseur paléo-babylonien de la serie *Šumma ālu*," 305–12.

62. Rochberg-Halton, *Aspects of Babylonian Celestial Divination*, 23–25.

63. Rutz, *Bodies of Knowledge*.

64. Parpola, LAS I and II, and SAA 10.

65. Hunger gives the range of dates for datable reports as -679 to -648, with a single report from the eight century (-708, SAA 8 No. 501), see SAA 8 introduction, p. xx.

66. Indicated by colophons from Assurbanipal's palace, see Hunger, *Babylonische und Assyrische Kolophone*, Nrs. 319–22 and 326. See the discussion of the Nineveh tablet collection and the idea of "Assurbanipal's library" in Robson, "Reading the Libraries of Assyria and Babylonia," 41–48. And justifying divination as a way to obtain knowledge, see Robson, "Empirical Scholarship in the Neo-Assyrian Court," 603–29.

67. Hunger, *Babylonische und Assyrische Kolophone*, 97, Nr. 317. Koch-Westenholz, *Babylonian Liver Omens*, 29, notes that Julian Reade's interpretation of the short identification of the tablet as belonging to Assurbanipal's palace in CRRAI 30 (1986) 220 indicates that these tablets were acquisitioned by, as opposed to written in, the palace.

68. See CAD s.v. *tapḫurtu*.

69. Hunger, *Babylonische und Assyrische Kolophone*, 97, Nr. 318.

70. See Wiseman and Black, *Literary Texts from the Temple of Nabû*, 3–7, concerning the library, or *girginakku*.

71. Gurney and Finkelstein, *Sultantepe Tablets*, I; also Gurney and Hulin, *Sultantepe Tablets*, II.

72. Rochberg, "Foresight in Ancient Mesopotamia."

73. Rochberg, "A Firm Yes," 5–12.

74. A survey of sites important for the recovery of scholarly tablets may be found in Pedersén, *Archives and Libraries in the Ancient Near East*. See also Lenzi, "Akkadian Scholarship: Kassite to Late Babylonian Periods." A discussion and comparison of Assyrian (Nineveh, Kalhu, and Huzirina) and Babylonian (Uruk) may be found in Robson, "Reading the Libraries of Assyria and Babylonia," 38–56. For the Neo-Babylonian to Seleucid periods, see De Breucker, "Berossos and the Mesopotamian Temple as Centre of Knowledge during the Hellenistic Period," 13–23.

75. Caplice, *Akkadian Namburbi Texts*; Maul, *Zukunftsbewältigung*; Koch, "Three Strikes and You're Out," 43–59.

76. Reiner, "Early Zodiologia and Related Matters," 421–27.

77. Reiner, *Astral Magic in Babylonia*.

78. Wee, "Discovery of the Zodiac Man," 217–33, with previous literature.

79. LBAT 1596–1598; BRM 4 19 and 20; and Geller, *Melothesia in Babylonia*.

80. Strassmaier, "Arsaciden-Inschriften," 129–58; Sachs, "Babylonian Horoscopes," 49–75; Neugebauer and van Hoesen, *Greek Horoscopes*; Rochberg, *Babylonian Horoscopes*.

81. Hunger and Pingree, *Astral Sciences*, 26–27.

82. Greenbaum and Ross, "Role of Egypt in the Development of the Horoscope," 146–82.

83. Rochberg, *Babylonian Horoscopes*, 66 rev. 10.

84. See Böck, "'An Esoteric Babylonian Commentary' Revisited," 615–20.

85. Rochberg, *In the Path of the Moon*, chapter 9.

86. SpTU 14, 159, see Koch-Westenholz, *Babylonian Liver Omens*, 24–25, and Koch, *Secrets of Extispicy*, 30–31.

87. TCL 6 14: 11, 12, 13 and 20; see Sachs, "Babylonian Horoscopes," 66.

88. BRM 4 20, with BRM 4 19 (and parallels with STT 300 although without the zodiacal references), see the discussion in Geller, *Melothesia*, 27–57.

89. Neugebuauer, *Astronomical Cuneiform Texts*.

90. See Maul, *Die Wahrsagekunst*, chapter 8, especially 270–72, for discussion of the ultimate development of astronomy as a discipline separate from divination, thereby lending celestial divination a special credibility. The divinatory and the astronomical are here understood more in terms of different aims of prediction, as discussed in chapter 8, below. I would emphasize, however, that the distinction between celestial divination and astronomy is purely modern.

91. Cooley, "An OB Prayer to the Gods of the Night," 75 line 7. See also Steinkeller, "Of Stars and Men," 11–47.

92. Sachs, "Latest Datable Cuneiform Tablets," 379–98, and Hunger and de Jong, "Almanac W 22340a from Uruk: The Last Datable Cuneiform Tablet," 186.

93. Reiner and Pingree, *Venus Tablet of Ammiṣaduqa*, and Walker, "Notes on the Venus Tablet of Ammiṣaduqa," 64–65.

94. Appearing as: mu giš.dúr.gar kù.sig$_{17}$.ga.kam "the year of the golden throne," which is an abbreviation of Ammiṣaduqa's eighth year name (YOS 13 408).

95. Reiner and Pingree, *Venus Tablet of Ammiṣaduqa*, 29.

96. De Jong, "Astronomical Fine-Tuning of the Chronology of the Hammurabi Age," 147–67.

97. Verderame, *La Tavole I–VI della serie astrologica Enuma Anu Enlil*, and Verderame, "*Enūma Anu Enlil* Tablets 1–13," 447–57.

98. Al-Rawi and George, "Enūma Anu Enlil XIV and Other Early Astronomical Tables," 52–73, and Hunger and Pingree, *Astral Sciences*, 44–50.

99. Rochberg-Halton, *Aspects of Babylonian Celestial Divination*.

100. Soldt, *Solar Omens of Enuma Anu Enlil: Tablets 23(24)–29(30)*.

101. Gehlken, *Weather Omens of Enūma Anu Enlil*.

102. Reiner and Pingree, *Babylonian Planetary Omens 2*.

103. Reiner and Pingree, *Babylonian Planetary Omens 3*; Reiner and Pingree, *Babylonian Planetary Omens 4*.

104. Frahm, *Babylonian and Assyrian Text Commentaries*, 136–55.

105. Koch-Westenholz, "Astrological Commentary Šumma Sîn ina Tāmartīšu Tablet I," 159–65; Gehlken, "Die Serie DIŠ Sîn ina tāmartīšu im Überblick," 3–5; Frahm, *Babylonian and Assyrian Text Commentaries*, 155–60; and Fincke, "Seventh Tablet of the *rikis gerri* Series of *enūma anu enlil*," 129–48.

106. Horowitz, "Astrolabes: Astronomy, Theology, and Chronology," 112, notes 7 and 8. See also Horowitz, *Three Stars Each*.

107. Reiner and Pingree, *Babylonian Planetary Omens 2*, 81, lines 19–24.

108. Rochberg, *In the Path of the Moon*, 294–302, and table 11.

109. Hunger and Pingree, *MUL.APIN*.

110. Frahm, *Babylonian and Assyrian Text Commentaries*, 166–67, suggests evidence for a third tablet of *MUL.APIN* on the basis of the catchline(?) "Jupiter is Šulpa'e" found in the colophon to Tablet II, see Hunger and Pingree, *MUL.APIN*, 123 Source D.

111. Strassmaier, *Inschriften von Cambyses*, No. 400; Britton, "Remarks on Strassmaier Cambyses 400," 7–33.

112. Huber and Steele, "Babylonian Lunar Six Tablets," 3–36.

113. Ibid., 15–16, Text B (N.2349).

114. An eclipse possibility occurs at a syzygy when the sun is within a half-month's progress in longitude from a lunar node. The thirty-eight eclipse possibilities per Saros cycle are spaced at six-month (thirty-three possibilities) and sometimes five-month (five possibilities) intervals. See

Aaboe et al., *Saros Cycle Dates and Related Babylonian Astronomical Texts*, and subsequently discussed in Steele, "Eclipse Prediction in Mesopotamia," and Britton, "Calendars, Intercalations, and Year-Lengths in Mesopotamian Astronomy," 88–89.

115. These were termed *Greek-letter phenomena* by Neugebauer. They are, for the outer planets (Jupiter, Saturn, Mars): Γ = IGI, reappearance in the east, or first visibility after conjunction with the sun, also known as heliacal rising; Φ = UŠ, UŠ IGI-*tú* is the first stationary point; Θ = E, E-ME, ME, ME-*a* is the acronychal rising or "opposition"; Ψ = UŠ, UŠ ÁR-*tú* second stationary point; Ω = ŠÚ last appearance, disappearance in the west, or last visibility before conjunction with the sun, also called heliacal setting. For the inner planets (Venus and Mercury) they are: Γ = IGI, IGI *ša* KUR first appearance in the east as morning star, morning rising; Φ = UŠ *ša* KUR morning station; Σ = ŠÚ, ŠÚ *ša* KUR morning setting, disappearance in the east, or last appearance in the east as a morning star before inferior conjunction; Χ = IGI, IGI *ša* ŠÚ is the reappearance in the west, or first appearance in the evening as evening star; Ψ = UŠ, UŠ *ša* ŠÚ evening station; Ω = ŠÚ, ŠÚ *ša* KUR last appearance in the west, evening setting.

116. The Lunar Six phenonena are: At the beginning of the month in the evening, NA = the time interval between sunset and moonset on the evening of the first lunar visibility after conjunction of sun and moon; in the middle of the month, ŠÚ = the interval from moonset to sunrise when the moon set for the last time before sunrise; *na* = in the middle of the month, the interval between sunrise and moonset when the moon set for the first time after sunrise; ME = interval between moonrise to sunset when the moon rose the last time before sunset; GE6 = the interval between sunset and moonrise when the moon rose the first time after sunset; at the end of the month, in the morning KUR = the interval between moonrise and sunrise when the moon was visible for the last time before conjunction.

117. Hunger and Pingree, *Astral Sciences*, 173–74.

118. Hunger, "Saturnbeobachtungen aus der Zeit Nebukadnezars II," 189–92.

119. Jones, "A Study of Babylonian Observations of Planets Near Normal Stars," 530–34.

120. Pingree and Reiner, "Observational Texts concerning the Planet Mercury," 175–80.

121. Ibid., 65. Rm 2, 303 lines Col.i 1'–3.'

122. Sachs, "Classification of the Babylonian Astronomical Texts of the Seleucid Period," 271–90.

123. Neugebauer, *A History of Ancient Mathematical Astronomy*, 351.

124. Ibid., 351–52.

125. A survey of the Non-Mathematical Astronomical Texts (nontabular, or GADEx texts) can be found in Hunger, "Non-Mathematical Astronomical Texts and Their Relationships," 77–96. The texts are available in Sachs-Hunger, ADRT 1–3; Hunger, ADRT 5; Hunger, ADRT 6; the copies are in Sachs, with the cooperation of J. Schaumberger, LBAT; the corpus of eclipse reports are in Huber and de Meis, *Babylonian Eclipse Observations from 750 BC to 1 BC*.

126. CAD s.v. *tērsītu*.

127. Ossendrijver, *Babylonian Mathematical Astronomy*.

128. Britton, "Studies in Babylonian Lunar Theory: Part I," 125; Ossendrijver, *Babylonian Mathematical Astronomy*, 116.

129. Steele, "Celestial Measurement in Babylonian Astronomy," 293–325.

130. Ossendrijver, "Exzellente Netzwerke: Die Astronomen von Uruk," 631–44; Ossendrijver, "Science in Action: Networks in Babylonian Astronomy," 213–21.

131. Parker, *A Vienna Demotic Papyrus on Eclipse- and Lunar Omina*.

132. Ross, "Horoscopic ostraca from Medinet Madi"; Ross, "An Introduction to the Horoscopic Ostraca of Medînet Mâdi," 147–63; Ross, "A Continuation of the Horocopic Ostraca of Medînet Mâdi," 153–71; Ross, "Further Horoscopic Ostraca from Medînet Mâdi," 61–91; Ross, "A Provisional Conclusion to the Horoscopic Ostraca from Medînet Mâdi," 47–80; Ross, "A Survey of Demotic Astrological Texts," 1–31; Stockhusen, "Babylonische astralwissenschaften in römerzeitlichen Ägypten: Das Beispiel Medînet Mâdi," 85–109.

133. Rochberg-Halton, *Aspects of Babylonian Celestial Divination*, 2–5; and Rochberg, *In the Path of the Moon*, 1–18.

134. For a discussion of the meaning of the terms *astrologia/astronomia*, see Bowen, "Three Introductions to Celestial Science in the First Century BC," 299.

135. Bowen and Goldstein, "Geminus and the Concept of Mean Motion in Greco-Latin Astronomy," 172, and Bowen, "Hupsiklēs of Alexandria (150–100 BCE)," 425.

136. Bowen and Goldstein, "Geminus and the Concept of Mean Motion," and Bowen, "Three Introductions to Celestial Science in the First Century BC," 322.

137. Jones, *Astronomical Papyri from Oxyrhynchus*.

138. Rochberg-Halton, "Elements of the Babylonian Contribution to Hellenistic Astrology," 51–62; Jones, *Astronomical Papyri from Oxyrhynchus*; Jones and Steele, "A New Discovery of a Component of Greek Astrology in Babylonian Tablets: The 'Terms.'"

139. Plofker and Knudson, "Paitāmahasiddhānta," 604–5.

140. Pingree, "Mesopotamian Astronomy and Astral Omens in Other Civilizations," 613–31; Pingree, "Venus Omens in India and Babylon," 293–315; Pingree, *From Astral Omens to Astrology*; Pingree, "Legacies in Astronomy and Celestial Omens," 125–37.

141. Pingree, *Yavanajātaka of Sphujidhvaja*.

142. Rochberg, *In the Path of the Moon*, 223–35, and Bhayro, "Book of the Signs of the Zodiac," 197–98.

143. See, for example, Boustan and Reed, *Heavenly Realms and Earthly Realities in Late Antique Religions*, or Hanegraaff, *Esotericism and the Academy*.

144. Stadhouders, "Pharmacopoeial Handbook *šammu šikinšu*—An Edition," 3–51; Scurlock, *Sourcebook for Ancient Mesopotamian Medicine*, 281–83.

145. Schuster-Brandis, *Steine als Schutz- und Heilmittel*, 24–47; Scurlock, *Sourcebook for Ancient Mesopotamian Medicine*, 283–89.

146. Landsberger and Krumbiegel, *Die Fauna des alten Mesopotamien*, 52–54 (transliteration and translation).

147. Ibid., p. 53 lines 1–2.

148. Reiner, *Astral Magic*, 29.

149. Duffin, Moody, and Gardner-Thorpe, *History of Geology and Medicine*.

150. Landsberger and Krumbiegel, *Die Fauna des alten Mesopotamien*, 55.

151. Böck, "Sourcing, Organizing, Administering Medicinal Ingredients," 693.

152. Ibid.

153. Unknown plant name.

154. A melon or cucumber-like plant, see CAD s.v. *qiššû*.

155. Stadhouders, "Pharmacopoeial Handbook *šammu šikinšu*: A Translation," 4 §23: 1–4.

156. See Landsberger, Kilmer, and Gordon MSL 8/1, 1960, and 8/2, 1962, and Landsberger, Reiner, and Civil, MSL 10, 1970.

157. See Heeßel, "Stein, Pflanze, und Holz," 1–22, and the remarks in Böck, "Sourcing, Organizing, Administering Medicinal Ingredients," 696–7.

158. Schuster-Brandis, *Steine als Schutz- und Heilmittel*, and Reiner, "Magic Figurines, Amulets, and Talismans," 27–36.

159. Schuster-Brandis, *Steine als Schutz- und Heilmittel*, 59–68.

160. Ibid., 27, for the edition see 24–47. The passage quoted here, according to Schuster-Brandis's sigla, is Texts A–D: 33–35.

161. Horowitz suggests *tikbušu* NU [TUKU] "it does not have spots"; see "Two Abnu šikinšu Fragments and Related Matters," 115, note 4.

162. The CAD s.v. *ḫa'attu* reads this passage, written *ḫa-ia-a'-tu*, as the substantive *ḫa'attu* meaning "panic" or "mortal terror." Schuster-Brandis translates "the *ḫajjattu*-demon," with comments on 46.

163. Schuster-Brandis, *Steine als Schutz- und Heilmittel*, edition Text E (= BAM 194 VII and KAR 185; VAT 9587): 11'–13,' 33 and 39.

164. Ibid., 8–10, especially 9 and note 39.

165. On the problems essentialism poses for defining natural kinds, see Magnus, *Scientific Enquiry and Natural Kinds*, 32–34.

166. Reiner, *Astral Magic*, 30 with note 117.

167. Postgate, "Mesopotamian Petrology," 205. Note that Scurlock also titled both series "The Nature of Plants" and "The Nature of Stones," although translated the texts using the phrase "the plant/stone that resembles . . . ," see her *Sourcebook for Ancient Mesopotamian Medicine*, 282–83 and 287–89.

168. Ibid., 220.

169. Huehnergard and Woods, "Akkadian and Eblaite," 109 Section 4.1.1.6.

170. CAD s.v. *šakānu*, meaning 3, 130.

171. Landsberger, *Fauna of Ancient Mesopotamia*, 53–54.

172. Oppenheim, *Ancient Mesopotamia*, 202–3.

173. See also Naddaf, *Greek Concept of Nature*, 13–14.

174. Ibid., 13.

175. Ibid., 13–14.

176. See Rochberg, *In the Path of the Moon*, 19–30, especially 28.

177. Oppenheim, *Ancient Mesopotamia*, 202.

178. For the edition, see van Dijk, *LUGAL UD ME-LÁM-bi NIR-GÁL*, and for relevant passages, see CAD s.v. *šīmtu*, lexical section. Note also Landsberger's comment on *nam*, which he said was "difficult to translate," but explained as "roughly 'the sum of effects of a thing on the outside world,' which is appended to every thing in the form of a formula of fate." See Landsberger, *Conceptual Autonomy of the Babylonian World*, 13.

179. Oppenheim, *Ancient Mesopotamia*, 203.

180. Steinkeller, "Luck, Fortune, and Destiny in Ancient Mesopotamia," ms p. 4.

181. Johnson, "Origins of Scholastic Commentary in Mesopotamia," 42.

182. See an early treatment in Maul, "Das Wort im Worte: Orthographie und Etymologie als hermeneutische Verfahren babylonischer Gelehrter," 1–18, and a more intensive study with respect to cuneiform commentaries in Frahm, *Babylonian and Assyrian Text Commentaries*. It is important to say that "etymography" itself is not so recent an invention. See, for example, Alinei, "Etymography and Etymothesis as Subfields of Etymology: A Contribution to the Theory of Diachronic Semantics," 41–56.

183. Johnson, "Origins of Scholastic Commentary," 43–44.

184. See Huehnergard, *A Grammar of Akkadian*, Section 13.3, 111–12, and 537 for a list of the most common determinatives.

185. Goldwasser, *Prophets, Lovers, and Giraffes: Wor(l)d Classification in Ancient Egypt*.

186. Goldwasser, "Determinative System as a Mirror of World Organization," 49.

187. Veldhuis, *History of the Cuneiform Lexical Tradition*, 49.

188. See Oppenheim, *Ancient Mesopotamia*, 48 and 248.

189. Veldhuis, *History of the Cuneiform Lexical Tradition*, 55. Cf. Hilgert, "Von 'Listenwissenschaft' und 'epistemischen Dingen.' Konzeptuelle Annäherungen an altorientalische Wissenspraktiken," 282.

190. For discussion of the relation between taxonomy and ontology from the perspective of Roman knowledge, see Lehoux, *What Did the Romans Know?* chapter 6, "The Trouble with Taxa."

191. Quine, "Natural Kinds," 42–43.

192. Goodman, "Seven Strictures on Similarity," 19.

193. Lehoux, *What Did the Romans Know?* 150.

194. Ibid.

195. In Rescher et al., *Essays in Honor of Carl G. Hempel*, 1–23.

196. Hacking, "A Tradition of Natural Kinds," 109.

197. Magnus, *Scientific Enquiry and Natural Kinds*, 4.

198. Churchland, "Conceptual Progress and Word/World Relations: In Search of the Essence of Natural Kinds," 2.

199. Wilson, Barker, and Brigandt, "When Traditional Essentialism Fails: Biological Natural Kinds," 189–215.

200. Lloyd, *Cognitive Variations*, 43. See further his discussion of the "deeply controversial" nature of taxonomic orders in Lloyd, *Being, Humanity and Understanding*, 104.

201. Lloyd, *Cognitive Variations*, 45, note 6.

202. Landsberger, *Conceptual Autonomy of the Babylonian World*, 11.

203. Wapnish, "Animal Names and Animal Classifications in Mesopotamia: An Interdisciplinary Approach Based on Folk Taxonomy," 92–93.

204. The terminology originated with Sokal and Sneath, *Principles of Numerical Taxonomy*. Cf. the somewhat different but related ideas of monotypic and polytypic classes as defined in Beckner, *Biological Way of Thought*.

205. Veldhuis, "How to Classify Pigs," 25–29.

206. Ibid., 26.

207. Ibid., 27.

208. Hacking, "Natural Kinds: Rosy Dawn, Scholastic Twilight," 203, emphasis in the original.

209. Goodman, *Ways of Worldmaking*, 10.

210. Reydon, "Natural Kinds No Longer Are What They Never Were," 263.

211. Ibid.

212. Borges, *Other Inquisitions*, 103.

213. Foucault, preface to *The Order of Things: An Archaeology of the Human Sciences*, xv.

214. Longxi, *Mighty Opposites*, 21.

CHAPTER FOUR

1. Radner and Robson, introduction to *The Handbook of Cuneiform Culture*, xxvii. The term "cuneiform culture" was also used in Hunger and Pingree, *Astral Sciences in Mesopotamia*, 270.

2. Allen, "Greek Philosophy and Signs," 29.

3. Deely, "From σημεῖον to 'signum' to 'sign,'" 130.

4. Ibid., 129.

5. Ibid., 134.

6. Harrison, *The Bible: Protestantism and the Rise of Natural Science*, 9.

7. Volk, "Aratus," 200; Netz, *Ludic Proof*, 186–87.

8. Volk, "Aratus," 200–201, my emphasis.

9. Beerden, *Worlds Full of Signs*, looks at some of the differences in how divinatory signs functioned in Mesopotamia, Greece, and Rome, focusing on the textual (cuneiform) or nontextual (Greek) nature of divination, and the various institutional or noninstitutional settings of the diviner.

10. Augustine, *City of God* 21.8; see also Daston and Park, *Wonders and the Order of Nature*, 40.

11. CAD s.v. *ittu*, meanings 1–4.

12. Veldhuis, "Theory of Knowledge and the Practice of Celestial Divination," 77–91.

13. On the difference between the omens from exta in the series versus those in the "extispicy reports," see Heeßel, "Hermeneutics of Mesopotamian Extispicy," 16–35, especially 33–35.

14. Koch-Westenholz, *Babylonian Liver Omens*, 13; Brown, *Mesopotamian Planetary Astronomy-Astrology*, 106 and passim.

15. De Zorzi, "Omen Series Summa Izbu: Internal Structure and Hermeneutic Strategies," 43–75.

16. Canguilhem, *Knowledge of Life*, 122.

17. David Brown first pinpointed the importance of the ideal as a norm in his *Mesopotamian Planetary Astronomy-Astrology*, 113–22, and 125–26.

18. Canguilhem, *Knowledge of Life*, 122.

19. Ibid.

20. Ibid., 90. From a completely different point of view, cf. Hanafi's notion of mechanical monsters in *The Monster in the Machine*, especially 76–96.

21. Canguilhem, *Knowledge of Life*, 90.

22. Koch-Westenholz, *Babylonian Liver Omens*, 72, text exemplars K 7, E 12, A 11,' and B 1,' and passim in the appendix to the introduction.

23. Ibid., 94, line 31, text exemplars A r 8' and B r 15.'

24. Starr, *Rituals of the Diviner*, 32; see also Koch-Westenholz, *Babylonian Liver Omens*, 52, note 139.

25. CAD s.v. *kajamānu*, usage a 2,' TCL 6 6 ii 3.

26. Koch-Westenholz, *Babylonian Liver Omens*, 40, CT 20 44 i 52–58.

27. Ibid., 40–41.

28. Note that the D-stem adjective *uzzubu*, attested only in lexical texts, according to the CAD entry s.v. means "freakish, anomalous, monstrous."

29. For an outline of this history, see Hanafi, *Monster in the Machine*.

30. Bates, *Emblematic Monsters: Unnatural Conceptions and Deformed Births in Early Modern Europe*, 12–13.

31. Isidore of Seville, *Etymologiarum sive originum libri* XX, II.3.1–4; see also Daston and Park, *Wonders and the Order of Nature*, 50.

32. CAD s.v. *izbu*, usage b, K. 2315:60ff., and for the reference to a malformed foal, see usage c 2.' For the series edition, see de Zorzi, *La Serie Teratomantica Šumma Izbu*.

33. CAD s.v. *izbu*, lexical section.

34. Leichty, *Omen Series Šumma Izbu*, 37.

35. Note the late *Izbu* commentary to the term *lillu*, cited in Frahm, *Babylonian and Assyrian Text Commentaries*, 209: lúlil: *sak-lu* "lúlil (means) imbecile," BM 77808 rev. 1.

36. CAD s.v. *akû* B, and note that the commentary text to *Izbu* explains *akû* as *enšu* "weak," see CAD *akû* B lexical section.

37. Leichty, *Omen Series Šumma Izbu*, 32–34, Tablet I 5–23.

38. Ibid., 34, *Izbu* I 28 and 29.

39. Ibid., 38, *Izbu* I 71.

40. Ibid., 39–44, *Izbu* I 83–131.

41. Ibid., 44, *Izbu* I 131.

42. De Zorzi, "Omen Series Šumma Izbu," 46.

43. Leichty, *Omen Series Šumma Izbu*, 39, *Izbu* I 82.

44. Ibid.

45. Ibid., 39, *Izbu* I 83 (identical twin boys), 86 (fraternal twins).

46. Ibid., 16.

47. Ibid., 62, *Izbu*, III 83 and 84.

48. De Zorzi, "Omen Series Summa Izbu," 52–53; Jeyes, "Divination as a Science in Ancient Mesopotamia," 35.

49. Rochberg, *In the Path of the Moon*, 135–42.

50. Canguilhem, *Knowledge of Life*, 90.

51. Leichty, *Omen Series Šumma Izbu*, 87, *Izbu* VI 28.

52. Ibid., 106, *Izbu* VIII 50.

53. Ibid., 126, *Izbu* X 58.'

54. Aristotle, *Physics* II.8.

55. Bates, *Emblematic Monsters*, 113.

56. Daston and Park say, "the contemporary French chronicler Johannes Multivallis related

its [the Ravenna monster's] deformities to particular moral failings: The horn indicates pride; the wings, mental frivolity and inconstancy; the lack of arms, a lack of good works; . . . the eye on the knee, a mental orientation solely toward earthly things; the double sex, sodomy. And on account of these vices, Italy is shattered by the sufferings of war, which the king of France has not accomplished by his own power, but only as the scourge of God." See *Wonders and the Order of Nature*, 182.

57. Ibid., 51.

58. Leichty, *Omen Series Šumma Izbu*, 8 on hermaphroditic *izbus*.

59. Daston and Park, *Wonders and the Order of Nature*, 177.

60. Ibid.

61. Jacobs, "Traces of the Omen Series Summa Izbu in Cicero, De Divinatione," 317–39.

62. Similarly, in the Treatise on Monsters of Fortunio Liceti of the early modern period, the possible generation of monsters is understood as "supernatural, infranatural, and natural productions." Although he speaks only about these natural causes, Liceti does not fail to mention that "the sole, efficient cause is Almighty God, that is, motive Intelligence and the Heavens." See Hanafi, *Monster in the Machine*, 35.

63. For two kinds of apotropaic rituals for extispicy, see Koch, "Sheep and Sky: Systems of Divinatory Interpretations," 465–56; Koch, "Three Strikes and You Are Out," 46.

64. Maul, *Zukunftsbewältigung*.

65. See CAD s.v.

66. Caplice, "Akkadian Text Genre Namburbi."

67. Ibid., 23.

68. Livingstone, *Hemerologies of Assryian and Babylonian Scholars*, 195–98.

69. Ibid., 195 lines 7–8.

70. Caplice, "Namburbi Texts in the British Museum, III," Text Nos. 25, 26, 27, and 28, republished in Maul, *Zukunftsbewältigung*, 234–48, 268–69, and 256–68.

71. For a *namburbi* against an eclipse, see Caplice, "Namburbi Texts in the British Museum, V," 166–68, Text No. 65, and republished in Maul, *Zukunftsbewältigung*, 458–60.

72. Caplice, "Namburbi Texts in the British Museum, I," 125–30, Text No. 10, republished in Maul, *Zukunftsbewältigung*, 336–43.

73. Caplice, "Akkadian Text Genre Namburbi," 29 where he cites Morgenstern, *Doctrine of Sin in the Babylonian Religion*.

74. Maul, *Zukunftsbewältigung*, 10.

75. Annus, "On the Beginnings and Continuities of Omen Sciences in the Ancient World," 7.

76. Ludlul I 41–42 and 49, translation from Annus and Lenzi, *Ludlul Bēl Nēmeqi*, 32.

77. Ibid., II 6 and 9.

78. Ibid., II 105–9.

79. Caplice, "Namburbi Texts in the British Museum, III," 273–74 (transliteration) and 275 (translation) Text No. 25: 14'–19'.

80. CAD s.v. *lumnu*, meaning 2.

81. Caplice, "Namburbi Texts in the British Museum, V," 168, rev. 1–8.

82. Ibid., rev. 9–11.

83. Rochberg-Halton, *Aspects of Babylonian Celestial Divination*, 48.

84. Stol, "Moon as Seen by the Babylonians," 257–58. For the text, see Civil, "Medical Commentaries from Nippur," 332 lines 17–20.

85. Stol, "Moon as Seen by the Babylonians," 257–58.

86. ACh Supp. 2 Šamas 40: 6.

87. LKA 109 lines 20–22.

88. CAD s.v. *minītu*, meaning 1d.

89. CAD s.v. *simanu*, usage c.

90. CAD s.v. *adannu*, meaning 2.

91. Verderame, *Le Tavole I–VI della serie astrologica Enuma Anu Enlil,* 10–11: šumma Sin ina la adannišu/ina la minâtišu/ina la simanišu innamir.

92. Brown, *Mesopotamian Planetary Astronomy-Astrology,* 113–22 for his description of "period schemes," and 146–55 for their impact on celestial divination.

93. Englund, "Administrative Timekeeping in Ancient Mesopotamia," 121–85; Brown, "Cuneiform Conception of Celestial Space and Time," 103–22; Brack-Bernsen, "The 360-Day Year in Mesopotamia," 83–100; Britton, "Calendars, Intercalations, and Year-Lengths in Mesopotamian Astronomy," 117–19. For month lengths and the Babylonian calendar, see Britton, "Calendars, Intercalations, and Year-Lengths," and Steele, "Length of the Month in Mesopotamian Calendars of the First Millennium BC," 133–48.

94. Hunger and Pingree, *MUL.APIN,* appendix, 163–64.

95. Text enclosed in square brackets is restored. Text in parentheses is a translator's gloss.

96. Hunger and Pingree, *MUL.APIN,* 102, Tablet II ii 44.

97. Al-Rawi and George, "Enūma Anu Enlil XIV and Other Early Astronomical Tables," 52–73.

98. *MUL.APIN* II i 11–12.

99. Al-Rawi and George, "Enūma Anu Enlil XIV," 58.

100. *šumma ūmu ana minâtišu erik palē ūmī arkūti,* see SAA 8 9 rev. 1, 10: 3, 28: 1–2, 37: 3–4, 86: 3–4, 87: 3–4. Note also the correspondence between the root 'rk (the verb *arāku* "to become/last long" and the adjective *arku* "to be long") in both the antecedent and consequent. I note as well that the CAD translates the antecedent clause as "if the day is longer than its normal measure," which is also possible.

101. SAA 8 87: 5–6.

102. For a discussion of the rule-based nature of astrological divination in the hellenistic Greek context, see Greenbaum, "Arrows, Aiming, and Divination: Astrology as a Stochastic Art," 179–210, especially 181.

103. Brown, *Mesopotamian Planetary Astronomy-Astrology,* 153, emphasis in the original.

104. Seneca, *Quaestiones naturales* 2.32.2: Hoc inter nos et Tuscos . . . interest: nos putamus, quia nubes collisae sunt, fulmina emitti; ipsi existimant nubes collidi, ut fulmina emittantur; nam cum omnia ad deum referant, in ea opinione sunt, tamquam non, quia facta sunt, significent, sed quia significatura sunt, fiant.

105. Brown, *Mesopotamian Planetary Astronomy-Astrology,* 155, emphasis in the original.

106. Daston, "Preternatural Philosophy," 15–41.

CHAPTER FIVE

1. Sigerist, *History of Medicine,* Vol. 1, *Primitive and Archaic Medicine,* 447, see also Geller, "West Meets East," 50.

2. A useful introduction with extensive bibliography from philosophy and psychology is Nickerson, *Aspects of Rationality.*

3. Of course the idea of knowledge, and with it rationality, is tied to a variety of ways of justifying a hierarchical relationship between knowledge and belief, knowledge once being defined as "justified true belief." I purposely do not include "belief" here, so as not to enter into a potential minefield of complexity in attributing to the Assyrians and Babylonians a "belief," or set of "beliefs" as against knowledge. Needham, *Belief, Language, and Experience,* shows how that subject demands a deft and concerted treatment all its own.

4. Betti and Jong, *Synthese.*

5. Ibid., 181.

6. Ibid., 186.

7. Ibid.

8. Nersessian, *Creating Scientific Concepts,* 11, however, notes that when it comes to theory building, as opposed to theoretical results expressed as laws and axioms, "many philosophers now

agree that the basic units for scientists in working with theories are most often not axiomatic systems or propositional networks, but models."

9. Lloyd, *Cognitive Variations*, 151–52.

10. Gellner, *Plough, Sword, and Book*, 39.

11. Anderson and Perrin, "Thinking with the Head: Race, Craniometry, Humanism," 83–98; Gould, *Mismeasure of Man*.

12. Ironically, craniometric analysis of Egyptian skulls was undertaken to prove that the founders of civilization in Africa, i.e., the ancient Egyptians, were a Caucasoid race native to the Nile Valley, and not related to black Africans; see Sanders, "Hamitic Hypothesis: Its Origins and Functions in Time Perspective," 528.

13. Stich, "Could Man Be an Irrational Animal? Some Notes on the Epistemology of Rationality," 15.

14. Then published as Dodds, *Greeks and the Irrational*.

15. Gellner, *Plough, Sword, and Book*, 111, and see Aya, "Devil in Social Anthropology," 560.

16. Gellner, *Spectacles and Predicaments*, 145–47; and Gellner, "Relativism and Universals," 188–200.

17. Latour, *We Have Never Been Modern*.

18. Ibid., 97.

19. Descola, "Beyond Nature and Culture," 146; Descola, "Cognition, Perception, and Worlding," 334–40; Descola, *Beyond Nature and Culture*.

20. Descola, *Beyond Nature and Culture*, 63.

21. Papineau, "Naturalism."

22. Pollock, *Language of the Gods in the World of Men*, 6.

23. Published in Dodds, *Greeks and the Irrational*.

24. Ibid., 1.

25. Ibid.

26. Ibid., 245.

27. Thorndike, *History of Magic and Experimental Science*, 1: 20–21.

28. Murray, *Five Stages of Greek Religion*, 139, quoted by Dodds, *Greeks and the Irrational*, 245 and note 50.

29. In this passage Strabo explained that Diogenes was from Seleucia, but that "as we call the country Babylonia, so also we call the men from there Babylonians, that is, not after the city, but after the country; but we do not call men after Seleuceia, if they are from there, as for example, Diogenes the Stoic philosopher."

30. Dodds, *Greeks and the Irrational*, 246. For Diogenes, see Lehoux, "Diogenēs of Babylōn," 253. He is also mentioned in Cicero, *De Natura Deorum*, 1.15 as following Chrysippus in turning myth to natural philosophy.

31. Dodds, *Greeks and the Irrational*, 246.

32. The famous passage is, "If we take in our hand any volume of divinity or school metaphysics (works on religion and philosophy) let us ask this question, does it contain any abstract reasoning concerning quantity or number? No. Does it contain any experimental reasoning concerning matter of fact or existence? No. Commit it then to the flames, for it can be nothing but sophistry and illusion." See Selby-Bigge, *Enquiries concerning Human Understanding and concerning the Principles of Morals by David Hume*, 165.

33. Dodds, *Greeks and the Irrational*, 189 with notes 63 and 64.

34. Ibid., with note 69.

35. Von Staden, "Galen's Daimon: Reflections on 'Irrational' and 'Rational,'" 15–43.

36. Dodds, *Greeks and the Irrational*, 121.

37. Neugebauer, *Exact Sciences in Antiquity*, 171.

38. Newman and Grafton, *Secrets of Nature: Astrology and Alchemy in Early Modern Europe*, 13–14.

39. Dodds, *Greeks and the Irrational*, 237.

40. Ibid., 240. For an expression of this idea, see Cicero, *De Natura Deorum*, I.15, on Chrysippus's notion of divine power, that "resides in reason and in the soul and mind of nature taken as a whole."

41. Rochberg, *Heavenly Writing*, 14–43.

42. Neugebauer, "Survival of Babylonian Methods in the Exact Sciences of Antiquity and the Middle Ages," 528–29.

43. Larsen, "Mesopotamian Lukewarm Mind: Reflections on Science, Divination, and Literacy," 203–25.

44. Bottéro, *Mesopotamia: Writing, Reasoning, and the Gods*, 125–37.

45. Rochberg, "Inference, Conditionals, and Possibility in Ancient Mesopotamian Science," 4–25; Rochberg, "If P, then Q: Form, Reasoning, and Truth in Babylonian Divination," 19–27; Rochberg, "Beyond Binarism in Babylon," 253–65. See further discussion of the issue in Geller, review of Heessel, *Divinatorische Texte I*, 118–21.

46. Watson and Horowitz, *Writing Science before the Greeks*.

47. Hilgert, "Von 'Listenwissenschaft' und 'epistemischen Dingen,'" 277–309.

48. Ibid., 298–99. For a description of the mechanics of Diri, see also Finkel, "Strange Byways in Cuneiform Writing," 14–16.

49. Larsen, "Mesopotamian Lukewarm Mind," 205. See also Salomon, "Collquiris Dam: The Colonial Re-Voicing of an Appeal to the Archaic," 270, which cites a paraphrase of Luc de Heusch by Jonathan Friedman about "hierarchical societies ('chiefdoms') and 'archaic' states—so-called lukewarm societies." See also Friedman, "Our Time, Their Time, World Time: The Transformation of Temporal Modes," 175.

50. Larsen, "Mesopotamian Lukewarm Mind," 209.

51. Koch, *Secrets of Extispicy*, 93–94.

52. Larsen, "Mesopotamian Lukewarm Mind," 215.

53. Ibid., 216.

54. YOS 10 31 ii 38–41.

55. Ibid., ii 24–30.

56. SAA 8 413:1.

57. Goody, *Domestication of the Savage Mind*, 16, and see chapter 8, "Grand Dichotomy Reconsidered," 146–62.

58. Cole and Cole, "Rethinking the Goody Myth," 306. Reactions from anthropologists to the technological determinism of the literacy thesis appeared in Halverson, "Goody and the Implosion of the Literacy Thesis," 301–17, and Probst, "Die Macht der Schrift. Zum ethnologischen Diskurs über eine populäre Denkfigur," 167–82.

59. Bottéro, *Mesopotamia*, 132. This argument from empiricism in omens, and others who have said the same, is critiqued in Rochberg, *Heavenly Writing*, 265–71.

60. Bottéro, *Mesopotamia*, 134.

61. Hunger, SAA 8 73 rev. 1.

62. Reiner and Pingree, *Babylonian Planetary Omens 4*, 181, K. 3780 iv 5.

63. Lloyd, *Polarity and Analogy: Two Types of Argumentation in Early Greek Thought*, showed the importance of the analogical argument, and analogical reasoning, in Greek philosophy and literature from the earliest Greek literary sources down to Aristotle.

64. Bottéro, *Mesopotamia*, 136.

65. Ibid.

66. Ibid., 136–37

67. Rochberg, "If P, then Q," 24–25.

68. Ibid., 19–27.

69. Horowitz, "Astral Tablets in the Hermitage," 203: lines 1–2 (*šumma šam*]*û uššu*[*šu ša*]*ttum lemn*[*at*).

70. Watson and Horowitz, *Writing Science*, 25.

71. Ibid., 44.

72. Ibid., 140.

73. Ibid., 155.

74. Gellner, *Plough, Sword, and Book*, 39.

75. Watson and Horowitz, *Writing Science*, 117.

76. Ossendrijver, *Babylonian Mathematical Astronomy*, 16 and note 90.

77. Watson and Horowitz, *Writing Science*, 116.

78. Ibid., 122.

79. For the Sargon Geography, for example, see Horowitz, *Mesopotamian Cosmic Geography*, 91–92.

80. Brown, *Mesopotamian Planetary Astronomy-Astrology*, 113–22.

81. Brack-Bernsen, "The 'Days in Excess' from MUL.APIN," 1–29, for a discussion of the use of *MUL.APIN* for astronomical calculation and the idea that the *MUL.APIN* schemes constituted "methods of astronomical modeling of nature."

82. Dodds, *Greeks and the Irrational*, 246.

83. Weill-Parot, "Astrology, Astral Influences, and Occult Properties in the Thirteenth and Fourteenth Centuries," 205.

84. Dodds, *Greeks and the Irrational*, 247.

85. Hallum, "Bōlos of Mendēs," 196–97.

86. Heeßel, "Astrological Medicine in Babyonia," 1–16; Heeßel, "Stein, Pflanze, und Holz: Ein neuer Text zur 'medizinischen Astrologie,'" 1–22.

87. BM 34035: 38–39, cited Livingstone, *Mystical and Mythological Explanatory Works of Assyrian and Babylonian Scholars*, 73. Livingstone takes *ṣītu* as the singular of *ṣâtu* "explanatory word list(s) (always in the plural)," which seems to be the only likely reading.

88. LBAT 1593 lines 17'–18,' see Reiner, "Early Zodiologia and Related Matters," 422. Reiner suggested that *umāmu* "animal" in LBAT 1593: 17' referred to the animals of the zodiacal signs, ibid., 424.

89. AMT 105: 21–24, also Hunger, *Babylonische und Assyrische Kolophone*, Nr. 533.

90. AMT 102: 20–25, and see Schuster-Brandis, *Steine als Schutz- und Heilmittel*, 133.

91. AMT 102: 1–6.

92. Examples can be found in VAT 7815 and 7816, see Weidner, *Gestirn-Darstellungen auf babylonischen Tontafeln*, 41–52; K 11151; ibid., 39–40 and Tf. 17; W 20030/133, see Hunger, "Noch ein Kalendertext," 43, with numbers alone; LBAT 1586 and 1587, Hunger, "Noch ein Kalendertext," 40–43, numbers alone; BM 96258 and BM 96293, see Brack-Bernsen and Steele, "Babylonian Mathemagics: Two Mathematical Astronomical-Astrological Texts," 96–97, only numbers; BM 36995 in Brack-Bernsen and Steele, ibid., 98–99, note 3, only numbers; LBAT 1593 in Reiner, "Early Zodiologia," 421–27; also mentioned in Boiy, *Late Achaemenid and Hellenistic Babylon*, 27; the Uruk tablets von Weiher SpTU III 104/ W. 22704 (Month IV) and SpTU105 /W.22619/6 + 22554/26 (Month VIII), and W. 20030/127 (Month II) in Hunger, "Die Tafeln des Iqīša," 163–65, and Oelsner, "Von Iqīša und eigenen anderen spätgeborenen Babyloniern," 797–813. Other related sources include BM 36326 where the *dodecatemoria* scheme is put into the same format as the Kalendertexte scheme, BM 47851, which has both schemes, in Hunger, "Ein astrologisches Zahlenschema," 191–96. The medical astrological BM 56605, edited by Heeßel in *Babylonisch-assyrische Diagnostik*, 112–30, 468–69 and pls. 1–2, and discussed in Wee, "Discovery of the Zodiac Man in Cuneiform," 225–33, belongs to this group of texts as well.

93. These texts and their number schemes are explicated in Brack-Bernsen and Steele, "Babylonian Mathemagics." See also Hunger and Pingree, *Astral Sciences in Mesopotamia*, 29–30.

94. Edited in Geller, "Look to the Stars," 54–56; Geller, *Melothesia*, 27–46 and 85–89. See also Texts 55 and 56 in Finkel, "On Late Babylonian Medical Training," 212–17, that assign the application of stone in a particular oil, a plant to be ingested in a particular liquid, and a colored wool to tie on an amulet according to the month (Text 55), and the application of stones, plants, and woods to certain days of the month (Text 56).

95. Translated by Geller, "Look to the Stars," 78; Geller, *Melothesia*, 87–88.

96. Scurlock, "Sorcercy in the Stars," 125–46.

97. *Šammu šikinšu* Text I, Source A (= STT 93): 43'–45' para. 18' in Stadhouders, "Pharmacopoeial Handbook *šammu šikinšu*: An Edition," 3–51, and Stadhouders, "Pharmacopoeial Handbook *šammu šikinšu*: A Translation," 3 for translation.

98. Notable examples are LBAT 1596–1598, BRM 4 19 and 20, K 11151 +, discussed in Heeßel, "Stein, Pflanze, und Holz"; Heeßel, "Astrological Medicine in Babyonia"; Geller, "Look to the Stars."

99. SpTU 14, 159, see Koch-Westenholz, *Babylonian Liver Omens*, 24–25.

100. BRM 4 20, with BRM 4 19 (and parallels with STT 300 although without the zodiacal references), see the discussion in Geller, "Look to the Stars," 25–54.

101. Neugebauer and Sachs, "The 'Dodecatemoria' in Babylonian Astrology," 65–66.

102. Reiner, "Early Zodiologia," 424.

103. Reiner, *Astral Magic in Babylonia*, 116.

104. See the contents of the text laid out in tabular form in Steele, "Astronomy and Culture in Late Babylonian Uruk," 337.

105. Reiner, *Astral Magic*, 115–16.

106. Heeßel, "Stein, Pflanze, und Holz," 19.

107. Ibid., 16–17.

108. Ibid., 18 and notes 35–37 for further references.

109. Pingree, "Some of the Sources of the Ghāyat al-Hakīm," 1–15, especially 2, 5–6.

110. Pliny, *Nat. Hist.* 37.169.

111. Scarborough, "Alexander of Tralleis," 58–59, and Reed, "Zachalias of Babylon," 843.

112. Rochberg, "Sudines," 767–68.

113. Ibid.

114. Bowen and Goldstein, "Meton of Athens and Astronomy in the Late Fifth Century B.C.," 75, my emphasis.

115. Holton, "On the Art of the Scientific Imagination," 192.

116. See note 56 above, and chapter 6 below.

117. Stich, "Could Man Be an Irrational Animal," 126.

118. Kahneman and Tversky, "On the Psychology of Prediction," 237.

119. Lloyd already noted the importance of analogy in Zande ideas about disease and treatment by magical means in *Polarity and Analogy*, 177–79.

120. Nisbett and Ross, *Human Inference: Strategies and Shortcomings of Social Judgement*, 115–16; see also Stich, "Could Man Be an Irrational Animal," 126.

121. Stich, "Could Man Be an Irrational Animal," 126.

122. Heeßel, "Hermeneutics of Mesopotamian Extispicy: Theory vs. Practice," 20.

123. De Zorzi, "Omen Series Summa Izbu: Internal Structure and Hermeneutic Strategies," 43–75.

124. Livingstone, *Mystical and Mythological Explanatory Works*, 2, and for *iNAMgišḫurankia*, 17–52.

125. Descola, "Cognition, Perception, and Worlding," 338.

126. Schwemer, "Magic Rituals: Conceptualizations and Performance," 418.

127. Tambiah, *Magic, Science, Religion, and the Scope of Rationality*.

128. Schwemer, "Magic Rituals," 420.

129. Ibid., 421–23.

130. Robson, "Mesopotamian Medicine and Religion," 463.

131. Henry, "Fragmentation of Renaissance Occultism and the Decline of Magic," 8.

132. Ibid., and see his note 29 for bibliography.

133. Descola, *Beyond Nature and Culture*, 82, and note 45 for Durkheim.

134. Ibid.

135. Heeßel, "Stein, Pflanze, und Holz," 17–18.

136. Schwemer, "Magic Rituals," 419, citing Versnel, "Some Reflections on the Relationship Magic-Religion," 181.

CHAPTER SIX

1. Papineau, "Naturalism."

2. Ibid.

3. Ibid.

4. Rochberg-Halton, *Aspects of Babylonian Celestial Divination*, 51–55.

5. Thorndike, "True Place of Astrology in the History of Science," 273.

6. Ibid., 274.

7. Cicero, *De Senectute*, 23, *De Divinatione*, i.3, ii.43, *De Natura Deorum*, i.15, *De Officiis*, iii.12, 13, 23; *De Finibus*, iii.10, 15. For Diogenes, see Lehoux, "Diogenēs of Babylōn," 253.

8. Salmon, *Introduction to the Philosophy of Science*, 35.

9. There are other sources for a wider investigation of causal thinking in Akkadian, primary among them the causative Š-stem of the verb, or the use of causal conjunctions such as *ištu*, *kīma*, or *aššum*.

10. Allen, *Inference from Signs: Ancient Debates about the Nature of Evidence*. Sextus Empiricus (*Outlines of Pyrrhonism* 2.104), for example, gives the Stoic definition of a sign as "a proposition antecedent in a sound conditional and revelatory of the consequent," a definition on which a Babylonian scribe could probably agree.

11. BRM 4 13:65.

12. YOS 10 56 i 11; Leichty, *Omen Series Šumma Izbu*, 202.

13. Reiner and Pingree, *Babylonian Planetary Omens 4*, 40:1.

14. Hunger, *Spätbabylonische Texte aus Uruk*, vol. I, Text 94:12–14.

15. Anscombe, "Causality and Determination," 88.

16. Note the lexical equivalents *ṭēmu* and *alaktu* for *a.rá*; see Abusch, "*Alaktu* and *Halakhah*," 19, and further discussion, 18–23.

17. Ibid., 17.

18. For further discussion, see Maul, "How the Babylonians Protected Themselves against Calamities Announced by Omens," 123–29, and Veldhuis on the nature of magical language in "Poetry of Magic," 35–48.

19. See Reiner's discussion of *purussâ parāsu* as a technical term in divination in "Fortune-Telling Mesopotamia," 25; Rochberg, *Heavenly Writing*, 194–96 and 266–67; Fincke, "Omina, die göttlichen 'Gesetze' der Divination," 131–47.

20. Rochberg-Halton, *Aspects of Babylonian Celestial Divination*, chapter 10 passim. See also ACh Suppl.I 1:1–8, and the discussion in Rochberg, "Lunar Data in Babylonian Horoscopes," 36 and note 7.

21. Brack-Bernsen and Hunger, "TU 11: A Collection of Rules for the Prediction of Lunar Phases and of Month Lengths," 12.

22. Gudea *Cyl.*A iv 26 and v 23, see Edzard, *Gudea and His Dynasty*, 72.

23. For "Blessing of Nisaba," see Hallo, "Cultic Setting of Sumerian Poetry," 125: 29–31; Sjöberg and Bergmann, *Collection of the Sumerian Temple Hymns*, 49: 538–39; Horowitz, *Mesopotamian Cosmic Geography*, 166–67. For MUL = *šiṭirtum*, see CAD s.v. lexical section.

24. Sjöberg and Bergmann, *Collection of the Sumerian Temple Hymns*, 138b, citing MSL II p. 132 VI 57 mul = *šiṭirtum*. Nisaba holds the "holy tablet of the heavenly star/writing (dub-mul-an-kù)" as well in the composition "Nisaba and Enki" lines 29–33; see Hallo, "Cultic Setting of Sumerian Poetry," 125, 129, and 131. In their discussion of the term *lumāšu* "constellation," used in the sense of a form of writing with astral pictographs or "astroglyphs," as they have been called, Roaf and Zgoll note that Sumerian mul "star" (or mul-an "heavenly star") "can refer both to a star in the

sky and to a cuneiform sign on a tablet"; see Roaf and Zgoll, "Assyrian Astroglyphs," 289. The notion of the god (often Šamaš) "writing" the signs on the exta of sheep is well-known, e.g., *ina libbi immeri tašaṭṭar šērē tašakkan dīnu* "you (Šamaš) write upon the flesh inside the sheep (i.e., the entrails), you establish (there) an oracular decision," OECT 6 pl. 30 K.2824:12.

25. In the following inscriptions of Nebuchadnezzar: VAB 4 187 i 39, also 74 ii 2, YOS 1 44 i 21; cf. BBSt. No. 5 ii 28. Also in the form *šiṭir burūmê*, literally "writing of the firmament," for which, see CAD s.v. *burūmû* usage b, predominantly in Neo-Assyrian royal inscriptions, but also in a hymn to Aššur, see Livingstone, *Court Poetry and Literary Miscellanea*, 4 Text No. 1: 21; Horowitz, *Mesopotamian Cosmic Geography*, 15, note 25, and 226.

26. Plotinus, *Ennead* 3.1.6. See also the remarks of Miller, "In Praise of Nonsense," 497, that "astrologers . . . found more and more correspondences between human writing and heavenly phenomena; when they contemplated the skies, they saw what one modern scholar has called 'Himmelsschrift,' a celestial text whose lights formed the moving script of divine order."

27. Translation from Foster, *Before the Muses*, 437 (*EnEl* IV 23–24).

28. *ṭēm ilūtiki rabīti* "your great divine decision," STT 73: 19, 33, and 41, in Reiner, "Fortune-Telling in Mesopotamia," 32.

29. Lambert, *Babylonian Oracle Questions*, 6.

30. The relationship between the specifically divinatory hermeneutical practice denoted by Akkadian *pišru* and the practice of pesharim in the Qumran community has been noted and discussed by Nissinen in "Pesharim as Divination: Qumran Exegesis, Omen Interpretation, and Literary Prophecy," 43–60.

31. CAD s.v. *pašāru*, meaning 8. Cf. the interpretation of Zgoll, *Traum und Welterleben im antiken Mesopotamien*, 383–96, in which *pašāru* means to be released from rather than to recount a dream.

32. See for example in Cicero, *De Divinatione* I.125–27, *De Fato* 13–14, and Sextus Empiricus *Adversus mathematicos* VIII.

33. Sorabji, *Necessity, Cause, and Blame*, chapter 4, "Stoic Embarrassment over Necessity."

34. Hume 2.3 1739 book I part III sections IV and XIV.

35. Sorabji, *Necessity, Cause, and Blame*, 37.

36. Hempel, *Aspects of Scientific Explanation and Other Essays in the Philosophy of Science*, chapter 12.

37. Taylor, *Explanation and Meaning: An Introduction to Philosophy*, 8.

38. Ibid.

39. Sorabji, *Necessity, Cause, and Blame*, 39.

40. STT I No. 71: 20, and see Lambert, "Sultantepe Tablets: A Review Article," 135.

41. SAA 10 278: 12–rev. 7.

42. Bobzien, *Determinism and Freedom*, 165, commented on this particular passage as follows: "we do not know how far this reflects early Stoic thought, but it comfortably fits with all we know about early Stoic theories of divination and causation." Cf. *De Div.* I.12–13, 16, 23, 25, II.47.

43. Bobzien, *Determinism and Freedom*, 6.

44. Augustine, *Civ.* V 1, 191.25–34, see also Bobzien, *Determinism and Freedom*, 166: "Now it could be said that the stars indicate those <human actions> rather than bring them about, so that their position is *some kind of speech* which foretells the future, and not an active power . . . but the astrologers do not usually say, for example, 'Mars in this position indicates a murderer,' but 'brings about a murderer.' However, let us concede that they do not express themselves as they should, and that they ought to take from the philosophers the rule of how to formulate their predictions of what they believe they find in the position of the stars" (my emphasis).

45. Cicero, *De Fato* VI 12–VII 14, and Sorabji, *Necessity, Cause, and Blame*, 72–78.

46. See Sharples, "Soft Determinism and Freedom in Early Stoicism," 266–67, where he discusses the Stoic chain of causes and its various identification or distinction from the idea of god as *pneuma*, the active principle in the universe. And despite philosophical argument and polemics among later Greek or Greco-Roman philosophers on the nature and effect of causes, divine cau-

sality remained a principal and pervasive conception throughout the ancient world of West Asia and the Mediterranean, as discussed in Lehoux, "Tomorrow's News Today: Astrology, Fate, and the Way Out," 109–11.

47. Zilsel, "Genesis of the Concept of Physical Law," 245–79, places the origins of the idea of nature as fully and independently nomological, and as emerging from the view of natural phenomena being law-like as a result of obeying God's commands, "in the period of Descartes, Hooke, Boyle and Newton" (p. 247). Ruby, "Origins of Scientific 'Law,'" 341–59, pushes the idea further back into the Middle Ages, saying, "although prima facie, the explanation of scientific 'law' as arising from the idea of divine legislation is highly plausible, it is for the most part mistaken. The idea of legislation by God or Nature does account for much of ancient use of 'law' for natural phenomena. However, the modern use emerged through different processes at different times in three distinct fields, in only one of which the idea of divine legislation had any part. In all three it appeared before 1543" (p. 342).

48. Needham, "Human Laws and Laws of Nature in China and the West," 3; Bodde, "Evidence for 'Laws of Nature' in Chinese Thought," 709, citing Needham, *Science and Civilization in China*. Vol. 2, *History of Scientific Thought*, 518.

49. Needham, "Human Laws and Laws of Nature in China and the West," 18.

50. Ibid., 20.

51. Striker said the idea that the natural law theory was invented by the Stoics "would be an exaggeration," but they "maintained that the reason which governs the universe can be described as a universal lawgiver—it prescribes what ought to be done, and prohibits what must not be done. In Cicero's solemn words, 'law is the highest reason, implanted in nature, which commands what ought to be done and forbids the opposite.'" See Striker, "Following Nature: A Study in Stoic Ethics," 35, citing Cicero, Leg. I.18.

52. Bodde, "Chinese 'Laws of Nature': A Reconsideration," 139–55.

53. Schofield, *Stoic Idea of the City*, 103.

54. See Lehoux, *What Did the Romans Know?* chapter 6, "Law in Nature, Nature in Law," and for the Stoics in particular, 53–59.

55. col. xl 1.

56. See CAD s.v. *kittu* A, meaning 1 usage b. See also Landsberger, "Die babylonischen Termini für Gesetz und Recht," 219–34.

57. Wilcke, *Early Ancient Near Eastern Law: A History of Its Beginnings; The Early Dynastic and Sargonic Periods*, 36.

58. See the address to the River (Euphrates), line 7: "You judge the case of mankind," in King STC I 200f. And KAR 294, see also King STC, 129.

59. Westbrook, "Judges in the Cuneiform Sources," 210.

60. The attribution of the power of judicial command and decree to gods is also a part of Sumerian religious discourse, the full exposition of which would take us far afield. To illustrate, see, for example, ETCSL 2.5.4.13, an adab hymn to Nanna for Išme-Dagan, Segment A 17–22: After you have extended yourself in the bright . . . , the daylight . . . , after you have established . . . on earth, on the day of the disappearance of the moon, as you have completed the month, you summon (?) the people, lord; and then in the nether world you decree great judgments, you decide sublime verdicts. Enki and Ninki, the great lords, the great princes, the lords who determine fates, await your utterances, father; they . . . the newborn (?) moon.

61. Craig ABRT 1 56:14, cited CAD s.v. *kittu* 1b.

62. Other gods can extend the rod and ring toward royal figures. See Wiggermann, "Ring und Stab (Ring and Rod)," 414–21. See also Rochberg, "Canon and Power in Cuneiform Scribal Scholarship."

63. Stolleis, "Legitimation of Law through God, Tradition, Will, Nature, and Constitution," 45, put it that the metaphor is "indicative of something that is not, in itself, law, but which underpins law, lending it validity and, as it were, making it 'right' or 'just.'"

64. For a summary and discussion of the debate concerning the legislative status of the cunei-

form law codes, see Westbrook, in *Law from the Tigris to the Tiber: The Writings of Raymond Westbrook*. Vol. 1, *The Shared Tradition*, 77–95; Westbrook, "Law Codes and Omen Series: Practical Application," 9–20.

65. Yoffee, "Context and Authority in Early Mesopotamian Law," 106, my emphasis.

66. Speiser, "Cuneiform Law and the History of Civilization," 537.

67. Goetze, "An Old Babylonian Prayer of the Divination Priest," 24, lines 9–12. See also Foster, *Before the Muses*, 209.

68. HSM 7494: 11–12; Starr, *Rituals of the Diviner*, 30 and 37. See also line 7 for "the case of so-and-so."

69. Lambert, *Babylonian Oracle Questions*, 5.

70. Bottéro, "Symptômes, signes, écritures," 139–42.

71. Rochberg, *Heavenly Writing*, 109 with note 44, and 266–67. Bottéro already described unprovoked divination as "judicial," see *Mesopotamia: Writing, Reasoning, and the Gods*, 141–42, where he says (p. 142), "each oracle [omen] was like a verdict against the interested parties on the basis of the elements of the omen, just as each sentence by a tribunal established the future of the guilty person based upon the dossier submitted to its judgment. The divinatory future, the predicted future, was what had to be expected at the moment that the gods publicized their decision by means of and in the omen."

72. ZA 43 306: 8. For discussion of the two senses of "heaven's interior," one located below the horizon and the other in the faraway (hence invisible) heavens, see Heimpel, "Sun at Night and the Doors of Heaven in Babylonian Texts," 130–32.

73. Horowitz, "Astral Tablets in the Hermitage," 194–206; Dossin, "Prières aux 'Dieux De La Nuit,'" 179–87, lines 5–8.

74. Steinkeller, "Of Stars and Men," 23–26.

75. Abusch, *Mesopotamian Witchcraft*, 124, note 23.

76. Ibid., 125.

77. Abusch, "*Alaktu* and *Halakhah*," 15–42.

78. LKA 139 rev. 25, cited in Abusch, "*Alaktu* and *Halakhah*," 21.

79. Burrows, "Hymn to Ninurta as Sirius (K 128)," 35:2–36:6 (pl. III).

80. Leichty, *Royal Inscriptions of Esarhaddon, King of Assyria (680–669 BC)*, 106, Esarhaddon 48: 58–59.

81. Ibid., 121, Esarhaddon 57 i 3'–8.'

82. SAA 3 25 ii 20'–27.'

83. The intertextual nature of *Enūma Eliš* and the Astrolabes raises the question of their respective dates of composition. The debate has not been resolved on the date of composition of *Enūma Eliš*, though arguments have been put forward for the latter part of the second millennium, perhaps the time of Nbk I (1125–1103), see Lambert, "Studies in Marduk," 4–6; for a later date, in the first millennium, see Abusch, s.v. "Marduk," in *Dictionary of Deities and Demons in the Bible*, 546–48. The earliest exemplar of the Astrolabe comes from a Near Assyrian copy datable to the time of TP I (1115–1077), though Horowitz has suggested that its composition is even Old Babylonian, in *Cosmic Geography*, 155. It is certain that this astronomical text underlies EnEl V 1–22.

84. For the emendation *a-<gu>-ú* in EnEl V 14, see Lambert in O. Kaiser, ed., *Texte aus der Umwelt des Alten Testament* (Gütersloh, 1982–), III/4, 588.

85. EnEl VII 130.

86. *ú-[ad-di*(?)] GISKIM, *EnEl* V 23, restoring a form of *idû*, with Glassner, "Droit et divination," 45, note 40, citing the parallel from Clay, YOS 1 45 I 7: *ú-ad-di it-ta-šu*.

87. EnEl 23–24.

88. CAD s.v. *milku* meaning 1 usage c.

89. See PSD s.v. galga, where [ĝalga] = [ĝá-al-ga] = [É×GAR] = *mi-el-ku* OB Aa 223:1.

90. For a.rá in this meaning, see OB Diri Nippur 316, and discussion in Abusch, "*Alaktu* and *Halakhah*"; Brown, "Astral Divination in the Context of Mesopotamian Divination," 77, note 22.

91. Laessoe, "A Prayer to Ea, Shamash, and Marduk, from Hama," 63. See duplicates, LKA 109: 1–8, and dupl. JRAS 1929 285: 1–6.

92. See "Enlil in the E-kur," ETCSL 4.05.1: 1–2: Enlil's commands are by far the loftiest, his words are holy, his utterances are immutable. The fate he decides is everlasting.

93. From a different point of view, Eichler, "Ethics and Law Codes, Mesopotamia," 518, sees these principles as transcending the divinities themselves, considering them a "moral cosmic standard" and "an innate force" of the universe.

94. The view that the stars form a decipherable language, written by the gods upon the sky as though on a tablet, is a trope that continues within the history of astrology well into the early modern period. Grafton, *Cardano's Cosmos*, 6, points out that, "Giovanni Gioviano Pontano, who published his treatise *On Celestial Things* in 1512, argued explicitly that the language of the stars conformed in all essential ways to the language of humans. . . . Stars and planets, Pontano argued, formed the letters of a cosmic alphabet. . . . Every planet . . . played the role of a letter with defined qualities. Every astrologically significant configuration of two or more planets—for example, when two of them met, or came into conjunction—resembled a word or a phrase, the sense of which the astrologer could determine." See also Rochberg, *In the Path of the Moon*, 304–5, and 417–18.

95. Cicero, *On the Republic* 3.33.

96. Cicero, *Laws*, vol. 16, I.3.

97. Philo, *On Moses* 2.48. In fact, Koester, "NOMOS PHYSEOS: The Concept of Natural Law in Greek Thought," 521–41, argued for Philo being the earliest philosopher to articulate the theory of natural law. Horsley, however, in his "Law of Nature in Philo and Cicero," 35–59, proposed an earlier voice on natural law in Cicero, which derived from a Stoic source on "universal law and right reason" (p. 36), namely, in Antiochus of Ascalon.

98. Armogathe, "Deus legislator," 266.

99. Ibid., 265, italics in the original.

100. Milton, "Laws of Nature," 684.

101. Quine, *From Stimulus to Science*, 25.

102. Ibid.

103. Ibid., 26.

104. Ibid.

105. Beebee, "Causes and Laws: Philosophical Aspects," 1574.

106. Zilsel, "Genesis of the Concept of Physical Law," 246.

107. Ibid., 247.

108. Rochberg, *Heavenly Writing*, 53, and Fincke, "Omina, die göttlichen 'Gesetze' der Divination," 131–47, both going back to Kraus, "Ein zentrales Problem des altmesopotamischen Rechtes: Was ist der Codex Hammu-rabi?" 288.

109. Chisholm, "Law Statements and Counterfactual Inference," 99.

110. Lewontin, review of Kay, 1263, said, "It seems impossible to do science without metaphors. . . . But the use of metaphor carries with it the consequence that we construct our view of the world, and formulate our methods for its analysis, as if the metaphor were the thing itself."

CHAPTER SEVEN

1. Daston and Lunbeck, introduction to *Histories of Scientific Observation*, 1.

2. Ibid.

3. Hanson, *Patterns of Discovery: An Inquiry into the Conceptual Foundations of Science*, 6.

4. Ibid., 19.

5. For the most recently available transliteration and translation of the Great Star List, see Koch-Westenholz, *Mesopotamian Astrology: An Introduction to Babylonian and Assyrian Celestial Divination*, appendix B, 187–205.

6. Ibid., 190–91, lines 83–113.

7. Ibid., 202–3, lines 281–86.

8. *kakkabāni tamšīli[šunu ēṣ]irū lumāši* AfO 17 89: 5, and see Rochberg-Halton, *Aspects of Babylonian Celestial Divination*, 271 and note 3. See further in Rochberg, "Stars Their Likenesses: Perspectives on the Relation between Celestial Bodies and Gods in Ancient Mesopotamia," 64–83.

9. The term may refer to the ritual lamentation that took place in the event of an eclipse. Discussion of the background for ritual weeping and lamentation can be found in Zgoll and Lämmerhirt, "Lachen und Weinen im antiken Mesopotamien," 453–67. For rituals associated with a lunar eclipse, see Brown and Linssen, "BM 134701 = 1965-10-14,1 and the Hellenistic Period Eclipse Ritual from Uruk," 147–66; Beaulieu and Britton, "Rituals for an Eclipse Possibility in the 8th Year of Cyrus," 73–86; Linssen, *Cults of Uruk and Babylon*, 109–17. See also the *namburbi* for an eclipse in Maul, *Zukunftsbewältigung*, 458–59.

10. The "Heaven of Anu" is discussed in Horowitz, *Mesopotamian Cosmic Geography*, 10, and 244–50, the *Abzu* (*apsû*) in ibid., passim.

11. Reiner and Pingree, *Babylonian Planetary Omens* 4, 28.

12. Steele, *Observations and Predictions of Eclipse Times by Early Astronomers*; Steele, "Eclipse Prediction in Mesopotamia," 421–54.

13. Cancik-Kirschbaum called them "thought experiments," in "Gegenstande und Methode: Sprachliche Erkenntnistechniken in der keilschriftlichen Überlieferung Mesopotamiens," 38.

14. Grant, *God and Reason in the Middle Ages*, 357–58.

15. Grant, *Nature of Natural Philosophy in the Late Middle Ages*, 222; Grant, *God and Reason in the Middle Ages*, 179–80. The quote from Murdoch is from "The Analytic Character of Late Medieval Learning: Natural Philosophy without Nature," 174. Note that in the ellipsis, Murdoch said: "in a very important way natural philosophy was not about nature," although this position has been countered by Park, in "Natural Particulars: Medical Epistemology, Practice and the Literature of Healing Springs," 347–67 and Park, "Observation in the Margins, 500–1500," 15–44.

16. Note the nominalization of *šumma* "if" in a tablet subscript to refer to medical omens, in the form ŠUM.MA.ME (*šummū*) "the ifs," in Finkel, "Adad-apla-iddina, Esagil-kīn-apli, and the Series SA.GIG," 152 and note 82.

17. Rochberg-Halton, *Aspects of Babylonian Celestial Divination*, 232.

18. Ibid., 197–98.

19. Ibid., 191.

20. Pongratz-Leisten, "King at the Crossroads between Divination and Cosmology," 41.

21. Hunger, "Uses of Enūma Anu Enlil for Chronology," 155–58.

22. CAD s.v. *u'iltu* meaning 1, and the description of the phenomena found in the reports (and the letters from the scholars to Esarhaddon and Assurbanipal) in Hunger and Pingree, *Astral Sciences*, 116–38.

23. SAA 10 76: 7-rev. 10.

24. SAA 10 114: 5–11.

25. SAA 8 126–42.

26. Reiner, "Constellation into Planet," 3–15. Cf. Frahm, *Babylonian and Assyrian Text Commentaries*, 79.

27. Kugler suggested that the Saros Tablet could be projected back to year 1 of Nabonassar; see Kugler, *Sternkunde und Sterndienst in Babel*, vol. 2, 364.

28. Grayson, *Assyrian and Babylonian Chronicles*, 8. The relationship of Diaries to Chronicles is discussed in Rochberg-Halton, "Babylonian Astronomical Diaries," 324–25.

29. Ibid., Chronicle 17, the "Religious Chronicle," which mentions a wolf, a badger, a panther and deer within city limits.

30. A single Diary from Uruk is extant from –463, see Sachs and Hunger, ADRT 1, 54–55.

31. Chalmers, *What Is This Thing Called Science?* 11.

32. Ibid., 29.

33. Neurath, "Foundations of the Social Sciences," 5 and 23.

34. Quine, "In Praise of Observation Sentences," 107–16.

35. Ibid., 109.

36. For a general effort to explore the empirical in cuneiform studies, see Selz and Wagensonner, *Empirical Dimension of Ancient Near Eastern Studies / Die empirische Dimension altorientalischer Forschung*.

37. Sahlins, *How "Natives" Think, about Captain Cook, For Example*, 153.

38. Obeyesekere, *Apotheosis of Captain Cook: European Mythmaking in the Pacific*.

39. Sahlins, preface to *How "Natives" Think*, ix.

40. Lukes, "Different Cultures Different Rationalities?" 7, quoting from Sahlins, *How "Natives" Think*, 6–7, 153, and 164.

41. Sahlins, *How "Natives" Think*, 162.

42. Armstrong, "Naturalism, Materialism, and First Philosophy," 261. See the extended discussion in Van Fraassen, *Empirical Stance*, 53–63.

43. For discussion of the "cattle pen" as a metaphorical expression for the horizon, see Rochberg, "Sheep and Cattle, Cows and Calves: The Sumero-Akkadian Astral Gods as Livestock," 347–59. On the TÙR/*tarbaṣu* meaning "halo (lunar and solar)," see Verderame, "Halo of the Moon," 91–104.

44. Brack-Bernsen, "Path of the Moon, the Rising Points of the Sun, and the Oblique Great Circle on the Celestial Sphere," 16–31.

45. Enūma Anu Enlil 50 Text III 24b rev. 16–17, in Reiner and Pingree, *Babylonian Planetary Omens* 2, 43.

46. SAA 10 363: 7–rev. 18.

47. SAA 8 456: 1–2, and cf. SAA 8 339: 1–2.

48. Hunger, ADRT 3, 315,-119 B_1 rev. 4.'

49. *MUL.APIN* I iv 31–37, and Hunger and Pingree, *Astral Sciences*, 67–69.

50. *MUL.APIN* II i 1–8; see Hunger and Pingree, *Astral Sciences*, 70–71; Brack-Bernsen, "Path of the Moon."

51. Transliterations and translations of the celestial omen reports are found in SAA 8 = Hunger, *Astrological Reports to Assyrian Kings*. The letters are edited in SAA 10 = Parpola, *Letters from Assyrian and Babylonian Scholars*.

52. SAA 8 216: bottom edge 2'–rev. 6.

53. SAA 10 149: 3–rev. 5.

54. On this scribe and his post, see Luukko, "Administrative Roles of the 'Chief Scribe' and the 'Palace Scribe' in the Neo-Assyrian Period," 227–56.

55. SAA 8 9 rev. 6–9: 4.

56. SAA 8 3.

57. SAA 10 45: 6–rev. 4.

58. For an edition of Tablet 1 of this at least seven-Tablet series, see Koch-Westenholz, "Astrological Commentary Šumma Sîn ina Tāmartīšu Tablet I," 159–65. See also Gehlken, "Die Serie DIŠ Sîn ina tāmartišu im Überblick," 3–5. For an argument that the reports made particular use of *Šumma Sin ina tāmartišu*, see Veldhuis, "Theory of Knowledge and the Practice of Celestial Divination," 77–91.

59. SAA 8 7: 5–rev. 4.

60. For the lunar omens, see Verderame, *La Tavole I–VI della serie astrologica Enuma Anu Enlil*; Rochberg-Halton, *Aspects of Babylonian Celestial Divination*; for discussion of eclipses in Enūma Anu Enlil, see further, Montelle, *Chasing Shadows: Mathematics, Astronomy, and the Early History of Eclipse Reckoning*, 54–58.

61. Rochberg, *In the Path of the Moon*, 257–70.

62. Livingstone, *Hemerologies of Assyrian and Babylonian Scholars*.

63. Schwemer, "Washing, Defiling, and Burning: Two Bilingual Anti-Witchcraft Incantations," 6, note to line 6.'

64. Cohen, *Cultic Calendars of the Ancient Near East*, 327 and 391–2.

65. SAA 8 342: 1–2.

66. SAA 8 207: 4–rev. 8, and cited in Hunger and Pingree, *Astral Sciences*, 121–22.

67. For discussion of the problem of the accuracy of measurement by the water clock, see Brown, Fermor, and Walker, "Water Clock in Mesopotamia," 142–44.

68. Sachs and Hunger, ADRT 1, -651: 6.

69. Ibid., Text -381: 7.

70. Rochberg-Halton, *Aspects of Babylonian Celestial Divination*, 77.

71. Ibid., 180.

72. Ibid., 188. For other examples, ibid., chapter 10 passim.

73. Brack-Bernsen and Hunger, "TU 11: A Collection of Rules for the Prediction of Lunar Phases and of Month Lengths," 26.

74. SAA 8 316: 3–5.

75. Rochberg-Halton, *Aspects of Babylonian Celestial Divination*, 16, and Rochberg, *Heavenly Writing*, 78 and 222–23.

76. Rochberg-Halton, *Aspects of Babylonian Celestial Divination*, 133.

77. Ibid., 233.

78. Ibid., 241.

79. SAA 8 46: 1–rev. 1.

80. SAA 8 321: 1–rev. 3.

81. SAA 8 316 rev. 3–5.

82. SAA 10 57: 4–9.

83. SAA 10 75: 8–rev. 7.

84. Verderame, *Le Tavole I–VI*, 9 and 13, I.4.1 § I.2.

85. SAA 8 147: 1–rev. 8.

86. Brack-Bernsen and Hunger, "TU 11," 11 and 16.

87. Soldt, *Solar Omens of Enuma Anu Enlil: Tablets 23(24)–29(30)*.

88. Ibid., 5, EAE 23(24) I 1.

89. Ibid., 29, EAE 24(25) III 29.

90. EAE 33 §I.1–3. See ACh Šamaš 10: 1–3 and dupls. Craig AAT 93–94 ii 28–31, Craig AAT 28: 1–4, and Craig AAT 29: 1–3.

91. EAE 32 §VIII.14. Sources include ACh Šamaš 9: 30, and dupls. ACh Šamaš 9–10 i 9,' K. 9225: 17' (unpub) and K.9233+ 14' (unpub), cited with the kind permission of the Trustees of the British Museum. See also EAE 33 §I.25 and §III.15, which assigns plague for the land.

92. EAE 33 §VI.15. Note that Jupiter is written ^{MUL}UD.AL.TAR. Sources include ACh Šamaš 10 rev. 20,' and dupls. Craig AAT 93–94 rev. iv 16,' Craig AAT 28 rev. 18,' Craig AAT 29 rev. 3,' and UCP 9 9 rev. 4.'

93. EAE 32 § X.14. Sources include ACh Šamaš 9–10:35,' and dupls. K. 9225+ rev. 8 (unpub), and K. 2208+ ii 6 (unpub, later join to Rm 2,136, in ACh Šamaš 8), cited with the kind permission of the Trustees of the British Museum.

94. EAE 33 § I.7. Sources include ACh Šamaš 10: 9–10, and dupls. Craig AAT 93–94 ii 37–38, Craig AAT 28: 8–9, and Craig AAT 29: 7–8.

95. SAA 8 5: 1–8.

96. SAA 8 83: 4–rev. 3.

97. SAA 8 27 rev. 3.

98. SAA 8 145: 3.

99. SAA 8 391: 5.

100. Reference to *kakkabū minâti* and *kakkabū la minâti* occurs in the Venus omens of *Enūma Anu Enlil*, translated by Reiner and Pingree in *Babylonian Planetary Omens 3*, 13 and passim, as "counted" and "uncounted" stars.

101. Epping, *Astronomisches aus Babylon*, 115.

102. Hunger and Pingree, *Astral Sciences*, 148–49. A detailed discussion of the Normal Stars

and their usage is found in Jones, "A Study of Babylonian Observations of Planets Near Normal Stars," 475–536.

103. BM 46083 was first published in Sachs, "A Late Babylonian Star Catalogue," 146–50. For both catalogs, see Roughton, Steele, and Walker, "A Late Babylonian Normal and *Ziqpu* Star Text," 537–72. See also the discussion in Steele, "Celestial Measurement in Babylonian Astronomy," 304–6. Equivalent ecliptical longitudes of the Normal Stars are discussed in Hunger and Pingree, *Astral Sciences*, 149–51, and Britton, "Studies in Babylonian Lunar Theory: Part III. The Introduction of the Uniform Zodiac," 617–63.

104. Comparison of ancient and modern positions of the planets shows that the cubit was somewhat larger than 2°; see Steele, "Celestial Measurement," 297–98.

105. Ibid., 308–10.

106. TCL 6 20: 13.

107. LBAT 1591: 1–4.

108. Neugebauer, "Alleged Babylonian Discovery of the Precession of the Equinoxes," 1–8, and Steele, "Celestial Measurement," 310.

109. Steele and Gray, "A Study of Babylonian Observations Involving the Zodiac," 443–58.

110. Slotsky, *Bourse of Babylon*; Boiy, *Late Achaemenid and Hellenistic Babylon*, 237–39; Geller, review of Slotsky, 409–12 on the identification of *kasû* and *saḫlû*.

111. The best summary treatment of this corpus is in Hunger and Pingree, *Astral Sciences*, 139–59.

112. Hunger and Pingree, *Astral Sciences*, 144.

113. Sachs and Hunger, ADRT 1, -651: 1–8.

114. Sachs and Hunger, ADRT 1, -567: 1–4.

115. Although this expression seems fairly localized to the Diaries, a fragmentary text from Uruk that mentions years 6–8 of Cambyses also uses it (NU PAP), perhaps in reference to observations of Mercury; see Hunger *Spätbabylonische Texte aus Uruk*, 1: 101–2, Text No. 100: 7' and rev. 9' and 10.'

116. Britton, "An Early Observation Text for Mars," 33–55.

117. Ibid., 52.

118. Ibid., 37, lines 1'–15.'

119. Neugebauer, *History of Ancient Mathematical Astronomy*, 363–65, and Sachs and Hunger, ADRT 1, 26–27.

120. Sachs and Hunger, ADRT 1, -651 i 6, iv 13' and -567 lines 4, 17, and rev. 16.'

121. Parpola, *Letters from Assyrian Scholars*, No. 62:3.

122. CAD s.v. *pitnu* A, meaning 3.

123. Reiner and Pingree, *Babylonian Planetary Omens* 2, EAE 50–51 III 8a, b, III 9a, and IV 4b, and p. 20 with Tablet VII.

124. Reiner and Pingree, *Babylonian Planetary Omens* 3, 6.

125. Steele, "Eclipse Prediction in Mesopotamia," 421–54.

126. Linssen, *Cults*, 16.

127. ACh Adad 19:49.

128. ACh Adad 6: 5.

129. TU 20 rev. 2–4, see Hunger, "Astrologische Wettervorhersagen," 239.

130. Brack-Bernsen and Hunger, "TU 11," 16.

131. TU 19, see Hunger, "Astrologische Wettervorhersagen," 246, lines 1–5.

132. Hunger, "Astrologische Wettervorgersagen," 248–49, lines 34–rev. 5.

133. Hunger, SpTU 1 94: 1–4, quoted in Koch-Westenholz, *Mesopotamian Astrology*, 171.

134. E.g., Diary -567: 7, "coughing and a little *rišūtu*-disease," Diary -273: 33' "much *ekketu*-disease."

135. E.g., Diary -567: 7 (a fox) and rev. 21' (a wolf that killed two dogs in Borsippa).

136. E.g., Diary -373 upper edge 1.

137. Swerdlow in *Babylonian Theory of the Planets*; Brown in *Mesopotamian Planetary Astronomy-Astrology*.

138. LBAT 1413–1432; Hunger, Sachs, and Steele, ADRT 5, 1–10 and 13–11; Huber and de Meis, *Babylonian Eclipse Observations from 750 BC to 1 BC*. See discussion in Hunger and Pingree, *Astral Sciences*, 181–82; Montelle, *Chasing Shadows*, 70–76.

139. LBAT 1413; see Huber and de Meis, *Babylonian Eclipse Observations*, 76–77.

140. Steele, *Observations and Predictions of Eclipse Times by Early Astronomers*, 43–45.

141. LBAT 1448+ 1–17; see Huber and de Meis, *Babylonian Eclipse Observations*, 150.

142. Rochberg-Halton, *Aspects of Babylonian Celestial Divination*, 57–60.

143. Ibid., 51–55.

144. EAE 31 §V, Craig AAT 70 rev. 12.'

145. EAE 32 § IX.4, Craig AAT 17:32,' and dupls Craig AAT 93–94 i 14,' K.9225+: 23 (unpub.) and K.9233+: 21 (unpub.), cited with the kind permission of the Trustees of the British Museum.

146. EAE 32 § X.11, AAT 93–94 i 32,' and dupls. K.9225+ rev. 5 (unpub.), and K.2208+ ii 3 (unpub.), cited with the kind permission of the Trustees of the British Museum.

147. EAE 32 § XI.11, Craig AAT 93–94 i 45, and dupls. K.9233+ rev. 21 (unpub.), K.2208+ ii 9 (unpub.), and ND 5490 ii 12–14 (unpub.), cited with the kind permission of the Trustees of the British Museum.

148. CAD s.v. *nalbašu*, meaning 3.

149. LBAT 1456: 1'–12'; see Huber and de Meis, *Babylonian Eclipse Observations*, 174–75.

150. Hunger and Pingree, *Astral Sciences*, 140.

151. Aaboe, "Mathematics, Astrology, and Astronomy," 278.

152. Swerdlow, *Babylonian Theory of the Planets*, 16 and 33.

153. For further discussion, see Hunger and Pingree, *Astral Sciences*, 139–40, and Rochberg, *Heavenly Writing*, 263–65.

154. Sachs and Hunger, ADRT 1, text B –375: 10.'

155. Sachs and Hunger, ADRT 2, –246: 7.

156. Rochberg, *Babylonian Horoscopes*, 53, Text 1 rev. 3–4.

157. Gray, "A Study of Babylonian Goal-Year Astronomy," 30.

158. Van Fraassen, *Empirical Stance*, 36.

159. Ibid., 31.

160. Dales, "De-Animation of the Heavens in the Middle Ages," 531–50, and the further discussion in Dales, "Some Effects of the Judeo-Christian Concept of Deity on Medieval Treatments of Classical Problems," 97–111. See also, Case, "'These Divine Animals': Physicality of the Stars in Platonic and Aristotelian Thought," 35–40.

161. Diodorus, *Bibl. Hist.* 2. 31.9.

162. Pliny, *Nat. Hist.* 7. 56.

163. See note 1 to this chapter.

CHAPTER EIGHT

1. Neugebauer, "From Assyriology to Renaissance Art," 392–93.
2. Ibid., 393.
3. Rochberg, *In the Path of the Moon*, 237–56.
4. Rochberg, "Observing and Describing the World," 633–34.
5. Bailer-Jones, *Scientific Models*, 160, emphasis in the original.
6. Bogen and Woodward, "Saving the Phenomena," 303–52.
7. Bailer-Jones, *Scientific Models*, 167.
8. Ibid.
9. Ossendrijver, *Babylonian Mathematical Astronomy*, 32.

10. Epping, *Astronomisches aus Babylon.*

11. Kugler, *Die Babylonische Mondrechnung, and Sternkunde und Sterndienst Babels,* vols. 1 and 2.

12. See Gray and Steele, "Studies on Babylonian Goal-Year Astronomy II," 624; Brack-Bernsen, "Ancient and Modern Utilization of the Lunar Data Recorded on the Babylonian Goal-Year Tablets," 13–39.

13. See chapter 3, note 92.

14. Kahneman and Tversky, "On the Psychology of Prediction," 237–51.

15. Ibid., 237–38.

16. Kahneman, *Thinking, Fast and Slow,* 13 and passim.

17. Ibid., 13.

18. Ibid., 24.

19. Ibid., 51–52. See Hume, *An Enquiry concerning Human Understanding,* 101. The principles, so defined, are a reiteration of the statement that the association of ideas is a property of human nature, and that "ideas are associated by resemblance, contiguity, and causation," from *A Treatise of Human Nature,* Book II, Section IV, 283.

20. Kahneman, *Thinking, Fast and Slow,* 52.

21. Tversky and Kahneman, "Judgment under Uncertainty: Heuristics and Biases," 1126, reprinted as appendix A in Kahneman, *Thinking, Fast and Slow,* 423.

22. Nersessian, *Creating Scientific Concepts,* passim, but especially 10–12, 55–60.

23. Kahneman and Tversky, "Intuitive Prediction: Biases and Corrective Procedures," 1.

24. See Lehoux, "Logic, Physics, and Prediction in Hellenistic Philosophy: x Happens, but y?" 125–42.

25. Especially chapter 6, "Anomaly and the Emergence of Scientific Discoveries," in Kuhn, *Structure of Scientific Revolutions,* 52–65.

26. Oppenheim, "A Babylonian Diviner's Manual," 204 lines 43–46.

27. Williams, "Signs from the Sky, Signs from the Earth: The Diviner's Manual Revisited," 473–85.

28. Hanson, "On the Symmetry between Explanation and Prediction," 349.

29. Hempel, "Function of General Laws in History," 38.

30. Hanson, "On the Symmetry between Explanation and Prediction," 351–52.

31. Hempel and Oppenheim, "Studies in the Logic of Explanation," 135–75. Following this, see Hempel, "Explanation in Science and Its History," 7–34.

32. Douglas, "Reintroducing Prediction to Explanation," 448.

33. Salmon, *Four Decades of Scientific Explanation,* 8 and 10.

34. Douglas, "Reintroducing Prediction to Explanation," 445.

35. Cf. Cryer, *Divination in Ancient Israel and Its Near Eastern Environment: A Socio-Historical Investigation,* especially 121–22, and 188, where he argues forcibly against the idea that divination was predictive by assimilating it entirely to magic. Geller convincingly critiqued that position in his review of Thomsen and Cryer, 148–49.

36. See the useful introduction in Dancygier, *Conditionals and Prediction: Time, Knowledge, and Causation in Conditional Constructions.*

37. Oppenheim, "Diviner's Manual," 205 lines 49–52, translation modified slightly.

38. Geller, review of Heeßel, *Divinatorische Texte I,* 119; Huehnergard, *A Grammar of Akkadian,* 99.

39. Brown, "Astral Divination in the Context of Mesopotamian Divination, Medicine, Religion, Magic, Society, and Scholarship," 109, emphases in the original.

40. Geller, review of Heeßel, *Divinatorische Texte,* 119.

41. Rochberg, "A Firm Yes," 5–12.

42. *têrētuja išara u ilum annam īpulanni* ARM 3 42:14.

43. YOS 1 45 i 16.

44. Leichty, 2011: 106–7, Esarhaddon 48: 57–61; ibid., 205, Esarhaddon 105 iii 24–28.

45. Ibid., 237, Esarhaddon 114 iv 1ff -iv 24; cf. ibid., 205, Esarhaddon 105 iii 24–29.

46. Oppenheim, "Diviner's Manual," 206, lines 72–80.

47. Kahneman and Tversky, "On the Psychology of Prediction," 237.

48. As defined in CAD s.v. *adannu*, meaning 2.

49. Geller, review of Heeßel, *Divinatorische Texte*, 119.

50. Heeßel, "Calculation of the Stipulated Term in Extispicy," 163–75.

51. Rochberg-Halton, *Aspects of Babylonian Celestial Divination*, 40–43; Montelle, *Chasing Shadows*, 57–58.

52. The most convincing explanation for which is found in Heeßel, "Calculation of the Stipulated Term in Extispicy," 166.

53. Brown, *Mesopotamian Planetary Astronomy-Astrology*; Brown, "Astronomy-Astrology in Mesopotamia," 41–59.

54. Brown, "Disenchanted with the Gods?" 11–27; Brown, "Scientific Revolution of 700 BC," 1–12.

55. Brown, "Disenchanted with the Gods?" 11.

56. Brown, "Scientific Revolution of 700 BC," 12, emphasis in the original. Raising the same questions, see Brown, *Mesopotamian Planetary Astronomy-Astrology*, 233.

57. This view was foreshadowed in Koch-Westenholz, *Mesopotamian Astrology: An Introduction to Babylonian and Assyrian Celestial Divination*, 52, as noted by Lehoux, "Observation and Prediction in Ancient Astrology," 242, note 41.

58. Lehoux, "Observation and Prediction in Ancient Astrology," 243.

59. Daston, "Nature of Nature in Early Modern Europe," 150–51.

60. Shea, review of Brooke and Cantor, *Reconstructing Nature*, 237.

61. See Holton, *Scientific Imagination: Case Studies*, 102–10.

62. Ibid., 108.

63. Nersessian, "Model-Based Reasoning in Conceptual Change," 5–22; Nersessian, "Model-Based Reasoning in Distributed Cognitive Systems," 699–709; Nersessian, *Creating Scientific Concepts*; Giere, "Using Models to Represent Reality," 41–57; Bailer-Jones, "Models, Metaphors, and Analogies," 108–27; Bailer-Jones, *Scientific Models*.

64. Nersessian, "Model-Based Reasoning in Conceptual Change," 8.

65. Nersessian, "Model-Based Reasoning in Distributed Cognitive Systems," 699.

66. Nersessian, *Creating Scientific Concepts*, 11.

67. Bailer-Jones, *Scientific Models*, 1; Bailer-Jones, "Models, Metaphors, and Analogies," 108.

68. Rochberg, *In the Path of the Moon*, 271–302.

69. Pinches, Strassmaier, and Sachs, introduction to LBAT, xxxv.

70. Huber and Steele, "Babylonian Lunar Six Tablets," 7. See also Huber, "A Lunar Six Text from 591 B.C.," 213–17.

71. See Brack-Bernsen, "Goal-Year Tablets: Lunar Data and Predictions," 149–77; Brack-Bernsen and Hunger, "TU 11: A Collection of Rules for the Prediction of Lunar Phases and of Month Lengths," 3–90; Brack-Bernsen and Hunger, "BM 42282+42294 and the Goal Year Method," 3–23; Brack-Bernsen, Hunger, and Walker, "KUR—When the Old Moon Can Be Seen a Day Later," 1–6; Brack-Bernsen, "Methods for Understanding and Reconstructing Babylonian Predicting Rules," 277–98; Gray and Steele, "Studies on Babylonian Goal-Year Astronomy I: A Comparison between Planetary Data in Goal-Year Texts, Almanacs, and Normal Star Almanacs," 553–600; Gray and Steele, "Studies on Babylonian Goal-Year Astronomy II: The Babylonian Calendar and Goal-Year Methods of Prediction," 611–33.

72. Sachs, "A Classification of Babylonian Astronomical Tablets of the Seleucid Period," 284–85.

73. Gray and Steele, "Studies on Babylonian Goal-Year Astronomy I," and "Studies on Babylonian Goal-Year Astronomy II." See also Hollywood and Steele, "Acronychal Risings in Babylonian Astronomy," 154–55; Steele, "Goal-Year Periods and Their Use in Predicting Planetary Phenomena," 101–10.

74. Ossendrijver, *Babylonian Mathematical Astronomy*, 17.

75. Britton, "Treatments of Annual Phenomena in Cuneiform Sources," 43.

76. Brack-Bernsen and Hunger, "TU 11: A Collection of Rules for the Prediction of Lunar Phases and Month Lengths," 3–90, and see 12, rev. 38 for the scribal name.

77. Ibid., 37–41.

78. Rochberg, *In the Path of the Moon*, 271–302.

79. Brack-Bernsen and Hunger, "TU 11: A Collection of Rules for the Prediction of Lunar Phases and Month Lengths," 31.

80. Ossendrijver, *Babylonian Mathematical Astronomy*, 50, suggests that the Akkadian term for "zone" is *babtu*, in the meaning "neighborhood" or "ward," as in a subdivision of the city, and expressed as "its ward" KÁ(*babtu*)-*tú šá* followed by the value for the zone associated with a particular planet, e.g., 30 and 36 for System A Jupiter.

81. See values for Z in Aaboe, "Period Relations in Babylonian Astronomy," 221.

82. Suárez, "Scientific Representation," 91.

83. Aaboe, "Period Relations," 225.

84. Pingree and Reiner, "A Neo-Babylonian Report on Seasonal Hours," 50–55, and reprinted in *Pathways into the Study of Ancient Sciences: Selected Essays by David Pingree*, 67–72.

85. Giere, *Science without Laws*, 25.

86. Borges, *Universal History of Infamy*, 141.

87. Cartwright, *How the Laws of Physics Lie*, 157–59. Cf. her later position in "Models: The Blueprints for Laws," S292–303.

88. Bailer-Jones, *Scientific Models*, 182, emphasis in the original.

89. Ibid., 183.

90. Ibid., 184.

91. Duhem, *Sauver les Phénomènes* and *To Save the Phenomena*.

92. Lloyd, "Saving the Appearances," 202–22.

93. Goldstein, "Saving the Phenomena," 1–12.

94. Ibid., 6.

95. Ibid., 6–7.

96. Freudenthal, "'Instrumentalism' and 'Realism' as Categories in the History of Astronomy," 270–71.

97. Ibid., citing Morgenbesser, "Realist-Instrumentalist Controversy," 200–218.

98. Ibid., 274.

99. Jones, *Astronomical Papyri from Oxyrhynchus*, 2 vols.

100. Evans and Berggren, *Geminos's Introduction to the Phenomena*, I 9 (on the Babylonian placement of the vernal equinox), VI 38 (on the constant differences of day lengths), and XVIII (using the Babylonian System B zigzag function for the moon's daily motion).

101. Bowen, "Demarcation of Physical Theory and Astronomy," 327–58; Bowen, *Simplicius on the Planets and Their Motions: In Defense of a Heresy*, 38–50.

102. Bowen, "Demarcation of Physical Theory and Astronomy," 328.

103. Evans and Berggren, *Geminos's Introduction to the Phenomena*, 51.

104. Lloyd, "Saving the Appearances," 214.

105. Goldstein, "Saving the Phenomena," 8.

106. Ibid.

107. The historiographical consequences of this have been neatly discussed in Bowen, "From Description to Prediction: An Unexamined Transition in Hellenistic Astronomy," 299–304.

108. Bowen, *Simplicius on the Planets*, 41; cf. Evans and Berggren, *Geminos's Introduction to the Phenomena*, 252–53.

109. Bowen, *Simplicius on the Planets*, 42; cf. Evans and Berggren, *Geminos's Introduction to the Phenomena*, 253.

110. Bowen, *Simplicius on the Planets*, 44.

111. For which, see the exposition of Barker and Goldstein, "Realism and Instrumentalism in Sixteenth Century Astronomy: A Reappraisal," 232–58.

112. Ibid., 240.

113. Aristotle, *Posterior Analytics* A 13, 78a 22, 78b 14–15 and 78b 33–34.

114. Ragep, "Copernicus and His Islamic Predecessors: Some Historical Remarks," 72, further clarifying how the "fact/reasoned fact (*quia/propter quid*) distinction of Aristotle's *Posterior Analytics*" played out within and was complicated by Islamic astronomy, see 72–73.

115. I thank Alan Bowen for his translation from Theon of Smyrna, *Expositio*, edition of Eduard Hiller, *Theonis Smyrnaei, philosophi platonici Expositio rerum mathematicarum ad legendum Platonem utilium* (Leipzig: Teubner, 1878), 177.9–178.2). Lloyd, "Saving the Appearances," 218, makes reference to this passage as well.

116. Hunger, *Babylonische und Assyrische Kolophone*, 42, Text No. 98, and Neugebauer, *Astronomical Cuneiform Texts*, 18 colophon U.

117. BM 42282+42294 lines: 1 [*tu*]*ppi niṣirtu šamê pirištu ilāni rabûti*, see Brack-Bernsen and Hunger, "BM 42282+42294 and the Goal-Year Method," 6.

118. Hankinson, *Cause and Explanation in Ancient Greek Thought*, 3, emphasis in the original.

119. Goldstein and Bowen, "Pliny and Hipparchus's 600-Year Cycle," 155–56.

120. Ibid., 158.

121. Hankinson, *Cause and Explanation in Ancient Greek Thought*, 3.

122. Faye, "Pragmatic-Rhetorical Theory of Explanation," 44.

123. Pingree, "Hellenophilia versus the History of Science," 560, my emphasis.

124. Frahm, *Babylonian and Assyrian Text Commentaries*, 23.

125. CT 20 1: 31 *šumma amūtu maṭṭalāt šamê*, see CAD s.v. *maṭṭalātu* "image."

126. Alasdair Livingstone discusses the meaning of the title in *Mystical and Mythological Explanatory Works of Assyrian and Babylonian Scholars*, 34–35.

127. Ibid., 28–29, line 33, taking the meaning of *mala bašmu* in accordance with the CAD s.v. *bašāmu*, meaning 1c as "pertaining to" instead of Livingtone's "as many as are designed," though both translations are justified.

128. An important early study of the characteristics of cuneiform hermeneutics is Lieberman, "A Mesopotamian Background for the So-Called *Aggadic* 'Measures' of Biblical Hermeneutics?" 157–225.

129. Frahm, *Babylonian and Assyrian Text Commentaries*, 22.

130. Ibid.

131. SAA 8 95 rev. 1–7.

132. SAA 8 8 rev. 1–7.

133. SAA 8 101: 1–5.

134. See CAD s.v. *ṭerû* C.

135. SAA 8 57: 5–rev. 4.

136. 11NT-3: 15–16, see Civil, "Medical Commentaries from Nippur," 331–36.

137. KAR 196 (VAT 8869).

138. KAR 196 (VAT 8869) iii 56, translated in Veldhuis, *A Cow of Sîn*, 14.

139. See the exposition of translation methodology in Crisostomo, "Bilingual Education and Innovations in Scholarship: The Old Babylonian Word List Izi."

140. Frahm, *Babylonian and Assyrian Text Commentaries*, 48.

141. Ibid., 49–50.

142. Hunger, *Babylonische und Assyrische Kolophone*, No. 321: 1–2, and see Frahm, *Babylonian and Assyrian Text Commentaries*, 49.

143. Frahm, *Babylonian and Assyrian Text Commentaries*, 50–55 collects the evidence.

144. Ossendrijver, *Babylonian Mathematical Astronomy*, 15.

145. Rochberg-Halton, *Aspects of Babylonian Celestial Divination*, 180.

146. Ibid., 288–89, No. 32 lines 1–4.

147. Ibid., No. 32 lines 5–6.

148. Britton, "Treatments of Annual Phenomena," appendix A, 59–61.

149. Several Labaši's are known from reports, letters, and later Uruk texts. This is probably not one of them, see Britton, "Treatments of Annual Phenomena," 61.

150. BM 45728: 5–14. Cf. translation in Ossendrijver, *Babylonian Mathematical Astronomy*, 23.

151. Steele, "Goal-Year Periods and Their Use," 105.

152. Ossendrijver, *Babylonian Mathematical Astronomy*, 36–37, 95, and passim.

153. Rochberg-Halton, *Aspects of Babylonian Celestial Divination*, 241. For further examples, see ibid., chapter 10 passim.

154. Ossendrijver, *Babylonian Mathematical Astronomy*, 329, No. 46 rev. ii 1–2.

155. Ibid., 12.

156. Ibid., 310, No. 42 rev. 7.

157. Ibid., 309, No. 41: 1–10.

CONCLUSION

1. Homans, *Nature of Social Science*, 4.

2. Merchant, "Scientific Revolution and the Death of Nature," 518–29, traces this trope about Bacon from the late seventeenth century to the middle of the 1970s.

3. Here Crombie cites Collingwood, *Idea of Nature*, giving no page number.

4. Crombie, "Some Reflections on the History of Science and Its Conception of Nature," 56.

5. See *Poems Translated from the German by Friedrich Schiller*, 74–77:

> a universe of gods must pass away!
> Mourning I search on yonder starry steeps,
> But thee no more, Selene, there I see! . . .
> Dull to the art that colors or creates,
> Like the dead timepiece, godless nature creeps.

6. No doubt underpinning Crombie's metaphor, consciously or unconsciously, is Plato's *Phaedrus* 264c, "that every discourse must be organized, like a living being, with a body of its own, as it were, so as not to be headless or footless, but to have a middle and members, composed in fitting relation to each other and to the whole."

7. Holton, *On the Thematic Origins of Scientific Thought*, 31–52; Holton, "On the Role of Themata in Scientific Thought," 328–34; Holton, *Scientific Imagination: Case Studies*.

8. Holton, "On the Role of Themata," 328.

9. Holton, *Thematic Origins*, 18.

10. See the University of California, Berkeley, project "Hellenistic Babylonia: Texts, Images, and Names, Dr. Laurie Pearce, Project Director: http://oracc.museum.upenn.edu/hbtin/.

11. Harrison, *Cambridge Companion to Science and Religion*, 11.

12. Daston, "Nature of Nature in Early Modern Europe," 154.

13. Fabian, *Memory against Culture*, 5 (emphasis in the original).

Bibliography

Aaboe, Asger. "Mathematics, Astrology, and Astronomy." In *The Cambridge Ancient History*. Vol. 3, part 2, *The Assyrian and Babylonian Empires and Other States of the Near East, from the Eighth to the Sixth Centuries BC*, edited by John Boardman, I. E. S. Edwards, E. Sollberger, and N. G. L. Hammond, 276–92. Cambridge: Cambridge University Press, 1992.

———. "Observation and Theory in Babylonian Astronomy." *Centaurus* 24 (1980): 14–35.

———. "On Period Relations in Babylonian Astronomy." *Centaurus* 10 (1964): 213–321.

Aaboe, Asger, J. P. Britton, J. A. Henderson, O. Neugebauer, and A. J. Sachs. *Saros Cycle Dates and Related Babylonian Astronomical Texts*. Transactions of the American Philosophical Society 81. Philadelphia: American Philosophical Society, 1991.

Abusch, Tzvi. "*Alaktu* and *Halakhah* Oracular Decision, Divine Revelation." *Harvard Theological Review* 80 (1987): 15–42.

———. "Marduk." In *Dictionary of Deities and Demons in the Bible*, edited by Karl van der Toorn, Bob Becking, and Pieter W. Van der Horst, 546–48. Leiden: Brill, 1999.

———. *Mesopotamian Witchcraft: Toward a History and Understanding of Babylonian Witchcraft Beliefs and Literature*. Leiden: Brill and Styx, 2002.

Abusch, Tzvi, and D. Schwemer. *The Corpus of Mesopotamian Anti-Witchcraft Rituals*. Leiden: Brill, 2011.

Achinstein, Peter. *The Nature of Explanation*. Oxford: Oxford University Press, 1983.

Adams, William Y. *The Philosophical Roots of Anthropology*. Stanford, CA: Center for the Study of Language and Information Publications, 1998.

Alinei, Mario. "Etymography and Etymothesis as Subfields of Etymology: A Contribution to the Theory of Diachronic Semantics." *Folia Linguistica* 16 (1982): 41–56.

Allen, James. "Greek Philosophy and Signs." In *Divination and Interpretation of Signs in the Ancient World*, edited by Amar Annus, 29–42. Oriental Institute Seminars 6. Chicago: Oriental Institute of the University of Chicago, 2010.

———. *Inference from Signs: Ancient Debates about the Nature of Evidence*. Oxford: Clarendon Press, 2001.

Al-Rawi, F. N. H., and Andrew George. "Enūma Anu Enlil XIV and Other Early Astronomical Tables." *Archiv für Orientforschung* 39 (1991–92): 52–73.

Anderson, Kay, and Colin Perrin. "Thinking with the Head: Race, Craniometry, Humanism." *Journal of Cultural Economy* 2 (2009): 83–98.

Annus, Amar. "On the Beginnings and Continuities of Omen Sciences in the Ancient World." In *Divination and Interpretation of Signs in the Ancient World*, edited by Amar Annus, 1–18. Oriental Institute Seminars 6. Chicago: Oriental Institute of the University of Chicago, 2010.

Annus, Amar, and Alan Lenzi. *Ludlul Bēl Nēmeqi: The Standard Babylonian Poem of the Righteous Sufferer*. State Archives of Assyria Cuneiform Texts 7. Helsinki: Neo-Assyrian Text Corpus Project, 2010.

Anscombe, G. E. M. "Causality and Determination." In *Causation*, edited by E. Sosa and M. Tooley, 88–104. Oxford: Oxford University Press, 1993, reprinted in 2005.

Aristotle. *Nicomachean Ethics*. Translated by H. Rackham. Loeb Classical Library 73. Cambridge, MA: Harvard University Press.

———. *Posterior Analytics*. Translated by H. Tredennick and E. S. Forster. Loeb Classical Library 391. Cambridge, MA: Harvard University Press.

Armogathe, Jean-Robert. "Deus legislator." In *Natural Law and Laws of Nature in Early Modern Europe: Jurisprudence, Theology, Moral and Natural Philosophy*, edited by L. Daston and M. Stolleis, 265–78. Translated by Ann T. Delehanty. Surrey: Ashgate, 2008.

Armstrong, A. H. *Plotinus Ennead III*. Loeb Classical Library 434. Cambridge, MA: Harvard University Press.

Armstrong, D. M. "Naturalism, Materialism, and First Philosophy." *Philosophia* 8 (1978): 261–76.

Arnaud, Daniel. *Recherches au pays d'Aštata, Emar VI*. Paris: Editions Recherche sur les Civilisations, 1985–87.

Assyrian Dictionary of the Oriental Institute of the University of Chicago (CAD). 21 vols. Chicago: Oriental Institute, 1956–2010.

Aya, Rod. "The Devil in Social Anthropology; or, the Empiricist Exorcist; or, the Case against Cultural Relativism." In *The Social Philosophy of Ernest Gellner*, edited by John A. Hall and Ian Jarvie, 553–64. Poznań Studies in the Philosophy of the Sciences and Humanities 48. Amsterdam: Rodopi, 1996.

Bailer-Jones, Daniela. "Models, Metaphors, and Analogies." In *The Blackwell Guide to the Philosophy of Science*, edited by Peter Machamer and Michael Silberstein, 108–27. Malden, MA: Blackwell, 2002.

———. *Scientific Models in Philosophy of Science*. Pittsburgh: University of Pittsburgh Press, 2009.

———. "Tracing the Development of Models in the Philosophy of Science." In *Model-Based Reasoning in Scientific Discovery*, edited by L. Magnani, N. J. Nersessian, and P. Thagard, 23–40. New York: Kluwer Academic, Plenum, 1999.

Barker, Peter, and Bernard R. Goldstein. "Realism and Instrumentalism in Sixteenth Century Astronomy: A Reappraisal." *Perspectives on Science* 6 (1999): 232–58.

Bates, Alan W. *Emblematic Monsters: Unnatural Conceptions and Deformed Births in Early Modern Europe*. Clio Medica 77. New York: Rodopi, 2005.

Beaulieu, Paul-Alain. "New Light on Secret Knowledge in Late Babylonian Culture." *Zeitschrift für Assyriologie* 82 (1992): 98–11.

———. "The Setting of Babylonian Wisdom Literature." In *Wisdom Literature in Mesopotamia and Israel*, edited by Richard J. Clifford, 3–19. Atlanta: Society of Biblical Literature, 2007.

Beaulieu, Paul-Alain, and John P. Britton. "Rituals for an Eclipse Possibility in the 8th Year of Cyrus." *Journal of Cuneiform Studies* 46 (1994): 73–86.

Beckner, Morton. *The Biological Way of Thought*. New York: Columbia University Press, 1959.

Beebee, Helen. "Causes and Laws: Philosophical Aspects." In *International Encyclopedia of the Social and Behavioral Sciences*, edited by Neil J. Smelser and Paul B. Baltes, 1572–78. Amsterdam: Elsevier, 2001.

Beebee, Helen, and Nigel Sabbarton-Leary, eds. *The Semantics and Metaphysics of Natural Kinds*. London: Routledge, 2010.

Beerden, Kim. *Worlds Full of Signs: Ancient Greek Divination in Context*. Leiden: Brill, 2013.

Bellah, Robert. *Religion in Human Evolution: From the Paleolithic to the Axial Age*. Cambridge, MA: Belknap Press of Harvard University Press, 2011.

Berggren, John Lennart. "Mathematics and Religion in Ancient Greece and Medieval Islam." In *The*

Alexandrian Tradition: Interactions between Science, Religion, and Literature, edited by Luis Arturo Guichard, Juan Luis García Alonso, and María Paz de Hoz, 11–34. Bern, New York: Peter Lang, 2014.

Betti, Arianna, and Willem R. de Jong, eds. "The Classical Model of Science I: A Millennia-Old Model of Scientific Rationality." Special issue, *Synthese* 174, no. 2 (May 2010).

Bhayro, Siam. "Book of the Signs of the Zodiac." In *Encyclopedia of Ancient Natural Scientists*, edited by Paul T. Keyser and Georgia L. Irby-Massie, 197–98. London: Routledge, 2008.

Bobzien, Susanne. *Determinism and Freedom in Stoic Philosophy*. Oxford: Clarendon Press, 1998.

Böck, Barbara. "'An Esoteric Babylonian Commentary' Revisited." *Journal of the American Oriental Society* 120, no. 4 (2000): 615–20.

———. "Sourcing, Organizing, Administering Medicinal Ingredients." In *The Oxford Handbook of Cuneiform Culture*, edited by Karen Radner and Eleanor Robson, 690–705. Oxford: Oxford University Press, 2011.

———. "When You Perform the Ritual of 'Rubbing': On Medicine and Magic in Ancient Mesopotamia." *Journal of Near Eastern Studies* 62 (2003): 1–16.

Bodde, Derk. "Chinese 'Laws of Nature': A Reconsideration." *Harvard Journal of Asiatic Studies* 39 (1979): 139–55.

———. "Evidence for 'Laws of Nature' in Chinese Thought." *Harvard Journal of Asiatic Studies* 20 (1957): 709–27.

Bogen, James, and James Woodward. "Saving the Phenomena." *Philosophical Review* 97 (1988): 303–52.

Boiy, T. *Late Achaemenid and Hellenistic Babylon*. Dudley, MA: Peeters, 2004.

Borger, Rykle. "Geheimwissen." In *Reallexikon der Assyriologie und Vorderasiatischen Archäologie* 3, edited by Erich Ebeling et al., 188–91. Berlin: Walter de Gruyter, 1971.

Borges, Jorge Luis. *Other Inquisitions, 1937–1952*. Translated by Ruth L. C. Simms. Introduction by James E. Irby. Austin: University of Texas Press, 1964.

———. *A Universal History of Infamy*. Translated by Norman Thomas di Giovanni. New York: E. P. Dutton, 1972.

Bottéro, Jean. *Mésopotamie: L'écriture, la raison et les dieux*. Paris: Gallimard, 1987. Translated 1992 by Zainab Bahrani and Marc Van de Mieroop as *Mesopotamia: Writing, Reasoning, and the Gods*. Chicago: University of Chicago Press, 1992.

———. "Symptômes, signes, écritures." In *Divination et Rationalité*, edited by J. P. Vernant, 139–42. Paris: Éditions du Seuil, 1974.

Boustan, Ra'anan S., and Annette Yoshiko Reed, eds. *Heavenly Realms and Earthly Realities in Late Antique Religions*. Cambridge: Cambridge University Press, 2004.

Bowen, Alan C. "The Demarcation of Physical Theory and Astronomy by Geminus and Ptolemy." *Perspectives on Science* 15 (2007): 327–58.

———. "From Description to Prediction: An Unexamined Transition in Hellenistic Astronomy." *Centaurus* 51 (2009): 299–304.

———. "Geminus and the Length of the Month: The Authenticity of Intro. Ast. 8.43–45." *Journal for the History of Astronomy* 37 (2006): 193–202.

———. "Hupsiklēs of Alexandria (150–100 BCE)." In *Encyclopedia of Ancient Natural Scientists*, edited by Paul T. Keyser and Georgia L. Irby-Massie, 425. London: Routledge, 2008.

———. *Simplicius on the Planets and Their Motions: In Defense of a Heresy*. Leiden: Brill, 2013.

———. "Three Introductions to Celestial Science in the First Century BC." In *Writing Science: Medical and Mathematical Authorship in Ancient Greece*, edited by Markus Asper, 299–327. Berlin: De Gruyter, 2013.

———. "Meton of Athens and Astronomy in the Late Fifth Century B.C." In *A Scientific Humanist: Studies in Memory of Abraham Sachs*, edited by E. Leichty, M. de J. Ellis, and P. Gerardi, 39–82. Occasional Publications of the Samuel Noah Kramer Fund 9. Philadelphia: University Museum, 1988.

Bowen, Alan C., and B. R. Goldstein. "Geminus and the Concept of Mean Motion in Greco-Latin Astronomy." *Archive for History of Exact Sciences* 50 (1996): 157–85.

Boyarin, Daniel. "Rethinking Jewish Christianity: An Argument for Dismantling a Dubious Category." *Jewish Quarterly Review* 99 (2009): 7-36.

Brack-Bernsen, Lis. "Ancient and Modern Utilization of the Lunar Data Recorded on the Babylonian Goal-Year Tablets." In *Actes de la Vème Conférence Annuelle de la SEAC*, edited by Arnold Lebeuf, Mariusz Ziółkowski, and Arkadiusz Soltysiak, 13–39. Warsaw: Warsaw University Institute of Archaeology, 1999.

———. "The 'Days in Excess' from MUL.APIN: On the 'First Intercalation' and 'Water Clock' Schemes from MUL.APIN." *Centaurus* 47 (2005): 1–29.

———. "Methods for Understanding and Reconstructing Babylonian Predicting Rules." In *Writings of Early Scholars in the Ancient Near East, Egypt, Rome, and Greece: Translating Ancient Scientific Texts*, edited by Annette Imhausen and Tanja Pommerening, 277–98. Beiträge zur Altertumskunde 286. Berlin: De Gruyter, 2010.

———. "The Path of the Moon, the Rising Points of the Sun, and the Oblique Great Circle on the Celestial Sphere." *Centaurus* 45 (2003): 16–31.

———. "The 360-Day Year in Mesopotamia." In *Calendars and Years: Astronomy and Time in Ancient Mesopotamia*, edited by John M. Steele, 83–100. Oxford: Oxbow Books, 2007.

———. "Methods for Understanding and Reconstructing Babylonian Predicting Rules." In *Writings of Early Scholars in the Ancient Near East, Egypt, Rome, and Greece*, edited by Annette Imhausen and Tanja Pommerening, 277–98. Berlin: De Gruyter, 2010.

———. "TU 11: A Collection of Rules for the Prediction of Lunar Phases and of Month Lengths." *SCIAMVS* 3 (2002): 3–90.

Brack-Bernsen, Lis, and Hermann Hunger. "BM 42282+42294 and the Goal-Year Method." *SCIAMVS* 9 (2008): 3–23.

Brack-Bernsen, Lis, Hermann Hunger, and Christopher Walker. "KUR—When the Old Moon Can Be Seen a Day Later." In *From the Banks of the Euphrates: Studies in Honor of Alice Louise Slotsky*, edited by Micah Ross, 1-6. Winona Lake, IN: Eisenbrauns, 2008.

Brack-Bernsen, Lis, and John M. Steele. "Babylonian Mathemagics: Two Mathematical Astronomical-Astrological Texts." In *Studies in the History of the Exact Sciences in Honour of David Pingree*, edited by Charles Burnett, J. P. Hogendijk, K. Plofker, and M. Yano, 95–125. Leiden: Brill, 2004.

———. "Eclipse Prediction and the Length of the Saros in Babylonian Astronomy." *Centaurus* 47 (2005): 181–206.

Britton, John P. "Calendars, Intercalations, and Year-Lengths in Mesopotamian Astronomy." In *Calendars and Years: Astronomy and Time in the Ancient Near East*, edited by John M. Steele, 115–32. Oxford: Oxbow Books, 2007.

———. "An Early Function for Eclipse Magnitudes in Babylonian Astronomy." *Centaurus* 32 (1989): 1–52.

———. "An Early Observation Text for Mars." In *Studies in the History of the Exact Sciences in Honour of David Pingree*, edited by Charles Burnett, Jan P. Hogendijk, Kim Plofker, and Michio Yano, 33–55. Leiden: Brill, 2004.

———. "Remarks on Strassmaier Cambyses 400." In *From the Banks of the Euphrates: Studies in Honor of Alice Slotsky*, edited by Michah Ross, 7–33. Winona Lake, IN: Eisenbrauns.

———. "Scientific Astronomy in Pre-Seleucid Babylon." In *Die Rolle der Astronomie in den Kulturen Mesopotamiens*, edited by Hannes D. Galter, 61–94. Grazer Morgenländische Studien 3. Graz: GrazKult, 1993.

———. "Studies in Babylonian Lunar Theory: Part I. Empirical Elements for Modeling Lunar and Solar Anomalies." *Archive for History of Exact Sciences* 61 (2007): 83–145.

———. "Studies in Babylonian Lunar Theory: Part II. Treatments of Lunar Anomaly." *Archive for History of Exact Sciences* 63 (2009): 357–431.

———. "Studies in Babylonian Lunar Theory: Part III. The Introduction of the Uniform Zodiac." *Archive for History of Exact Sciences* 64 (2010): 617–63.

———. "A Tale of Two Cycles." *Centaurus* 33 (1990): 57–69.

———. "Treatments of Annual Phenomena in Cuneiform Sources." In *Under One Sky: Astronomy and Mathematics in the Ancient Near East*, edited by John M. Steele and Annette Imhausen, 21–78. Münster: Ugarit Verlag, 2002.

Broadie, Sarah. *Nature and Divinity in Plato's Timaeus*. Cambridge: Cambridge University Press, 2012.

Brown, David. "Astral Divination in the Context of Mesopotamian Divination, Medicine, Religion, Magic, Society, and Scholarship." In *Special Issue in Honor of Prof. Ho Peng Yoke's 80th Birthday*, edited by H. U. Vogel, 69–126. East Asian Science, Technology, and Medicine 25. Tübingen: International Society for the History of East Asian Science, Technology, and Medicine, 2006.

———. "Astronomy-Astrology in Mesopotamia." *Bibliotheca Orientalis* 58 (2001): 41–59.

———. "The Cuneiform Conception of Celestial Space and Time." *Cambridge Archaeological Journal* 10 (2000): 103–22.

———. "Disenchanted with the Gods? The Advent of Accurate Prediction and Its Influence on Scholarly Attitudes towards the Supernatural in Ancient Mesopotamia and Ancient Greece." In *Your Praise Is Sweet: A Memorial Volume for Jeremy Black from Students, Colleagues, and Friends*, edited by Heather Baker, Eleanor Robson, and Gábor Zólyomi, 11–27. London: British Institute for the Study of Iraq, 2010.

———. *Mesopotamian Planetary Astronomy-Astrology*. Cuneiform Monographs 18. Groningen: Styx, 2000.

———. "The Scientific Revolution of 700 BC." In *Learned Antiquity: Scholarship and Society in the Near-East, the Greco-Roman World, and the Early Medieval West*, edited by Alasdair A. MacDonald, Michael W. Twomey, and Gerrit J. Reinink, 1–12. Dudley, MA: Peeters, 2003.

Brown, David, J. Fermor, and C. Walker. "The Water Clock in Mesopotamia." *Archiv für Orientforschung* 46–47 (1999–2000): 130–48.

Brown, David, and Marc Linssen. "BM 134701 = 1965-10-14,1 and the Hellenistic Period Eclipse Ritual from Uruk." *Revue d'Assyriologie* 91 (1997): 147–66.

Buber, Martin. *I and Thou*. Translated by Walter Kaufmann. New York: Macmillan, 1970.

———. *Ich und Du*. Leipzig: Insel-Verlag, 1923.

Buffon, Georges Louis Leclerc, comte de. *Oeuvres choisies de Buffon: Contenant les Discours Académiques, des Extraits de la Théorie de la Terre, les Époques de la Nature, La Génésie des Minéraux, L'Histoire Naturelle de l'Homme et des Animaux*. Paris: Chez Firmin Didot Frères, Libraires, imprimeurs de l'Institut, Rue Jacob, 56, 1843.

Burrows, Eric. "Hymn to Ninurta as Sirius (K 128)." *Journal of the Royal Asiatic Society* n.s. 56 (1924): 33–40.

Burstein, Stanley Mayer, ed. and trans. *The Babyloniaca of Berossus*. Malibu, CA: Undena, 1978.

Campbell, Joseph Keim, Michael O'Rourke, and Matthew H. Slater, eds. *Carving Nature at Its Joints: Natural Kinds in Metaphysics and Science*. Cambridge, MA: MIT Press, 2011.

Cancik-Kirschbaum, Eva. "Gegenstande und Methode: Sprachliche Erkenntnistechniken in der keilschriftlichen Überlieferung Mesopotamiens." In *Writings of Early Scholars in the Ancient Near East, Egypt, Rome, and Greece: Translating Ancient Scientific Texts*, edited by Annette Imhausen and Tanja Pommerening, 13–45. Beiträge zur Altertumskunde 286. Berlin: De Gruyter, 2010.

Canguilhem, Georges. *Knowledge of Life*. 1965; New York: Fordham University Press, 2008.

Caplice, Richard I. *The Akkadian Namburbi Texts: An Introduction*. Los Angeles: Undena, 1974.

———. "The Akkadian Text Genre Namburbi." PhD diss., University of Chicago, 1963.

———. "Namburbi Texts in the British Museum, I." *Orientalia* 34 (1965): 105–31, plates 15–18.

———. "Namburbi Texts in the British Museum, III." *Orientalia* 36 (1967): 1–38, plates 1–4.

———. "Namburbi Texts in the British Museum, V." *Orientalia* 40 (1971): 133–83, plates 2–18.

Carroll, John W. *Laws of Nature*. Cambridge Studies in Philosophy. Cambridge: Cambridge University Press, 1994.

Cartwright, Nancy. *How the Laws of Physics Lie*. Oxford: Oxford University Press, 1983.

———. "Models: The Blueprints for Laws." *Philosophy of Science* 64 (Proceedings) (1997): S292–303.

Case, Stephen. "'These Divine Animals': Physicality of the Stars in Platonic and Aristotelian Thought." In *Selected Proceedings of the Newberry Center for Renaissance Studies, 2013 Multidisciplinary Graduate Student Conference*, edited by Carla Zecher and Karen Christianson, 35–40. Newberry Essays in Medieval and Early Modern Studies 7. Chicago: Newberry Library, 2013.

Cassirer, Ernst. *An Essay on Man: An Introduction to a Philosophy of Human Culture*. New Haven, CT: Yale University Press, 1944.

Castro, Eduardo Viveiros de. "Cosmological Deixis and Amerindian Perspectivism." *Journal of the Royal Anthropological Institute of Great Britain and Ireland* 4 (1998): 469–88.

Catagnoti, Amalia, and Marco Bonechi. "Magic and Divination at IIIrd Millennium Ebla, 1: Textual Typologies and Preliminary Lexical Approach." *Studi Epigrafici e Linguistici su Vicino Oriente Antico* 15 (1998): 17–39.

Chalmers, A. F. *What Is This Thing Called Science?* 2nd ed. Queensland: University of Queensland Press, 1982.

Chang, Hasok. *Is Water H_2O? Evidence, Realism, and Pluralism*. Boston Studies in the Philosophy of Science 293. Dordrecht: Springer, 2012.

———. "Putting Science Back in History of Science." BSHS Presidential Address, 24th International Congress of the History of Science, Technology, and Medicine, at the University of Manchester, July 22, 2013.

Chisholm, Roderick M. "Law Statements and Counterfactual Inference." *Analysis* 15 (1955): 97–105.

Churchland, Paul. "Conceptual Progress and Word/World Relations: In Search of the Essence of Natural Kinds." *Canadian Journal of Philosophy* 15 (1985): 1–17.

Cicero. *De Divinatione*. Translated by W. A. Falconer. Loeb Classical Library 154. Cambridge, MA: Harvard University Press; and London: William Heinemann, 1979.

———. *De Natura Deorum*. Translated by H. Rackham. Loeb Classical Library 268. Cambridge, MA: Harvard University Press, and London: William Heinemann, 1933, reprinted in 2000.

Civil, Miguel. "Medical Commentaries from Nippur." *Journal of Near Eastern Studies* 33 (1974): 329–38.

Cohen, Mark E. *The Cultic Calendars of the Ancient Near East*. Bethesda, MD: CDL Press, 1983.

Cole, David, and Jennifer Cole. "Rethinking the Goody Myth." In *Technology, Literacy, and the Evolution of Society: Implications of the Work of Jack Goody*, edited by David R. Olson and Michael Cole, 305–24. Mahwah, NJ: Lawrence Erlbaum Associates, 2006.

Collingwood, R. G. *The Idea of Nature*. Oxford: Clarendon Press, 1945.

Conant, James, and John Haugeland, eds. *The Road since Structure: Thomas S. Kuhn*. Chicago: University of Chicago Press, 2000.

Cooke, Roger. *The History of Mathematics: A Brief Course*. New York: Wiley, 1997.

Cooley, Jeffrey L. "An OB Prayer to the Gods of the Night." In *Reading Akkadian Prayers and Hymns*, edited by Alan Lenzi, 71–84. Atlanta: Society of Biblical Literature, 2011.

Cornwell, John. *Explanations: Styles of Explanation in Science*. Oxford: Oxford University Press, 2004.

Craig, James A. *Astrological-Astronomical Texts*. Leipzig: J. C. Hinrichs, 1899.

Crawley, Ernest. *The Mystic Rose: A Study of Primitive Marriage*. London: Macmillan, 1902.

Crisostomo, C. Jay. "Bilingual Education and Innovations in Scholarship: The Old Babylonian Word List Izi." PhD diss., University of California, Berkeley, 2014.

Crombie, A. C. "Some Reflections on the History of Science and Its Conception of Nature." *Annals of Science* 6 (1948): 54–75.

Cryer, Frederick H. *Divination in Ancient Israel and Its Near Eastern Environment: A Socio-Historical Investigation. Journal for the Study of the Old Testament* Supplement Series 142. Sheffield: Journal for the Study of the Old Testament Press, 1994.

Cunningham, Andrew. "Getting the Game Right: Some Plain Words on the Identity and Invention of Science." *Studies in History and Philosophy of Science* 19 (1988): 365–89.

Dales, Richard C. "The De-Animation of the Heavens in the Middle Ages." *Journal of the History of Ideas* 41 (1980): 531–50.

———. "Some Effects of the Judeo-Christian Concept of Deity on Medieval Treatments of Classical Problems." In *Man and Nature in the Middle Ages*, edited by Susan J. Ridyard and Robert G. Benson, 97–111. Sewanee, TN: University of the South Press, 1995.

Dalley, Stephanie. *Myths from Mesopotamia: Creation, the Flood, Gilgamesh, and Others*. Oxford: Oxford University Press, 1989.

Dancygier, Barbara. *Conditionals and Prediction: Time, Knowledge, and Causation in Conditional Constructions*. Cambridge Studies in Linguistics 87. Cambridge: Cambridge University Press, 1999.

Daston, Lorraine. "The Nature of Nature in Early Modern Europe." *Configurations* 6 (1998): 149–72.

———. "Preternatural Philosophy." In *Biographies of Scientific Objects*, edited by Lorraine Daston, 15–41. Chicago: University of Chicago Press, 2000.

———, ed. *Biographies of Scientific Objects*. Chicago: University of Chicago Press, 2000.

Daston, Lorraine, and Peter Galison. *Objectivity*. New York: Zone Books, 2007.

Daston, Lorraine, and Elizabeth Lunbeck, eds. 2011. *Histories of Scientific Observation*. Chicago: University of Chicago Press.

Daston, Lorraine, and Katharine Park. *Wonders and the Order of Nature, 1150–1750*. New York: Zone Books, 1998.

Davidson, Basil. *The African Genius: An Introduction to African Social and Cultural History*. Boston: Little, Brown, 1969.

Davidson, Donald. *Problems of Rationality*. Oxford: Oxford University Press, 2004.

Dear, Peter. "Intelligibility in Science." *Configurations* 11 (2003): 145–61.

De Breucker, Geert. "Berossos and the Mesopotamian Temple as Centre of Knowledge during the Hellenistic Period." In *Learned Antiquity: Scholarship and Society in the Near-East, the Greco-Roman World, and the Early Medieval West*, edited by Alaisdair A. MacDonald, Michael W. Twomey, and Gerrit J. Reinink, 13–23. Dudley, MA: Peeters, 2003.

Deely, John. "From σημεῖον to 'signum' to 'sign': Translating Sign from Greek to Latin to English." In *Essays in Translation, Pragmatics, and Semiotics*, edited by Esmeli Helin, 129–72. Helsinki: Multilingual Communication Programme, Helsinki University Press, 2002.

De Jong, Teije. "Astronomical Fine-Tuning of the Chronology of the Hammurabi Age." *Jaarbericht "Ex Oriente Lux"* 44 (2012–13): 147–67.

Delany, David. *Law and Nature*. Cambridge: Cambridge University Press, 2003.

Dentan, Robert C. *The Idea of History in the Ancient Near East*. New Haven, CT: Yale University Press, 1955.

Descola, Phillipe. *Beyond Nature and Culture*. Translated by Janet Lloyd. Chicago: University of Chicago Press, 2013.

———. "Beyond Nature and Culture." *Proceedings of the British Academy* 139 (2006): 137–55.

———. "Cognition, Perception, and Worlding." *Interdisciplinary Science Reviews* 35 (2010): 334–40.

———. "Who Owns Nature?" http://www.laviedesidees.fr/IMG/pdf/20080121_descola_en.pdf.

De Zorzi, Nicla. "Bird Divination in Mesopotamia: New Evidence from BM 108874." *KASKAL, Rivista di storia, ambienti e culture del Vicino Oriente Antico* 6 (2009): 85–135.

———. "The Omen Series Summa Izbu: Internal Structure and Hermeneutic Strategies." *KASKAL, Rivista di storia, ambienti e culture del Vicino Oriente Antico* 8 (2011): 43–75.

———. *La Serie Teratomantica Šumma Izbu: Testo, Tradizione, Orizzonti Culturali*. 2 vols. Padua: S.A.R.G.O.N. Editrice e Libreria, 2014.

Dodds, E. R. *The Greeks and the Irrational*. Berkeley: University of California Press, 1951.

Dolansky, Shawna. *Now You See It, Now You Don't: Biblical Perspectives on the Relationship between Magic and Religion*. Winona Lake, IN: Eisenbrauns, 2008.

Dossin, Georges. "Prières aux 'Dieux De La Nuit.'" *Revue d'Assyriologie et d'archéologie orientale* 32, no. 4 (1935): 179–87.

Douglas, Heather E. "Reintroducing Prediction to Explanation." *Philosophy of Science* 76 (2009): 444–63.

Douglas, Mary, and David Hull. *How Classification Works: Nelson Goodman among the Social Sciences*. Edinburgh: Edinburgh University Press, 1992.

Duffin, C. J., R. T. J. Moody, and C. Gardner-Thorpe, eds. *A History of Geology and Medicine*. Geological Society Special Publication 375. Bath: Geological Society Publishing House, 2013.

Duhem, Pierre. *Sauver les phénomènes: Essai sur la notion de théorie physique de Platon à Galilée*. Paris: A. Hermann, 1908.

———. *To Save the Phenomena: An Essay on the Idea of Physical Theory from Plato to Galileo*. Translated by Edmund Doland and Stanley L. Jaki. Chicago: University of Chicago Press, 1969.

Edzard, D. O. *Gudea and His Dynasty*. Royal Inscriptions of Mesopotamia Early Periods 3/1. Toronto: University of Toronto Press, 1997.

Eichler, Barry L. "Ethics and Law Codes, Mesopotamia." In *Religions of the Ancient World*, edited by Sarah Iles Johnston, 516–19. Cambridge, MA: Belknap Press of Harvard University Press, 2004.

Englund, R. K. "Administrative Timekeeping in Ancient Mesopotamia." *Journal of the Economic and Social History of the Orient* 31 (1988): 121–85.

Epping, Joseph. *Astronomisches aus Babylon*. Ergänzungsheft zu den Stimmen aus Maria Laach 44. Freiburg im Breisgau: Herder, 1889.

Evans, James, and J. Lennart Berggren. *Geminos's Introduction to the Phenomena: A Translation and Study of a Hellenistic Survey of Astronomy*. Princeton, NJ: Princeton University Press, 2006.

Evans-Pritchard, E. E. *Theories of Primitive Religion*. Oxford: Clarendon Press, 1965.

Fabian, Johannes. *Memory against Culture: Arguments and Reminders*. Durham, NC: Duke University Press, 2007.

Fales, F. M., and J. N. Postgate. *Imperial Administrative Records, Part I: Palace and Temple Administration*. State Archives of Assyria 7. Helsinki: Helsinki University Press, 1992.

Falkenstein, A. "'Wahrsagung' in der sumerische Überlieferung." In *La divination en Mésopotamie ancienne*, CRRA XIV, edited by Jean Nougayrol, 45–68. Paris: Presses Universitaires de France, 1966.

Faye, Jan. "Explanation and Interpretation in the Sciences of Man." In *Explanation, Prediction, and Confirmation*, edited by Dennis Dieks, Wenceslao J. Gonzalez, Stephan Hartmann, Thomas Uebel, and Marcel Weber, 269–79. Philosophy of Science in a European Perspective 2. Dordrecht, London: Springer, 2011.

———. "Explanation Explained." *Synthese* 120 (1999): 61–75.

———. *The Nature of Scientific Thinking: On Interpretation, Explanation, and Understanding*. New York: Palgrave Macmillan, 2014.

———. "The Pragmatic-Rhetorical Theory of Explanation." In *Rethinking Explanation*, edited by Johannes Persson and Petri Ylikoski, 43–68. Boston Studies in the Philosophy of Science 252. Dordrecht, London: Springer, 2007.

Fincke, Jeanette C. "The Babylonian Texts of Nineveh." *Archiv für Orientforschung* 50 (2003–4): 111–49.

———. "Omina, die göttlichen 'Gesetze' der Divination." *Jaarbericht "Ex Oriente Lux"* 40 (2006–7): 131–47.

———. "The Seventh Tablet of the *rikis gerri* Series of *enūma anu enlil*." *Journal of Cuneiform Studies* 66 (2014): 129–48.

Finkel, I. L. "Adad-apla-iddina, Esagil-kīn-apli, and the Series SA.GIG." In *A Scientific Humanist:*

Studies in Memory of Abraham Sachs, edited by Erle Leichty and Maria Dejong Ellis, 143–59. Philadelphia: University Museum, 1988.

———. "On Late Babylonian Medical Training." In *Wisdom, Gods, and Literature: Studies in Assyriology in Honour of W. G. Lambert*, edited by A. R. George and Irving L. Finkel, 137–224. Winona Lake, IN: Eisenbrauns. 2000.

———. "Strange Byways in Cuneiform Writing." In *The Idea of Writing: Play and Complexity*, edited by Alex de Voogt and Irving Finkel, 9–25. Leiden: Brill, 2010.

Finkelstein, J. J. "Law in the Ancient Near East." *Encyclopedia Biblica* 5 (1968), cols. 588–614. Reprinted in *Jewish Law and Decision-Making: A Study through Time*, edited by Aaron M. Schreiber. Philadelphia: Temple University Press, 1979.

———. "Mesopotamian Historiography." *Proceedings of the American Philosophical Society* 107 (1963): 461–72.

Fiske, Shanyn. "From Ritual to the Archaic in Modernism: Frazer, Harrison, Freud, and the Persistence of Myth." In *A Handbook of Modernism Studies*, edited by Jean-Michel Rabaté, 173–91. London: John Wiley and Sons, 2013.

Foster, Benjamin R. *Before the Muses: An Anthology of Akkadian Literature*. 3rd ed. Bethesda, MD: CDL Press, 2005.

Foucault, Michel. *The Order of Things: An Archaeology of the Human Sciences*. New York: Vintage Press, 1973.

Frahm, Eckart. *Babylonian and Assyrian Text Commentaries: Origins of Interpretation*. Guides to the Mesopotamian Textual Record 5. Münster: Ugarit-Verlag, 2011.

Frankfort, Henri, H. A. Groenewegen-Frankfort, John A. Wilson, Thorkild Jacobsen, and William A. Irwin, eds. *The Intellectual Adventure of Ancient Man: An Essay of Speculative Thought in the Ancient Near East*. Chicago: University of Chicago Press, 1946.

Frankfort, Henri, H. A. Groenewegen-Frankfort, John A. Wilson, and Thorkild Jacobsen. *Before Philosophy: The Intellectual Adventure of Ancient Man*. Baltimore, MD: Penguin, 1949.

Freedman, S. M. *If a City Is Set on a Height*. Occasional Papers of the Samuel Noah Kramer Fund 17. Philadelphia: University of Pennsylvania Museum, 1998.

Freudenthal, Gad. "'Instrumentalism' and 'Realism' as Categories in the History of Astronomy: Duhem vs. Popper, Maimonides vs. Gerson." *Centaurus* 45 (2003): 227–48; revised in *The Significance of the Hypothetical in the Natural Sciences*, edited by Michael Heidelberger and Gregor Schiemann, 269–94. Berlin: Walter de Gruyter, 2009.

Friedman, Jonathan. "Our Time, Their Time, World Time: The Transformation of Temporal Modes." *Ethnos* 50 (1985): 167–83.

Funkenstein, Amos. "The Disenchantment of Knowledge: The Emergence of the Ideal of Open Knowledge in Ancient Israel and Classical Greece." *Aleph* 3 (2003): 15–95.

Galton, Eugene. *English Men of Science: Their Nature and Nurture*. London: Macmillan, 1874.

Geertz, Clifford. *The Interpretation of Cultures*. 1973; New York: Basic Books, 2000.

———. "1983 Distinguished Lecture: Anti Anti-Relativism." *American Anthropologist* 82 (1984): 263–78.

Gehlken, Erlend. "Die Adad-Tafeln der Omenserie Enūma Anu Enlil. Teil 1: Einführung." *Baghdader Mitteilungen* 36 (2005): 235–73.

———. "Die Serie DIŠ Sîn ina tāmartīšu im Überblick." *Nouvelles Assyriologiques Brèves et Utilitaires* 4 (2007): 3–5.

———. *Weather Omens of Enūma Anu Enlil: Thunderstorms, Wind, and Rain (Tablets 44–49)*. Cuneiform Monographs. Leiden: Brill, 2012.

Geller, Markham J. *Ancient Babylonian Medicine: Theory and Practice*. London: Wiley-Blackwell, 2010.

———. *Evil Demons: Canonical Utukkū Lemnūtu Incantations*. State Archives of Assyria Cuneiform Texts 5. Helsinki, 2007.

———. *Forerunners to Udug-Hul: Sumerian Exorcistic Incantations*. Wiesbaden: F. Steiner Verlag, 1985.

———. *Look to the Stars: Babylonian Medicine, Magic, Astrology, and Melothesia*. Preprint 401. Berlin: Max-Planck-Inst. für Wissenschaftsgeschichte, 2010.

———. *Melothesia in Babylonia: Medicine, Magic, and Astrology in the Ancient Near East*. Science, Technology, and Medicine in Ancient Cultures 2. Berlin: De Gruyter, 2014.

———. "A New Piece of Witchcraft." In *DUMU-E2-DUB-BA-A: Studies in Honor of Ake W. Sjöberg*, edited by Hermann Behrens, Darlene Loding, and Martha T. Roth, 193–205. Occasional Publications of the Samuel Noah Kramer Fund 11. Philadelphia: University Museum, 1989.

———. Review of Alice Louise Slotsky, *The Bourse of Babylon*. *Orientalische Literaturzeitung* 95 (2000): 409–12.

———. Review of Marie-Louise Thomsen and Frederick H. Cryer, *Witchcraft and Magic in Europe: Biblical and Pagan Societies*. Athlone History of Witchcraft and Magic in Europe 1. London: Athlone Press. 2001, in *Medical History* 48 (2004): 148–49.

———. Review of Nils P. Heessel, *Divinatorische Texte I: Terrestrische, teratologische, physiognomische und oneiromantische Texte*. *Keilschrifttexte aus Assur literarischen Inhalts*, 1, *Welt des Orients* 41 (2011): 118–21.

———. "West Meets East: Early Greek and Babylonian Diagnosis." *Archiv für Orientforschung* 48–49 (2001–2): 50–75.

Geller, Markham J., and Klaus Geus. "Esoteric Knowledge in Antiquity—Some Thoughts." In *Esoteric Knowledge in Antiquity*, TOPOI-Dahlem Seminar for the History of Ancient Sciences, II, edited by Klaus Geus and Mark Geller, 3–6. Preprint 454. Berlin: Max-Planck-Institut für Wissenschaftsgeschichte, 2014.

Geller, Markham J., in collaboration with Ludek Vacin. *Healing Magic and Evil Demons: Canonical Udug-Hul Incantations*. Die babylonisch-assyrische Medizin in Texten und Untersuchungen 8. Berlin: De Gruyter, 2015.

Gellner, Ernest. *Plough, Sword, and Book: The Structure of Human History*. Chicago: University of Chicago Press, 1988.

———. "Relativism and Universals." In *Rationality and Relativism*, edited by Martin Hollis and Steven Lukes, 188–200. Cambridge, MA: MIT Press, 1982.

———. *Spectacles and Predicaments: Essays in Social Theory*. Cambridge: Cambridge University Press, 1979.

———. *Words and Things: A Critical Account of Linguistic Philosophy and a Study in Ideology*. Boston: Beacon Press, 1959.

George, Andrew. *The Babylonian Gilgamesh Epic: Introduction, Critical Edition, and Cuneiform Texts*. Vol. 1. Oxford: Oxford University Press, 2003.

Giere, Ronald N. *Explaining Science: A Cognitive Approach*. Chicago: University of Chicago Press, 1988.

———. *Science without Laws*. Chicago: University of Chicago Press, 1999.

———. "Using Models to Represent Reality." In *Model-Based Reasoning in Scientific Discovery*, edited by L. Magnani, N. J. Nersessian, and P. Thagard, 41–57. New York: Kluwer Academic, Plenum, 1999.

Glacken, Clarence J. *Traces on the Rhodian Shore: Nature and Culture in Western Thought from Ancient Times to the End of the Eighteenth Century*. Berkeley: University of California Press, 1967.

Glassner, Jean-Jacques. "Droit et divination: Deux manières de rendre la justice; à Propos de *dīnum, uṣurtum et awatum*." *Journal of Cuneiform Studies* 64 (2012): 39–56.

Goetze, Albrecht. "An Old Babylonian Prayer of the Divination Priest." *Journal of Cuneiform Studies* 22 (1968): 25–29.

———. "Reports on Acts of Extispicy from Old Babylonian and Kassite Times." *Journal of Cuneiform Studies* 11 (1957): 89–105.

Goldstein, B. R. "Babylonian Solar Theory Reconsidered." *Archives internationale d'histoire des sciences* 30 (1980): 189–91.

———. "On the Babylonian Discovery of the Periods of Lunar Motion." *Journal for the History of Astronomy* 33 (2002): 1–13.

———. "Saving the Phenomena: The Background to Ptolemy's Planetary Theory." *Journal for the History of Astronomy* 28 (1997): 1–12.

Goldstein, B. R., and A. C. Bowen. "On Early Hellenistic Astronomy: Timocharis and the First Callippic Calendar." *Centaurus* 32 (1989): 272–93.

———. "Pliny and Hipparchus's 600-Year Cycle." *Journal for the History of Astronomy* 26 (1995): 155–58.

Goldwasser, Orly. "The Determinative System as a Mirror of World Organization." *Göttinger Miszellen* 170 (1999): 49–68.

———. *Prophets, Lovers, and Giraffes: Wor(l)d Classification in Ancient Egypt*. With an appendix by Matthias Müller. Wiesbaden: Harrassowitz Verlag, 2002.

Goodman, Nelson. *Fact, Fiction, and Forecast*. 4th ed. Cambridge, MA: Harvard University Press, 1983.

———. "Seven Strictures on Similarity." In *How Classification Works: Nelson Goodman among the Social Sciences*, edited by Mary Douglas and David Hull, 13–23. Edinburgh: Edinburgh University Press, 1992.

———. *Ways of Worldmaking*. Indianapolis: Hackett, 1978.

Goody, Jack. *Domestication of the Savage Mind*. Cambridge: Cambridge University Press, 1977.

Gould, Stephen Jay. *The Mismeasure of Man*. New York: Norton, 1996.

Grafton, Anthony. *Cardano's Cosmos*. Cambridge, MA: Harvard University Press, 1999.

Grant, Edward. *God and Reason in the Middle Ages*. Cambridge: Cambridge University Press, 2001.

———. *A History of Natural Philosophy: From the Ancient World to the Nineteenth Century*. Cambridge: Cambridge University Press, 2007.

———. *The Nature of Natural Philosophy in the Late Middle Ages*. Washington, DC: Catholic University of America Press, 2010.

———. "Scientific Imagination in the Middle Ages." *Perspectives on Science* 12 (2004): 394–423.

Grant, Robert M. *Miracle and Natural Law in Graeco-Roman and Early Christian Thought*. Amsterdam: North-Holland Publishing, 1952.

Gray, Jennifer M. K. "A Study of Babylonian Goal-Year Astronomy." PhD diss., Durham University, 2009.

Gray, Jennifer M. K., and J. M. Steele. "Studies on Babylonian Goal-Year Astronomy I: A Comparison between Planetary Data in Goal-Year Texts, Almanacs, and Normal Star Almanacs." *Archive for History of Exact Sciences* 62 (2008): 553–600.

———. "Studies on Babylonian Goal-Year Astronomy II: The Babylonian Calendar and Goal-Year Methods of Prediction." *Archive for History of Exact Sciences* 63 (2009): 611–33.

Grayson, Kirk. *Assyrian and Babylonian Chronicles*. Texts from Cuneiform Sources 5. Locust Valley, NY: J. J. Augustin, 1975.

Greenbaum, Dorian Gieseler. "Arrows, Aiming, and Divination: Astrology as a Stochastic Art." In *Divination: Perspectives for a New Millennium*, edited by Patrick Curry, 179–210. Burlington, VT: Ashgate Publishing, 2010.

Greenbaum, Dorian Gieseler, and Micah Ross. "The Role of Egypt in the Development of the Horoscope." In *Egypt in Transition: Social and Religious Development of Egypt in the First Millennium BCE*, edited by L. Bareš, F. Coppens, and K. Smoláriková, 146–82. Prague: Czech Institute of Egyptology, Faculty of Arts, Charles University in Prague, 2010.

Gregersen, F., and Simo Køppe. "Against Epistemological Relativism." *Studies in History and Philosophy of Science* 19 (1988): 447–87.

Groneberg, Brigitte R. M., *Syntax, Morphologie und Stil der jungbabylonischen "hymnischen" Literatur*. Teil 1: *Grammatik*. Teil 2: *Belegsammlung und Textkatalog*. Stuttgart: Franz Steiner Verlag Wiesbaden, 1987.

Guinan, A. "A Severed Head Laughed: Stories of Divinatory Interpretation." In *Magic and Divination in the Ancient World*, edited by L. Ciraolo and J. Seidel, 7–40. Leiden: Brill and Styx, 2002.

Gurney, O. R., and J. J. Finkelstein. *The Sultantepe Tablets*, I. Occasional Publications of the British Institute of Archaeology at Ankara 3. London: British Institute of Archaeology at Ankara, 1957.

Gurney, O. R., and P. Hulin. *The Sultantepe Tablets*, II. Occasional Publications of the British Institute of Archaeology at Ankara 7. London: British Institute of Archaeology at Ankara, 1964.

Hacking, Ian. "Almost Zero." *London Review of Books* 29 (2007): 17–19.

———. "Natural Kinds: Rosy Dawn, Scholastic Twilight." *Royal Institute of Philosophy Supplement* 82 (2007): 203–39.

———. "A Tradition of Natural Kinds." *Philosophical Studies* 61 (1991): 109–26.

Hadot, Pierre. *The Veil of Isis: An Essay on the History of the Idea of Nature*. Cambridge, MA: Belknap Press of Harvard University Press, 2006.

Hallo, William W. "The Cultic Setting of Sumerian Poetry." In *Actes de la XVIIe Rencontre Assyriologique Internationale*. Université Libre de Bruxelles 30 Juin–4 Juillet 1969. Publications de Comité belge de recherches historiques, épigraphiques et archéologiques en Mésopotamie 1, edited by André Finet, 117–34. Brussels: Ham-sur-Heure, 1970.

———. "On the Antiquity of Sumerian Literature." *Journal of the American Oriental Society* 83 (1963): 167–76.

Hallum, Bink. "Bōlos of Mendēs." In *Encyclopedia of Ancient Natural Scientists: The Greek Tradition and Its Many Heirs*, edited by Paul T. Keyser and Georgia Irby-Massie, 196–97. London: Routledge, 2008.

Halverson, John. "Goody and the Implosion of the Literacy Thesis." *Man* 27 (1992): 301–17.

Hanafi, Zakiya. *The Monster in the Machine: Magic, Medicine, and the Marvelous in the Time of the Scientific Revolution*. Durham, NC: Duke University Press, 2000.

Hanegraaff, Wouter J. *Esotericism and the Academy: Rejected Knowledge in Western Culture*. Cambridge: Cambridge University Press, 2012.

Hankinson, R. J. *Cause and Explanation in Ancient Greek Thought*. Oxford: Clarendon Press, 1998.

Hanson, Norwood Russell. *Observation and Explanation: A Guide to the Philosophy of Science*. New York: Harper and Row, 1971.

———. "On the Symmetry between Explanation and Prediction." *Philosophical Review* 68 (1959): 349–58.

———. *Patterns of Discovery: An Inquiry into the Conceptual Foundations of Science*. London: Cambridge University Press, 1958.

Harris, Marvin. "History and Significance of the Emic/Etic Distinction." *Annual Review of Anthropology* 5 (1976): 329–50.

Harrison, Peter. *The Bible: Protestantism and the Rise of Natural Science*. Cambridge: Cambridge University Press, 1998.

———, ed. *The Cambridge Companion to Science and Religion*. Cambridge: Cambridge University Press, 2010.

Heeßel, Nils P. "Astrological Medicine in Babyonia." In *Astro-Medicine: Astrology and Medicine, East and West*, edited by Anna Akasoy, Charles Burnett, and Ronit Yoeli-Tlalim, 1–16. Micrologus Library 25. Florence: Sismel, Edizioni del Galluzzo, 2008.

———. *Babylonisch-assyrische Diagnostik*. Alter Orient und Altes Testament 43. Münster: Ugarit Verlag, 2000.

———. "Bibliographie zur altorientalischen Medizin 2000 bis August 2005 (mit Nachträgen aus früheren Jahren)." *Journal des Médecines Cunéiformes* 6 (2005): 34–40.

———. "The Calculation of the Stipulated Term in Extispicy." In *Divination and Interpretation of Signs in the Ancient World*, edited by Amar Annus, 163–75. Oriental Institute Seminars 6. Chicago: Oriental Institute of the University of Chicago, 2010.

———. "The Hands of the Gods: Disease Names, and Divine Anger." In *Disease in Babylonia*, edited by Irving L. Finkel and Markham J. Geller, 120–30. Cuneiform Monographs 36. Leiden: Brill, 2007.

———. "The Hermeneutics of Mesopotamian Extispicy: Theory vs. Practice." In *Mediating between Heaven and Earth: Communication with the Divine in the Ancient Near East*, edited by C. L. Crouch, Jonathan Stökl, and Anna Elise Zernecke, 16–35. New York: T. and T. Clark, 2012.

——. "Stein, Pflanze, und Holz: Ein neuer Text zur 'medizinischen Astrologie.'" *Orientalia* n.s. 74 (2005): 1–22.

Heimpel, Wolfgang. "The Sun at Night and the Doors of Heaven in Babylonian Texts." *Journal of Cuneiform Studies* 38 (1986): 127–51.

Hempel, Carl G. *Aspects of Scientific Explanation and Other Essays in the Philosophy of Science*. New York: Free Press, 1965.

——. "Explanation in Science and Its History." In *Frontiers of Science and Philosophy*, edited by Robert G. Colodny, 7–33. Pittsburgh: University of Pittsburgh Press, 1962.

——. "The Function of General Laws in History." *Journal of Philosophy* 39 (1942): 35–48.

Hempel, Carl G., and P. Oppenheim. "Studies in the Logic of Explanation." *Philosophy of Science* 15 (1948): 135–75.

Henry, John. "The Fragmentation of Renaissance Occultism and the Decline of Magic." *History of Science* 46 (2008): 1–48.

Hesse, Mary. *Revolutions and Reconstructions in the Philosophy of Science*. Brighton: Harvester Press, 1980.

Hilgert, Markus. "Von 'Listenwissenschaft' und 'epistemischen Dingen.' Konzeptuelle Annäherungen an altorientalische Wissenspraktiken." *Journal of General Philosophy of Science* 40 (2009): 277–309.

Hollywood, Louise, and John M. Steele. "Acronycal Risings in Babylonian Astronomy." *Centaurus* 46 (2004): 145–62.

Holton, Gerald. "On the Art of the Scientific Imagination." *Daedalus* 125 (1996): 183–208.

——. "On the Role of Themata in Scientific Thought." *Science* 188 (1975): 328–34.

——. *On the Thematic Origins of Scientific Thought: Kepler to Einstein*. Cambridge, MA: Harvard University Press, 1973; 1988 rev. ed.

——. *The Scientific Imagination: Case Studies*. London: Cambridge University Press, 1978.

Homans, George C. *The Nature of Social Science*. New York: Harcourt, Brace, and World, 1967.

Horowitz, Wayne. "Astral Tablets in the Hermitage." *Zeitschrift für Assyrologie und Vorderasiatische Archäologie* 90 (2000): 194–206.

——. "The Astrolabes: Astronomy, Theology, and Chronology." In *Calendars and Years: Astronomy and Time in the Ancient Near East*, edited by John M. Steele, 101–13. Oxford: Oxbow Books, 2007.

——. *Mesopotamian Cosmic Geography*. Mesopotamian Civilizations 8. Winona Lake, IN: Eisenbrauns, 1998.

——. *The Three Stars Each: The Astrolabes and Related Texts*. *Archiv für Orientforschung* Suppl. 33. Vienna: Institut für Orientalistik der Universität Wien, 2014.

Horsley, Richard A. "The Law of Nature in Philo and Cicero." *Harvard Theological Review* 71 (1978): 35–59.

Huber, Peter J. "A Lunar Six Text from 591 B.C." *Wiener Zeitschrift für die Kunde des Morgenlandes* 97 (2007): 213–17.

Huber, Peter J., and Salvo de Meis. *Babylonian Eclipse Observations from 750 BC to 1 BC*. Rome: IsIAO; Milan: Mimesis, 2004.

Huber, Peter J., and John M. Steele. "Babylonian Lunar Six Tablets." *SCIAMVS* 8 (2007): 3–36.

Huehnergard, John. *A Grammar of Akkadian*. Atlanta: Scholars Press, 1997.

Huehnergard, John, and Christopher Woods. "Akkadian and Eblaite." In *The Ancient Languages of Mesopotamia, Egypt, and Aksum*, edited by Roger D. Woodard, 218–80. Cambridge: Cambridge University Press, 2008.

Hume, David. *An Enquiry concerning Human Understanding* (1740). Oxford Philosophical Texts, edited by Tom L. Beauchamp. Oxford: Oxford University Press, 1999.

——. *A Treatise of Human Nature* (1738), edited by L. A. Selby-Bigge. Oxford: Clarendon Press, 1888, reprinted in 1964.

Hunger, Hermann. *Astrological Reports to Assyrian Kings*. State Archives of Assyria 8. Helsinki: Helsinki University Press, 1992.

——. "Astrologische Wettervorhersagen." *Zeitschrift für Assyrologie und Vorderasiatische Archäologie* 66 (1976): 234–60.

——. "Ein astrologisches Zahlenschema." *Wiener Zeitschrift für die Kunde des Morgenlandes* 86 (1996): 191–96.

——. *Astronomical Diaries and Related Texts from Babylon*. Vol. 6, *Goal Year Texts*. Vienna: Österreichische Akademie der Wissenschaften, 2006.

——. *Babylonische und Assyrische Kolophone*. Alter Orient und Altes Testament 2. Kevelaer: Butzon and Bercker; Neukirchen-Vluyn: Neukirchener Verlag des Erziehungsvereins, 1968.

——. "Noch ein Kalendertext." *Zeitschrift für Assyrologie und Vorderasiatische Archäologie* 64 (1974): 40–43.

——. "Saturnbeobachtungen aus der Zeit Nebukadnezars II." In *Assyriological et Semitica: Festschrift für Joachim Oelsner*, edited by Joachim Marzahn and Hans Neumann, 189–92. Münster: Ugarit Verlag, 2000.

——. *Spätbabylonische Texte aus Uruk*. Vol. 1. Berlin: Mann, 1976.

——. "Die Tafeln des Iqīša." *Welt des Orients* 6 (1971): 163–65.

——. "Uses of Enuma Anu Enlil for Chronology." *Akkadica* 119–20 (2000): 155–58.

——. "Die Wissenschaft der babylonischen Astronomen." In *Wissenskultur im Alten Orient: Weltanschauung, Wissenschaften, Techniken, Technologien*, edited by Hans Neumann, 95–103. Colloquien der Deutschen Orient-Gesellschaft 4. Wiesbaden: Harrassowitz Verlag, 2012.

Hunger, Hermann, and Teije de Jong. "Almanac W22340a from Uruk: The Latest Datable Cuneiform Tablet." *Zeitschrift für Assyrologie und Vorderasiatische Archäologie* 104 (2014): 182–94.

Hunger, Hermann, and David Pingree. *Astral Sciences in Mesopotamia*. Handbuch der Orientalistik. Leiden: Brill, 1999.

——. *MUL.APIN: An Astronomical Compendium in Cuneiform*. Archiv für Orientforschung Suppl. 24. Horn: Verlag Ferdinand Berger and Söhne, 1989.

Hunger, Hermann, and Abraham J. Sachs. *Astronomical Diaries and Related Texts from Babylonia*. Vol. 5, *Lunar and Planetary Texts*. Vienna: Österreichischen Akademie der Wissenschaften, 2001.

Jacobs, John. "Traces of the Omen Series Summa Izbu in Cicero, De Divinatione." In *Divination and Interpretation of Signs in the Ancient World*, edited by Ammar Annus, 317–39. Oriental Institute Seminars 6. Chicago: Oriental Institute of the University of Chicago, 2010.

Jeyes, Ulla. "Divination as a Science in Ancient Mesopotamia." *Jaarbericht "Ex Oriente Lux"* 32 (1991–92): 23–41.

Joannès, F. "Un precurseur paléo-babylonien de la serie *Šumma ālu*." In *Cinquante-deux reflections sur le proche-orient ancient: Offertes en homage à Léon De Meyer*, edited by Herman Gasche et al., 305–12. Leuven: Peeters, 1994.

Johnson, J. Cale. "The Origins of Scholastic Commentary in Mesopotamia: Second-Order Schemata in the Early Dynastic Exegetical Imagination." In *Visualizing Knowledge and Creating Meaning in Ancient Writing Systems*, edited by Shai Gordin, 11–55. Berliner Beiträge zum Vorderen Orient 23. Berlin: PeVe Verlag, 2014.

Jolly, Karen Louise. "Magic and Science." In *Encyclopaedia of the History of Science, Technology, and Medicine in Non-Western Cultures*, edited by Helaine Selin, 1246–51. Berlin: Springer-Verlag, 2008.

Jones, Alexander. *Astronomical Papyri from Oxyrhynchus*, 2 vols. Memoirs of the American Philosophical Society 233. Philadelphia: American Philosophical Society, 1999.

——. "A Study of Babylonian Observations of Planets Near Normal *Stars*." *Archive for History of Exact Sciences* 58 (2004): 475–536.

Jones, Alexander, and John M. Steele. 2011. "A New Discovery of a Component of Greek Astrology in Babylonian Tablets: The 'Terms.'" *ISAW Papers* 1 (2011), online at http://dlib.nyu.edu/awdl/isaw/isaw-papers/1/.

Kahneman, Daniel. "Intuitive Prediction: Biases and Corrective Procedures." *Advanced Decision*

Technology. Technical Report PTR 1042-77-6. Eugene, OR: Decision Research, a Division of Perceptronics, 1977.

——. *Thinking, Fast and Slow*. New York: Farrar, Straus and Giroux, 2013.

Kahneman, Daniel, and Amos Tversky. "On the Psychology of Prediction." *Psychological Review* 80 (1973): 237–51.

Kelley, Donald R. *The Human Measure: Social Thought in the Western Legal Tradition*. Cambridge, MA: Harvard University Press, 1990.

Khalidi, Muhammad Ali. *Natural Categories and Human Kinds: Classification in the Natural and Social Sciences*. Cambridge: Cambridge University Press, 2013.

Kirk, G. S. *The Nature of Greek Myths*. New York: Overlook Press, 1974.

Koch, Ulla Susanne. *Secrets of Extispicy: The Chapter Mutābiltu of the Babylonian Extispicy Series and Niṣirti bārûti, Texts Mainly from Assurbanipal's Library*. Alter Orient und Altes Testament 326. Münster: Ugarit Verlag, 2005.

——. "Sheep and Sky: Systems of Divinatory Interpretations." In *The Oxford Handbook of Cuneiform Culture*, edited by Karen Radner and Eleanor Robson, 447–69. Oxford: Oxford University Press, 2011.

——. "Three Strikes and You Are Out!" In *Divination and Interpretation of Signs in the Ancient World*, edited by Amar Annus, 43–60. Chicago: Oriental Institute of the University of Chicago, 2010.

Koch-Westenholz, Ulla. "The Astrological Commentary Šumma Sîn ina Tāmartīšu Tablet I." In *La science des cieux: Sages, mages, astrologues*, edited by R. Gyselen, 159–65. Res Orientales 12. Bures-sur-Yvette: Groupe pour l'étude de la civilisation du Moyen-Orient, 1999.

——. *Babylonian Liver Omens: The Chapters Manzāzu, Padānu, Pān tākalti of the Babylonian Extispicy Series Mainly from Assurbanipal's Library*. Carsten Niebuhr Institute of Near Eastern Studies Publications 25. Copenhagen: Museum Tusculanum Press, 2000.

——. *Mesopotamian Astrology: An Introduction to Babylonian and Assyrian Celestial Divination*. Carsten Niebuhr Institute of Near Eastern Studies Publications 19. Copenhagen: Museum Tusculanum Press, 1995.

Koester, Helmut. "Nomos Physeos: The Concept of Natural Law in Greek Thought." In *Religions in Antiquity*, edited by Jacob Neusner, 521–41. Leiden: Brill, 1968.

Kramer, S. N. "New Literary Catalogue from Ur." *Revue d'Assyriologie* 55 (1961): 169–76.

——. "The Oldest Literary Catalogue: A Sumerian List of Literary Compositions Composed about 2000 B.C." *Bulletin of the Schools of Oriental Research* 88 (1942): 10–19.

Kraus, F. R. "Ein zentrales Problem des altmesopotamischen Rechtes: Was ist der Codex Hammurabi?" *Genava* 8 (1960): 283–96.

Kugler, F. X. *Die Babylonische Mondrechnung: Zwei Systeme der Chaldäer über den Lauf des Mondes und der Sonne*. Freiburg im Breisgau: Herder, 1900.

——. *Sternkunde und Sterndienst in Babel: Babylonische Planetenkunde*. Vol. 1. Münster in Westfalen: Aschendorff, 1907.

——. *Sternkunde und Sterndienst in Babel: Assyriologische, astronomische und astral-mythologische Untersuchungen*. Vol. 2. Münster in Westfalen: Aschendorff, 1909–10.

Kuhn, Thomas S. *The Structure of Scientific Revolutions*. 3rd ed. Chicago: University of Chicago Press, 1996.

Kuukkanen, Jouni-Matti. "I Am Knowledge: Get Me Out of Here! On Localism and the Universality of Science." *Studies in History and Philosophy of Science* 42 (2011): 590–601.

Labat, René. *Un Calendrier babylonien des travaux, des signes et des mois-séries iqqur îpuš*. Bibliothéque de l'École des Hautes Études 321. Paris: H. Champion, 1965.

Laessoe, J. "A Prayer to Ea, Shamash, and Marduk, from Hama." *Iraq* 18 (1956): 60–67.

——. *Studies on the Assyrian Ritual and Series bīt rimki*. Copenhagen: Munksgaard, 1955.

Lambert, W. G. 1959. *Babylonian Oracle Questions*. Winona Lake, IN: Eisenbrauns, 2007.

——. "Enmeduranki and Related Matters." *Journal of Cuneiform Studies* 21 (1967): 126–38.

——. "The Laws of Hammurabi in the First Millennium." In *Reflets des deux flueves: Volume de Mélanges offerts à André Finet*, edited by M. Lebeau and Ph. Talon, 95–98. Akkadica Supplementum 6. Leuven: Peeters, 1989.

——. "The Qualifications of Babylonian Diviners." In *Festschrift für Rykle Borger zu seinem 65. Geburtstag am 24. Mai 1994: Tikip Santakki Mala Bašmu*, edited by Stefan M. Maul, 141–58. Cuneiform Monographs 10. Groningen: Styx, 1998.

——. "Studies in Marduk." *Bulletin of the School of Oriental and African Studies* 47 (1984): 1–9.

——. "The Sultantepe Tablets: A Review Article." *Revue d'Assyriologie* 53 (1959): 119–38.

Landsberger, Benno. *The Conceptual Autonomy of the Babylonian World*. Translated by T. Jacobsen, B. Foster, and H. von Siebenthal. Sources and Monographs on the Ancient Near East 1, fasc. 4. Malibu, CA: Undena, 1976.

——. "Die Eigenbegrifflichkeit der babylonischen Welt." *Islamica* 2 (1926): 355–72.

——. "Die Babylonischen Termini für Gesetz und Recht." In *Symbolae ad Iura Orientis Antiqui Pertinentes Paulo Koschaker Dedicatae*, edited by J. Friedrich, J. G. Lautner, and J. Miles, 219–34. Studia et Documenta ad Iura Orientis Antiqui Pertinentia 2. Leiden: Brill, 1939.

Landsberger, Benno, in cooperation with Anne Draffkorn Kilmer and Edmond I. Gordon. *The Fauna of Ancient Mesopotamia: First Part, Tablet XIII*. Materialien zum Sumerischen Lexikon 8/1. Rome: Pontificium Institutum Biblicum, 1960.

Landsberger, Benno, with the collaboration of I. Krumbiegel. *Die Fauna des alten Mesopotamien nach der 14. Tafel der Serie Har-ra = hubullu*. XLII Bd. der Ahbandungen der philologisch-historischen Klasse der Sächsischen Akademie der Wissenschaften 6. Leipzig: Verlag von S. Hirzel, 1934.

Landsberger, Benno, E. Reiner, and M. Civil. *The Series HAR-ra = hubullu: Tablets XVI, XVII, XIX and Related Texts*. Materialien zum sumerischen Lexikon 10. Rome: Pontificium Institutum Biblicum, 1970.

Langdon, Stephen. *Babylonian Menologies*. Schweich Lectures of the British Academy. London: Oxford University Press, 1935.

Larsen, Mogens Trolle. "The Mesopotamian Lukewarm Mind: Reflections on Science, Divination, and Literacy." In *Language, Literature, and History: Philological and Historical Studies Presented to Erica Reiner*, edited by F. Rochberg-Halton, 203–25. New Haven, CT: American Oriental Society, 1987.

Latour, Bruno. *We Have Never Been Modern*. Translated by Catherine Porter. Cambridge, MA: Harvard University Press, 1993.

Lehoux, Daryn. "Diogenēs of Babylōn." In *Encyclopedia of Ancient Natural Scientists: The Greek Tradition and Its Many Heirs*, edited by Paul T. Keyser and Georgia L. Irby-Massie, 253. London: Routledge, 2008.

——. "Logic, Physics, and Prediction in Hellenistic Philosophy: x Happens, but y?" In *Foundations of the Formal Sciences IV: History of the Concept of the Formal Sciences*, edited by Benedikt Löwe, Volker Peckhaus, and Thoralf Räsch, 125–42. Studies in Logic 3. London: Strand, 2006.

——. "Observation and Prediction in Ancient Astrology." *Studies in History and Philosophy of Science* 35 (2004): 227–46.

——. "Tomorrow's News Today: Astrology, Fate, and the Way Out." *Representations* 95 (2006): 105–22.

——. *What Did the Romans Know? An Inquiry into Science and Worldmaking*. Chicago: University of Chicago Press, 2012.

Leichty, Erle. *The Omen Series Šumma Izbu*. Texts from Cuneiform Sources 4. Locust Valley, NY: J. J. Augustin Publishers, 1970.

——. *The Royal Inscriptions of Esarhaddon, King of Assyria (680–669 BC)*. Royal Inscriptions of the Neo-Assyrian Period 4. Winona Lake, IN: Eisenbrauns, 2011.

Lenzi, Alan. "Akkadian Scholarship: Kassite to Late Babylonian Periods." *Journal of Ancient Near Eastern History* 2 (2015), in press.

——. *Secrecy and the Gods: Secret Knowledge in Ancient Mesopotamia and Biblical Israel*. State Archives of Assyria Studies 19. Helsinki: Neo-Assyrian Text Corpus Project, 2008.

——. "The Uruk List of Kings and Sages and Late Mesopotamian Scholarship." *Journal of Ancient Near Eastern Religions* 8 (2008): 137–69.

Lenzi, Alan, and Jonathan Stökl, eds. 2014. *Divination, Politics, and Ancient Near Eastern Empires*. Ancient Near Eastern Monographs 7. Atlanta: Society of Biblical Literature.

Lévy-Bruhl, Lucien. *Les Carnets de Lévy-Bruhl*, published posthumously. Paris: Presses Universitaires de France, 1949.

——. *How Natives Think*. Translated by Lilian A. Clare. Princeton, NJ: Princeton University Press, 1985.

——. *The Notebooks on Primitive Mentality*. Translated by Peter Rivière. Oxford: Basil Blackwell, 1975.

Lewontin, R. C. Review of Lily E. Kay, *Who Wrote the Book of Life? A History of the Genetic Code*. Stanford, CA: Stanford University Press, 2000, in *Science* 291 (2001): 1263–64.

Lieberman, Stephen J. "A Mesopotamian Background for the So-Called *Aggadic* 'Measures' of Biblical Hermeneutics?" *Hebrew Union College Annual* 58 (1987): 157–225.

Lienhardt, Godfrey. *Divinity and Experience: The Religion of the Dinka*. Oxford: Clarendon Press, 1961.

Linssen, Marc. *The Cults of Uruk and Babylon: The Temple Ritual Texts as Evidence for Hellenistic Cult Practice*. Leiden: Brill, 2004.

Liverani, Mario. *Myth and Politics in Ancient Near Eastern Historiography*. London: Equinox Publishing, 2004.

Livesey, Steven J., and Richard H. Rouse. "Nimrod the Astronomer." *Traditio* (1981): 203–66.

Livingstone, Alasdair. *Court Poetry and Literary Miscellanea*. State Archives of Assyria 3. Helsinki: University of Helsinki, 1989.

——. *Hemerologies of Assyrian and Babylonian Scholars*. Cornell University Studies in Assyriology and Sumerology 25. Bethesda, MD: CDL Press, 2013.

——. *Mystical and Mythological Explanatory Works of Assyrian and Babylonian Scholars*. Oxford: Oxford University Press, 1986. Reprinted by Eisenbrauns, 2007.

Lloyd, G. E. R. *Being, Humanity, and Understanding*. Oxford: Oxford University Press. 2012.

——. *Cognitive Variations: Reflections of the Unity and Diversity of the Human Mind*. Oxford: Oxford University Press, 2007.

——. *Methods and Problems in Greek Science*. Cambridge: Cambridge University Press, 1991.

——. *Polarity and Analogy: Two Types of Argumentation in Early Greek Thought*. 1966; Cambridge: Cambridge University Press, 1992.

——. "Saving the Appearances." *Classical Quarterly* 28 (1978): 202–22.

Löhnert, Anne. "Scribes and Singers of Emesal Lamentations in Ancient Mesopotamia." In *Papers on Ancient Literatures: Greece, Rome, and the Near East*, edited by E. Cingano and Lucio Milano, 421–47. Quaderni del Dipartimento di scienze dell'antichità e del Vicino oriente 4. Padova: Sargon, 2008.

Longino, Helen. *The Fate of Knowledge*. Princeton, NJ: Princeton University Press, 2001.

Longxi, Zhang. *Mighty Opposites: From Dichotomies to Differences in the Comparative Study of China*. Stanford, CA: Stanford University Press, 1998.

Lukes, Steven. "Different Cultures Different Rationalities?" *History of the Human Sciences* 13 (2000): 3–18.

Lüthy, Christoph. "What to Do with Seventeenth-Century Natural Philosophy? A Taxonomic Problem." *Perspectives in Science* 8 (2000): 164–95.

Luukko, Mikko. "The Administrative Roles of the 'Chief Scribe' and the 'Palace Scribe' in the Neo-Assyrian Period." *State Archives of Assyria Bulletin* 16 (2007): 227–56.

Magnani, Lorenzo, Nancy J. Nersessian, and Paul Thagard. *Model-Based Reasoning in Scientific Discovery*. New York: Kluwer Academic, Plenum, 1999.

Magnus, P. D. *Scientific Enquiry and Natural Kinds: From Planets to Mallards*. New York: Palgrave Macmillan, 2012.

Marchetti, Nicolò. "Divination at Ebla during the Old Syrian Period: The Archaeological Evidence." In *Exploring the Longue Durée: Essays in Honor of Lawrence E. Stager*, edited by J. D. Schloen, 279–95. Winona Lake, IN: Eisenbrauns, 2009.

Mauer, Gerlinde. "Ein Schülerexzerpt aus Enūma Anu Enlil." *Baghdader Mitteilungen* 18 (1987): 239–42.

Maul, Stefan M. "Der Assyrische König—Hüter der Weltordnung." In *Gerechtigkeit: Richten und Retten in der abendländischen Tradition und ihren altorientalischen Ursprüngen*, edited by Jan Assmann, Bernd Janowski, and Michael Welker, 65–77. Munich: Wilhelm Fink Verlag, 1998.

———. "How the Babylonians Protected Themselves against Calamities Announced by Omens." In *Mesopotamian Magic: Textual, Historical, and Interpretive Perspectives*. Studies in Ancient Magic and Divination 1, edited by T. Abusch and K. Van der Toorn, 123–29. Groningen: Styx, 1999.

———. *Die Wahrsagekunst im Alten Orient*. Munich: C. H. Beck, 2013.

———. "Das Wort im Worte: Orthographie und Etymologie als hermeneutische Verfahren babylonischer Gelehrter." In *Commentaries—Kommentare*, edited by Glen W. Most, 1–18. Aporemata: Kritische Studien zur Philologiegeschichte 4. Göttingen: Vandenhoeck and Ruprecht, 1999.

———. *Zukunftsbewältigung: Eine Untersuchung altorientalischen Denkens anhand der babylonisch-assyrischen Löserituale (Namburbi)*. Mainz am Rhein: Verlag Philipp von Zabern. 1994.

Mayhoff. C., and L. Jan, eds. *Naturalis Historiae*. Vol. 5, *Libri XXXI–XXXVII / Plinii secundi naturalis historiae libri xxxvii*. 5 vols. Bibliotheca scriptorum Graecorum et Romanorum Teubneriana (Book 1654). K. G. Saur Verlag, 1998 / Leipzig: Teubner, 1892–1909.

Medin, D. L., and A. Ortony. "Psychological Essentialism." In *Similarity and Analogical Reasoning*, edited by S. Vosniadou and A. Ortony, 179–95. Cambridge: Cambridge University Press, 1989.

Meijer, Diederik J. W., ed. *Natural Phenomena: Their Meaning, Depiction, and Description in the Ancient Near East*. North-Holland, Amsterdam: Royal Netherlands Academy of Arts and Sciences, 1992.

Merchant, Carolyn. "The Scientific Revolution and the Death of Nature." *Isis* 97 (2006): 513–33.

Merlan, P. "The Old Academy, Section G: Theology and Demonology." In *The Cambridge History of Later Greek and Early Medieval Philosophy*, edited by A. H. Armstrong, 7th printing, 14–38. 1967; Cambridge: Cambridge University Press, 2007.

Michalowski, Piotr. "Presence at the Creation." In *Lingering over Words: Studies in Ancient Near Eastern Literature in Honor of William L. Moran*, edited by T. Abusch, J. Huehnergard, and P. Steinkeller, 381–96. Atlanta: Scholars Press, 1990.

Miller, Patricia Cox. "In Praise of Nonsense." In *Classical Mediterranean Spirituality: Egyptian, Greek, Roman*, edited by A. H. Armstrong, 481–505. New York: Crossroad Publishing, 1986.

Milton, J. R. "Laws of Nature." In *The Cambridge History of Seventeenth-Century Philosophy*, vol. 1, edited by Daniel Garber and Michael Ayers, 680–701. Cambridge: Cambridge University Press, 1998.

———. "The Origin and Development of the Concept of the 'Laws of Nature.'" *Archives Européennes de Sociologie* 22 (1981): 173–95.

Montelle, Clemency. *Chasing Shadows: Mathematics, Astronomy, and the Early History of Eclipse Reckoning*. Baltimore: Johns Hopkins University Press, 2011.

Morgenbesser, Sidney. "The Realist-Instrumentalist Controversy." In *Essays in Honor of Ernest Nagel: Philosophy, Science, and Method*, edited by Sidney Morgenbesser, Patrick Suppes, and Morton White, 200–218. New York: St. Martin's Press, 1969.

Morgenstern. Julian. *The Doctrine of Sin in the Babylonian Religion*. Berlin: W. Peiser, 1905.

Müller, Friedrich Max. *Chips from a German Workshop*. Vol. 2. New York: Longmans, Green, 1898.

Murdoch, John E. "The Analytic Character of Late Medieval Learning: Natural Philosophy without Nature." In *Approaches to Nature in the Middle Ages*, edited by Lawrence D. Roberts, 171–213. Binghamton, NY: Center for Medieval and Early Renaissance Studies, 1982.

Murray, Gilbert. *Five Stages of Greek Religion*. Garden City, NY: Doubleday Anchor, 1951.

Naddaf, Gerard. *The Greek Concept of Nature*. Albany: State University of New York Press, 2005.

Nagel, Thomas. "What Is It Like to Be a Bat?" *Philosophical Review* 83 (1974): 435–50.

Needham, Joseph. "Human Laws and Laws of Nature in China and the West." *Journal of the History of Ideas* 12 (1951): 3–30 (pt. 1) and 194–230 (pt. 2).

———. *Science and Civilization in China*. Vol. 2, *History of Scientific Thought*. Cambridge: Cambridge University Press, 1956.

Needham, Rodney. *Belief, Language, and Experience*. Chicago: University of Chicago Press, 1972.

Nersessian, Nancy J. *Creating Scientific Concepts*. Cambridge, MA: MIT Press, 2008.

———. "Model-Based Reasoning in Conceptual Change." In *Model-Based Reasoning in Scientific Discovery*, edited by Lorenzo Magnani, Nancy J. Nersessian, and Paul Thagard, 5–22. New York: Kluwer Academic, Plenum, 1999.

———. "Model-Based Reasoning in Distributed Cognitive Systems." *Philosophy of Science* 73 (2006): 699–709.

Netz, Reviel. *Ludic Proof: Greek Mathematics and the Alexandrian Aesthetic*. Cambridge: Cambridge University Press, 2009.

Neugebauer, Otto. "The Alleged Babylonian Discovery of the Precession of the Equinoxes." *Journal of the American Oriental Society* 70 (1950): 1–8.

———. *Astronomical Cuneiform Texts*. 3 vols. London: Lund Humphries, 1955.

———. *Astronomy and History Selected Essays*. Berlin: Springer Verlag, 1983.

———. *The Exact Sciences in Antiquity*. 2nd ed. New York: Dover, 1969.

———. "From Assyriology to Renaissance Art." *Proceedings of the American Philosophical Society* 133 (1989): 391–403.

———. "The History of Ancient Astronomy: Problems and Methods." *Journal of Near Eastern Studies* 4 (1945): 1–38, also published in 1946, *Proceedings of the Astronomical Society of the Pacific* 58, and in *Astronomy and History Selected Essays*.

———. *A History of Ancient Mathematical Astronomy*. 3 vols. Berlin: Springer Verlag, 1975.

———. "The Survival of Babylonian Methods in the Exact Sciences of Antiquity and the Middle Ages." *Proceedings of the American Philosophical Society* 107 (1963): 528–35.

Neugebauer, Otto, and H. B. van Hoesen. *Greek Horoscopes*. Memoirs of the American Philosophical Society 48. Philadelphia: American Philosophical Society, 1959.

Neugebauer, Otto, and A. J. Sachs. "The 'Dodecatemoria' in Babylonian Astrology." *Archiv für Orientforschung* 16 (1953): 65–66.

Neurath, Otto. "Foundations of the Social Sciences." In *Foundations of the Unity of Science*, vol. 2, edited by O. Neurath, R. Carnap and C. Morris, 1–52. Chicago: University of Chicago Press, 1971.

Newman, William R., and Anthony Grafton. *Secrets of Nature: Astrology and Alchemy in Early Modern Europe*. Cambridge, MA: MIT Press, 2001.

Nickerson, Raymond S. *Aspects of Rationality: Reflections on What It Means to Be Rational and Whether We Are*. New York: Psychology Press, 2008.

Nisbett, Richard E., and Lee Ross. *Human Inference: Strategies and Shortcomings of Social Judgement*. Upper Saddle River, NJ: Prentice Hall, 1980.

Nissinen, Marttii. "Pesharim as Divination: Qumran Exegesis, Omen Interpretation, and Literary Prophecy." In *Prophecy after the Prophets? The Contribution of the Dead Sea Scrolls to the Understanding of Biblical and Extra-Biblical Prophecy*, edited by K. De Troyer and A. Lange, with the assistance of Lucas L. Schulte, 43–60. Leuven: Peeters, 2009.

Oakley, Francis. "Christian Theology and the Newtonian Science: The Rise of the Concept of the Laws of Nature." *Church History* 30 (1961): 433–57.

Obeyesekere, Gananath. *The Apotheosis of Captain Cook: European Mythmaking in the Pacific*. Princeton, NJ: Princeton University Press, 1992.

Oelsner, J. "Von Iqīša und eigenen anderen spätgeborenen Babyloniern." In *Studi su vicino oriente antico dedicati alla memoria di Luigi Cagni*, edited by S. Graziani, 797–813. Naples: Instituto Universitario Orientale, 2000.

Oppenheim, A. L. *Ancient Mesopotamia: Portrait of a Dead Civilization*, rev. ed., Erica Reiner. Chicago: University of Chicago Press, 1977.

———. "A Babylonian Diviner's Manual." *Journal of Near Eastern Studies* 33 (1974): 197–220.

———. "Man and Nature in Mesopotamian Civilization." In *Dictionary of Scientific Biography* 15, edited by C. C. Gillispie, 634–66. Oxford: Oxford University Press, 1978.

Oshima, Takayoshi. *The Babylonian Theodicy*. State Archives of Assyria Cuneiform Texts 9. Helsinki: Neo-Assyrian Text Corpus Project, 2013.

Ossendrijver, Mathieu. *Babylonian Mathematical Astronomy: Procedure Texts*. Berlin: Springer Verlag, 2012.

———. "Exzellente Netzwerke: Die Astronomen von Uruk." In *The Empirical Dimension of Ancient Near Eastern Studies*, edited by G. J. Selz and K. Wagensonner, 631–44. Wiener Offene Orientalistik 8. Vienna: LIT Verlag, 2011.

———. "Science in Action: Networks in Babylonian Astronomy." In *Babylon: Wissenskultur zwischen Orient und Okzident*, edited by E. Cancik-Kirschbaum, Margarete van Ess, and Joachim Marzahn, 213–21. Topoi: Berlin Studies of the Ancient World. Berlin: De Gruyter, 2011.

Papineau, David. "Naturalism." *Stanford Encyclopedia of Philosophy*, 2013 edition, edited by Edward N. Zalta, http://plato.stanford.edu/entries/naturalism/.

Park, Katharine. "Natural Particulars: Medical Epistemology, Practice, and the Literature of Healing Springs." In *Natural Particulars: Nature and the Disciplines in Renaissance Europe*, edited by Anthony Grafton and Nancy Siraisi, 347–67. Cambridge, MA: MIT Press, 1999.

———. "Observation in the Margins, 500–1500." In *Histories of Scientific Observation*, edited by Lorraine Daston and Elizabeth Lunbeck, 15–44. Chicago: University of Chicago Press, 2011.

Parker, Richard A. *A Vienna Demotic Papyrus on Eclipse- and Lunar-Omina*. Providence, RI: Brown University Press, 1959.

Parpola, Simo. *Letters from Assyrian and Babylonian Scholars*. State Archives of Assyria 10. Helsinki: Helsinki University Press, 1993.

———. *Letters from Assyrian Scholars to the Kings Esarhaddon and Assurbanipal*. Part 1, *Texts*. Alter Orient und Altes Testament. Neukirchen-Vluyn: Verlag Butzon and Bercker Kevelaer, 1970.

Pearce, Laurie E. "Materials of Writing and Materiality of Knowledge." In *Gazing on the Deep: Ancient Near Eastern and Other Studies in Honor of Tzvi Abusch*, edited by Jeffery Stackert, Barbara Nevling Porter, and David P. Wright, 167–79. Winona Lake, IN: Eisenbrauns, 2010.

Pedersén, Olof. *Archives and Libraries in the Ancient Near East, 1500–300 B.C.* Bethesda, MD: CDL Press, 1998.

Pingree, David. "From Alexandria to Baghdād to Byzantium. The Transmission of Astrology." *International Journal of the Classical Tradition* 8 (2001): 3–37.

———. *From Astral Omens to Astrology; From Babylon to Bikaner*. Rome: Istituto Italiano Per L'Africa E L'Oriente, 1997.

———. "Hellenophilia versus the History of Science." *Isis* 83 (1992): 554–63.

———. "Legacies in Astronomy and Celestial Omens." In *The Legacy of Mesopotamia*, edited by Stephanie Dalley, 125–37. Oxford: Oxford University Press, 1998.

———. "The Logic of Non-Western Science: Mathematical Discoveries in Medieval India." *Daedalus* 132 (2003): 45–53.

———. "Mesopotamian Astronomy and Astral Omens in Other Civilizations." In *Mesopotamien und seine Nachbarn. Politische und kulturelle Wechselbezeihungen im Alten Vorderasien vom 4. bis 1. Jahrtausend v. Chr. bis 1 Jahrtausend v. Chr*, edited by H. Nissen and J. Renger, 613–31. RAI 25. Berlin: Reimer, 1982.

———. *Pathways into the Study of Ancient Sciences: Selected Essays by David Pingree*, edited by Isabelle Pingree and John M. Steele. Philadelphia: American Philosophical Society Press, 2014.

———. "Some of the Sources of the Ghāyat al-Hakīm." *Journal of the Warburg and Courtauld Institutes* 43 (1980): 1–15.

———. "Venus Omens in India and Babylon." In *Language, Literature, and History: Philological and Historical Studies Presented to Erica Reiner*, edited by F. Rochberg-Halton, 293–315. AOS 67. New Haven, CT: American Oriental Society, 1987.

———. *The Yavanajātaka of Sphujidhvaja*. Harvard Oriental Series 48. Cambridge, MA: Harvard University Press, 1978.

Pingree, David, and Erica Reiner. "Observational Texts concerning the Planet Mercury." *Revue d'Assyriologie* 69 (1975): 175–80.

Plato. *Phaedrus*. Translated by Harold North Fowler. Vol. 1. Loeb Classical Library 36. Cambridge, MA: Harvard University Press, 1982.

Pliny. *Natural History*, Books 1–7. Translated by H. Rackham. Vol. 2. Loeb Classical Library 352. Cambridge, MA: Harvard University Press, 1942.

———. *Natural History*, Books 36–37. Translated by D. E. Eichholz. Vol. 10. Loeb Classical Library 419. Cambridge, MA: Harvard University Press, 1962.

Plofker, Kim, and Toke Lindegaard Knudson. "Paitāmahasiddhānta." In *Encyclopedia of Ancient Natural Scientists*, edited by Paul T. Keyser and Georgia L. Irby-Massie, 604–5. London: Routledge, 2008.

Pollock, Sheldon. *The Language of the Gods in the World of Men: Sanskrit, Culture, and Power in Premodern India*. Berkeley: University of California Press, 2009.

Pongratz-Leisten, Beate. *Herrschaftswissen in Mesopotamien: Formen der Kommunikation zwischen Gott und König im 2. und 1. Jahrtausend v. Chr.* State Archives of Assyria Studies 10. Helsinki: Neo-Assyrian Text Corpus Project, 1999.

———. "The King at the Crossroads between Divination and Cosmology." In *Divination, Politics, and Ancient Near Eastern Empires*, edited by Alan Lenzi and Jonathan Stökl, 33–48. Atlanta: Society of Biblical Literature, 2014.

Postgate, Nicholas. "Mesopotamian Petrology: Stages in the Classification of the Material World." *Cambridge Archaeological Journal* 7 (1997): 205–24.

Prescott, F. C. *Poetry and Myth*. New York: Macmillan, 1927.

Probst, P. "Die Macht der Schrift. Zum ethnologischen Diskurs über eine populäre Denkfigur." *Anthropos* 87 (1992): 167–82.

Putnam, Hilary. *Reason, Truth, and History*. Cambridge: Cambridge University Press, 1981.

Quine, Willard van Ormand. *From Stimulus to Science*. Cambridge, MA: Harvard University Press, 1996.

———. "In Praise of Observation Sentences." *Journal of Philosophy* 90 (1993): 107–16.

———. "Natural Kinds." In *Essays in Honor of Carl G. Hempel*, edited by Nicholas Rescher et al., 1–23. Dordrecht: D. Reidel, 1970.

Radner, Karen, and Eleanor Robson. *The Handbook of Cuneiform Culture*. Oxford: Oxford University Press, 2011.

Ragep, Jamil. "Copernicus and His Islamic Predecessors: Some Historical Remarks." *History of Science* 45 (2007): 65–81.

Reed, Annette Yoshiko. "Zachalias of Babylon." In *Encyclopedia of Ancient Natural Scientists: The Greek Tradition and Its Many Heirs*, edited by Paul T. Keyser and Georgia L. Irby-Massie, 843. London: Routledge, 2008.

Reiner, E. *Astral Magic in Babylonia*. Transactions of the American Philosophical Society 85. Philadelphia: American Philosophical Society, 1995.

———. "Constellation into Planet." In *Studies in the History of the Exact Sciences in Honour of David Pingree*, edited by Charles Burnett, Jan P. Hogendijk, Kim Plofker, and Michio Yano, 3–15. Leiden: Brill, 2004.

———. "Early Zodiologia and Related Matters." In *Wisdom, Gods, and Literature: Studies in Assyriology in Honour of W. G. Lambert*, edited by A. R. George and Irving L. Finkel, 421–27. Winona Lake, IN: Eisenbrauns, 2000.

———. "The Etiological Myth of the 'Seven Sages.'" *Orientalia* n.s. 30, no. 1 (1961): 1–11.

———. "Fortune-Telling in Mesopotamia." *Journal of Near Eastern Studies* 19 (1960): 23–35.

——. "Magic Figurines, Amulets, and Talismans." In *Monsters and Demons in the Ancient and Medieval Worlds: Papers Presented to Edith Porada*, edited by Ann E. Farkas, Prudence O. Harper, and Evelyn B. Harrison, 27–36. Mainz on Rhein: Verlag Philipp von Zabern, 1987.

——. *Šurpu: A Collection of Sumerian and Akkadian Incantations. Archiv für Orientforschung* Suppl. 11. Graz: Selbstverlag des Herausgebers, 1958.

——. *Your Thwarts in Pieces, Your Mooring Rope Cut: Poetry from Babylonia and Assyria*. Michigan Studies in the Humanities. Ann Arbor: Michigan Slavic Publications, 1985.

Reiner, E., and David Pingree. *Babylonian Planetary Omens 2: Enūma Anu Enlil Tablets 50–51*. Bibliotheca Mesopotamica 2. Malibu, CA: Undena, 1981.

——. *Babylonian Planetary Omens 3*. Groningen: Styx, 1998.

——. *Babylonian Planetary Omens 4*. Cuneiform Monographs 30. Leiden: Brill, 2005.

——. *The Venus Tablet of Ammiṣaduqa*. Babylonian Planetary Omens 1. Malibu, CA: Undena, 1975.

Renger, Johannes. "Untersuchungen zum Priestertum der altbabylonischen Zeit 2. Teil." *Zeitschrift für Assyrologie und Vorderasiatische Archäologie* 59 (1969): 104–230.

Rescher, Nicholas, et al., eds. *Essays in Honor of Carl G. Hempel*. Dordrecht: D. Reidel, 1970.

Reydon, Thomas A. C. "Natural Kinds No Longer Are What They Never Were." *Metascience* 24 (2015): 259–64.

Richardson, Seth F. C. "On Seeing and Believing: Liver Divination and the Era of Warring States." In *Divination and the Interpretation of Signs in the Ancient World*, edited by Amar Annus, 225–66. Oriental Institute Seminars 6. Chicago: Oriental Institute Publications, 2010.

Roaf, Michael, and Annette Zgoll. "Assyrian Astroglyphs: Lord Aberdeen's Black Stone and the Prisms of Esarhaddon." *Zeitschrift für Assyrologie und Vorderasiatische Archäologie* 91 (2001): 264–95.

Robson, Eleanor. *Ancient Knowledge Networks* (forthcoming).

——. "Empirical Scholarship in the Neo-Assyrian Court." In *The Empirical Dimension of Ancient Near Eastern Studies = Die empirische Dimension altorientalischer Forschungen*, edited by Gebhard J. Selz and Klaus Wagensonner, 603–29. Berlin: LIT Verlag, 2011.

——. "Mesopotamian Medicine and Religion." *Religion Compass* 2/4 (2008): 455–83.

——. "Numeracy." In *The Princeton Companion to Mathematics*, edited by T. Gowers and J. E. Barrow-Green, 983–91. Princeton, NJ: Princeton University Press, 2008.

——. "The Production and Dissemination of Scholarly Knowledge." In *The Oxford Handbook of Cuneiform Culture*, edited by Karen Radner and Eleanor Robson, 557–76. Oxford: Oxford University Press, 2011.

——. "Reading the Libraries of Ancient Assyria and Babylonia." In *Ancient Libraries*, edited by Jason König, Katerina Oikonomopolou, and Greg Woolf, 38–56. Cambridge: Cambridge University Press, 2013.

——. "The Uses of Mathematics in Ancient Iraq (6000–600 BC)." In *Mathematics across Cultures: The History of Non-Western Mathematics*, edited by H. Selin, 93–114. Dordrecht: Kluwer Academic, 2001.

Robson, Eleanor, and K. R. Stevens. "Scholarly Tablet Collections in First Millennium Assyria and Babylonia." In *The Earliest Libraries: Library Tradition in the Ancient Near East*, edited by G. Barjamovic and K. Ryholt. Oxford University Press, in press.

Rochberg, Francesca. *Babylonian Horoscopes*. Transactions of the American Philosophical Society 88. Philadelphia: American Philosophical Society, 1998.

——. "Beyond Binarism in Babylon." *Interdisciplinary Science Reviews* 35 (2010): 253–65.

——. "Canon and Power in Cuneiform Scribal Scholarship." In *Problems of Canonicity and Identity Formation in Ancient Egypt and Mesopotamia*, edited by Kim Ryholt and Gojko Barjamovic, 217–29. Copenhagen: Carsten Niebuhr Institute, in press.

——. "Conceiving the History of Science Forward." In *The Frontiers of Ancient Science: Essays in Honor of Heinrich von Staden*, 515–31, edited by Brooke Holmes and Klaus-Dietrich Fischer. Beiträge zur Altertumskunde 338. Berlin: De Gruyter, 2015.

———. "A Critique of the Cognitive-Historical Thesis of *The Intellectual Adventure of Ancient Man*." In *The Adventure of the Human Intellect: Self, Society, the Divine in Ancient World Cultures*, edited by Kurt A. Raaflaub, 16–28. The Ancient World: Comparative Histories. Malden, MA: Wiley, 2016.

———. "Divine Causality and Cuneiform Divination." In *A Common Cultural Heritage: Studies in Mesopotamia and the Biblical World in Honor of Barry L. Eichler*, edited by G. Frame, Erle Leichty, Jeffery Tigay, and Steve Tinney, 189–203. Bethesda, MD: CDL Press, 2011.

———. "'A Firm Yes': Certainty in the Assyrian Culture of Knowledge." *Bulletin for the Canadian Society for Mesopotamian Studies* 7 (2014): 5–12.

———. "Foresight in Ancient Mesopotamia." In *Foresight*, edited by Lawrence W. Sherman. Darwin College Lecture Series. Cambridge: Cambridge University Press, forthcoming.

———. *The Heavenly Writing: Divination, Horoscopy, and Astronomy in Mesopotamian Culture*. Cambridge: Cambridge University Press, 2004.

———. "The History of Science and Ancient Mesopotamia." *Journal of Ancient Near Eastern History* 1 (2013): 37–60.

———. "If P, then Q: Form, Reasoning, and Truth in Babylonian Divination." In *Divination and Interpretation of Signs in the Ancient World*, edited by Amar Annus, 19–27. Oriental Institute Seminars 6. Chicago: Oriental Institute of the University of Chicago, 2010.

———. *In the Path of the Moon: Babylonian Celestial Divination and Its Legacy*. Leiden: Brill, 2010.

———. "Inference, Conditionals, and Possibility in Ancient Mesopotamian Science." *Science in Context* 22 (2009): 4–25.

———. "Lunar Data in Babylonian Horoscopes." In *Astronomy and Astrology from the Babylonians to Kepler: Essays Presented to Bernard R. Goldstein on the Occasion of His 65th Birthday*, edited by P. Barker, A. C. Bowen, J. Chabas, G. Freudenthal, and T. Langermann. *Centaurus* 45 (2003): 32–45.

———. "Observing and Describing the World through Divination and Astronomy." In *The Oxford Handbook of Cuneiform Culture*, edited by Karen Radner and Eleanor Robson, 618–36. Oxford: Oxford University Press, 2011.

———. "Sheep and Cattle, Cows and Calves: The Sumero-Akkadian Astral Gods as Livestock." In *Opening the Tablet Box: Near Eastern Studies in Honor of Benjamin R. Foster*, edited by S. Melville and A. Slotsky, 347–59. Culture and History of the Ancient Near East 42. Leiden: Brill, 2010.

———. "The Stars Their Likenesses: Perspectives on the Relation between Celestial Bodies and Gods in Ancient Mesopotamia." In *What Is a God? Anthropomorphic and Non-Anthropomorphic Aspects of Deity in Ancient Mesopotamia*, edited by Barbara Nevling Porter, 1–52. Winona Lake, IN: Eisenbrauns, 2009.

———. "Sudines." In *Encyclopedia of Ancient Natural Scientists: The Greek Tradition and Its Many Heirs*, edited by Paul T. Keyser and Georgia L. Irby-Massie, 767–68. London: Routledge, 2008.

———. "Where Were the Laws of Nature before There Was Nature?" In *Laws of Heaven—Laws of Nature: The Legal Interpretation of Cosmic Phenomena in the Ancient World*, edited by Konrad Schmid and Christoph Uehlinger, 21–39. Orbis Biblicus et Orientalis 276. Fribourg: Academic Press, and Göttingen: Vandenhoeck and Ruprecht, 2016.

Rochberg-Halton, Francesca. *Aspects of Babylonian Celestial Divination: The Lunar Eclipse Tablets of Enuma Anu Enlil. Archiv für Orientforschung* Suppl. 22. Horn: Ferdinand Berger und Söhne, 1988.

———. "The Babylonian Astronomical Diaries." Review of Sachs and Hunger, *Astronomical Diaries and Related Texts from Babylon*, vol. 1. *Journal of the American Oriental Society* 111 (1991): 323–32.

———. "Elements of the Babylonian Contribution to Hellenistic Astrology." *Journal of the American Oriental Society* 108, no. 1 (1988): 51–62.

Rorty, Richard. *Philosophy and the Mirror of Nature*. Princeton, NJ: Princeton University Press, 1979.

Ross, Micah. "A Continuation of the Horocopic Ostraca of Medînet Mâdi." *Egitto e Vicino Oriente* 30 (2007): 153–71.

———. "Further Horoscopic Ostraca from Medînet Mâdi." *Egitto e Vicino Oriente* 32 (2009): 61–91.

———. "Horoscopic ostraca from Medinet Madi." PhD diss., Brown University, 2006.

———. "An Introduction to the Horoscopic Ostraca of Medînet Mâdi." *Egitto e Vicino Oriente* 29 (2006): 147–63.

———. "A Provisional Conclusion to the Horoscopic Ostraca from Medînet Mâdi." *Egitto e Vicino Oriente* 34 (2011): 47–80.

———. "A Survey of Demotic Astrological Texts." *Culture and Cosmos* 11 (2007): 1–31.

Roth, Martha T. *Law Collections from Mesopotamia and Asia Minor*. Atlanta: Scholars Press, 2003.

Roughton, N. A., J. M. Steele, and C. B. F. Walker. "A Late Babylonian Normal and *Ziqpu* Star Text." *Archive for History of Exact Sciences* 58 (2004): 537–72.

Ruby, Jane E. "The Origins of Scientific 'Law.'" *Journal of the History of Ideas* 47 (1986): 341–59.

Rutz, Matthew. *Bodies of Knowledge in Ancient Mesopotamia: The Diviners of Late Bronze Age Emar and Their Tablet Collection*. Ancient Magic and Divination 9. Leiden: Brill, 2013.

Sachs, A. J. "Babylonian Horoscopes." *Journal of Cuneiform Studies* 6, no. 2 (1952): 49–75.

———. "Classification of the Babylonian Astronomical Texts of the Seleucid Period." *Journal of Cuneiform Studies* 2 (1948): 271–90.

———. *Late Babylonian Astronomical and Related Texts*. With the co-operation of J. Schaumberger, copied by T. G. Pinches and J. N. Strassmaier. Providence, RI: Brown University Press, 1955.

———. "A Late Babylonian Star Catalogue." *Journal of Cuneiform Studies* 6 (1952): 146–50.

———. "The Latest Datable Cuneiform Tablets." In *Kramer Anniversary Volume: Cuneiform Studies in Honor of Samuel N. Kramer*, edited by B. Eichler, 379–98. Alter Orient und Altes Testament 25. Kevelaer: Butzon and Bercker, 1976.

Sachs, A. J., and Hermann Hunger. *Astronomical Diaries and Related Texts from Babylon*. Vol. 1, *Diaries from 652 BC to 262 BC*. Vienna: Österreichische Akademie der Wissenschaften,1988.

———. *Astronomical Diaries and Related Texts from Babylon*. Vol. 2, *Diaries from 261 BC to 165 BC*. Vienna: Österreichische Akademie der Wissenschaften, 1989.

———. *Astronomical Diaries and Related Texts from Babylon*. Vol. 3, *Diaries from 164 BC to 61 BC*. Vienna: Österreichische Akademie der Wissenschaften, 1996.

Sahlins, Marshall. *Culture in Practice: Selected Essays*. New York: Zone Books, 2000.

———. *How "Natives" Think, about Captain Cook, for Example*. Chicago: University of Chicago Press, 1995.

———. *What Kinship Is—and Is Not*. Chicago: University of Chicago Press, 2013.

Salmon, Wesley C. "Four Decades of Scientific Explanation." In *Scientific Explanation*, edited by Philip Kitcher and Wesley C, Salmon, 3–219. Minnesota Studies in the Philosophy of Science 13. Minneapolis: University of Minnesota Press, 1989.

———. *Four Decades of Scientific Explanation*. Pittsburgh: University of Pittsburgh Press, 2006.

———. *Introduction to the Philosophy of Science*. Indianapolis: Hackett, 1999.

Salomon, Frank. "Collquiris Dam: The Colonial Re-Voicing of an Appeal to the Archaic." In *Native Traditions in the Postconquest World*, edited by Elizabeth Hill Boone and Tom Cummins, 265–93. Washington, DC: Dumbarton Oaks Research Library and Collection, 1998.

Sanders, Edith R. "The Hamitic Hypothesis: Its Origins and Functions in Time Perspective." *Journal of African History* 10 (1969): 521–32.

Scafi, Alessandro. *Mapping Paradise: A History of Heaven on Earth*. Chicago: University of Chicago Press, 2006.

Scarborough, John. "Alexander of Tralleis." In *Encyclopedia of Ancient Natural Scientists: The Greek Tradition and Its Many Heirs*, edited by Paul T. Keyser and Georgia L. Irby-Massie, 58–59. London: Routledge, 2008.

Schiller, Friedrich. *Poems Translated from the German by Friedrich Schiller*. New York: New York Book, 1898.

Schofield, Malcolm. *The Stoic Idea of the City*. Cambridge: Cambridge University Press, 1991.

Schuster-Brandis, Anais. *Steine als Schutz- und Heilmittel: Untersuchung zu ihrer Verwendung in der Beschwörungskunst Mesopotamiens im 1. Jt. v. Chr*. Münster: Ugarit-Verlag, 2008.

Schwemer, Daniel. "Magic Rituals: Conceptualizations and Performance." In *The Oxford Handbook of Cuneiform Culture*, edited by Karen Radner and Eleanor Robson, 418–42. Oxford: Oxford University Press, 2011.

———. "Washing, Defiling, and Burning: Two Bilingual Anti-Witchcraft Incantations." *Orientalia* n.s. 78 (2009): 44–68.

Scurlock, Joann. "Sorcery in the Stars: STT 300, BRM 4 19–20 and the Mandaean Book of the Zodiac." *Archiv für Orientforschung* 51 (2005–6): 125–46.

———. *Sourcebook for Ancient Mesopotamian Medicine*. Writings from the Ancient World 36. Atlanta: SBL Press, 2014.

Segal, Robert. Review of Frankfort et al., in *Journal of the American Academy of Religion* 47 (1979): 662–64.

Selby-Bigge, L. A., ed. *Enquiries concerning Human Understanding and concerning the Principles of Morals by David Hume*. 3rd edition revised by P. H. Nidditch. Oxford: Clarendon Press, 1975.

Selin, Helaine, ed. *Encyclopedia of the History of Science, Technology, and Medicine in Non-Western Cultures*. 2nd ed. Berlin: Springer, 2008.

Selz, Gebhard J., with the collaboration of Klaus Wagensonner. *The Empirical Dimension of Ancient Near Eastern Studies / Die empirische Dimension altorientalischer Forschung*. Berlin: LIT Verlag, 2011.

Seneca. *Questiones Naturales*, II.32. translated by Thomas Corcoran, 1971.

Shapere, Dudley. "External and Internal Factors in the Development of Science." *Science and Technology Studies* 4 (1986): 1–9.

Shapin, Steven. "History of Science and Its Sociological Reconstructions." *History of Science* 20 (1982): 157–211.

Sharples, R. W. "Soft Determinism and Freedom in Early Stoicism." *Phronesis* 31 (1986): 266–79.

Shea, William. Review of John Brooke and Geoffrey Cantor, *Reconstructing Nature: The Engagement of Science and Religion*. Edinburgh: T. and T. Clark, 1998, in *British Journal for the History of Science* 32 (1999): 237–39.

Shweder, Richard A. "Anthropology's Romantic Rebellion against the Enlightenment, or There's More to Thinking Than Reason and Evidence." In *Culture Theory: Essays on Mind, Self, and Emotion*, edited by Richard A. Shweder and Robert A. LeVine, 27–66. Cambridge: Cambridge University Press, 1984.

Sigerist, Henry. *History of Medicine*. Vol. 1, *Primitive and Archaic Medicine*. Oxford: Oxford University Press, 1951.

Sjöberg, Åke W., and E. Bergmann. "The Collection of the Sumerian Temple Hymns." In *The Collection of the Sumerian Temple Hymns*, edited by Åke W. Sjöberg, E. Bergmann, and Gene B. Gragg. Texts from Cuneiform Sources 3. Locust Valley, NY: J. J. Augustin, 1969.

Slotsky, Alice. *The Bourse of Babylon: Market Quotations in the Astronomical Diaries of Babylonia*. Bethesda, MD: CDL Press, 1997.

Snyder, Gary. *The Practice of the Wild*. San Francisco: North Point Press, 1990.

Soden, Wolfram von. "Leistung und Grenze sumerischer und babylonischer Wissenschaft." *Die Welt als Geschichte* 2 (1936): 411–64 and 509–57.

Sokal, Robert R., and Peter H. A. Sneath. *Principles of Numerical Taxonomy*. San Francisco: W. H. Freeman, 1963.

Soldt, W. H. van. *Solar Omens of Enuma Anu Enlil: Tablets 23 (24)–29 (30)*. Publications de l'Institut historique-archéologique néerlandais de Stamboul 73. Istanbul: Nederlands Historisch-Archaeologisch Instituut Te Istanbul, 1995.

Sorabji, R. N. *Necessity, Cause, and Blame: Perspectives on Aristotle's Theory*. Chicago: University of Chicago Press, 1980.

Speiser, E. A. "Authority and Law in Ancient Mesopotamia." *Journal of the American Oriental Society Supplement* 17 (1954): 8–15.

———. "Cuneiform Law and the History of Civilization." *Proceedings of the American Philosophical Society* 107 (1963): 536–41.

Špelda, Daniel. "The Search for Antediluvian Astronomy: Sixteenth- and Seventeenth-Century Astronomers' Conceptions of the Origin of the Science." *Journal for the History of Astronomy* 44 (2013): 337–62.

Stadhouders, H. "The Pharmacopoeial Handbook *šammu šikinšu*: An Edition." *Journal des médecines cunéiformes* 18 (2011): 3–51.

———. "The Pharmacopoeial Handbook *šammu šikinšu*: A Translation." *Journal des médecines cunéiformes* 19 (2012): 1–20.

Starr, Ivan. *Queries to the Sun-God: Divination and Politics in Sargonid Assyria*. State Archives of Assyria 4. Helsinki: Helsinki University Press, 1990.

———. *The Rituals of the Diviner*. Bibliotheca Mesopotamica 12. Malibu, CA: Undena, 1983.

Steele, John M. "Astronomy and Culture in Late Babylonian Uruk." In *"Oxford IX" International Symposium on Archaeoastronomy Proceedings IAU Symposium No. 278*, edited by Clive L. N. Ruggles, 331–41. Cambridge: Cambridge University Press, 2011.

———. "Celestial Measurement in Babylonian Astronomy." *Annals of Science* 64 (2007): 293–325.

———. "Eclipse Prediction in Mesopotamia." *Archive for History of Exact Sciences* 54 (2000): 421–54.

———. "Goal-Year Periods and Their Use in Predicting Planetary Phenomena." In *The Empirical Dimension of Ancient Near Eastern Studies / Die empirische Dimension altorientalischer Forschung*, edited by Gebhard J. Selz with the collaboration of Klaus Wagensonner, 101–10. Berlin: LIT Verlag, 2011.

———. "The Length of the Month in Mesopotamian Calendars of the First Millennium BC." In *Calendars and Years: Astronomy and Time in the Ancient Near East*, edited by John M. Steele, 133–48. Oxford: Oxbow Books, 2007.

———. *Observations and Predictions of Eclipse Times by Early Astronomers*. Archimedes 4. Dordrecht: Kluwer Academic, 2000.

Steele, John M., and J. M. K. Gray. "A Study of Babylonian Observations Involving the Zodiac." *Journal for the History of Astronomy* 38 (2007): 443–58.

Steinkeller, Piotr. "Luck, Fortune, and Destiny in Ancient Mesopotamia—Or How the Sumerians and Babylonians Thought of Their Place in the Flow of Things." Paper presented at the sixtieth Rencontre Assyriologique Internationale, "Fortune and Misfortune in the Ancient Near East," Warsaw, July 21–25, 2014.

———. "Of Stars and Men: The Conceptual and Mythological Setup of Babylonian Extispicy." In *Biblical and Oriental Essays in Memory of William L. Moran*, edited by A. Gianto, 11–47. Biblica et Orientalia 48. Rome: Pontifical Biblical Institute, 2005.

Stich, Stephen P. "Could Man Be an Irrational Animal? Some Notes on the Epistemology of Rationality." *Synthese* 64 (1985): 115–35.

Stockhusen, Marco. "Babylonischen Astralwissenschaften im römerzeitlichen Ägypten: Das Beispiel Medînet Mâdi." *Welt des Orients* 42 (2012): 85–109.

Stol, Marten. "The Moon as Seen by the Babylonians." In *Natural Phenomena: Their Meaning, Depiction, and Description in the Ancient Near East*, edited by Diederik J. W. Meijer, 245–75. North-Holland, Amsterdam; New York: Royal Netherlands Academy of Arts and Sciences, 1992.

Stolleis, Michael. "The Legitimation of Law through God, Tradition, Will, Nature, and Constitution." In *Natural Law and Laws of Nature in Early Modern Europe: Jurisprudence, Theology, Moral and Natural Philosophy*, edited by Lorraine Daston and Michael Stolleis, 45–55. Surrey: Ashgate, 2008.

Strabo. *Geography*. Translated by H. L. Jones. Vol. 7. Loeb Classical Library 241. Cambridge, MA: Harvard University Press.

Strassmaier, J. N. "Arsaciden-Inschriften." *Zeitschrift für Assyrologie und Vorderasiatische Archäologie* 3 (1888): 129–58.

———. *Inschriften von Cambyses*. Vol. 9 of *Babylonische Texte*. Leipzig: Eduard Pfeiffer, 1890.

Strathern, Marilyn. "No Nature No Culture: The Hagen Case." In *Nature, Culture, and Gender*, edited by Carol P. MacCormack and Marilyn Strathern, 174–222. 1980. Reprint, Cambridge: Cambridge University Press, 1998.

Strawson, P. F. *Analysis and Metaphysics: An Introduction to Philosophy*. Oxford: Oxford University Press, 1992.

Streck, Michael P. Review of *The Assyrian Dictionary of the University of Chicago, Ill: Oriental Institute*. Vol. 19, in *Zeitschrift für Assyrologie und Vorderasiatische Archäologie* 99 (2009): 136–40.

Striker, Gisela. "Following Nature: A Study in Stoic Ethics." *Oxford Studies in Ancient Philosophy* 9 (1991): 1–74.

———. "Origins of the Concept of Natural Law." In *Proceedings of the Boston Area Colloquium in Ancient Philosophy*, vol. 2, edited by John J. Cleary, 79–94. Lanham, MD: University Press of America, 1987.

Suárez, Mauricio. "Scientific Representation." *Philosophy Compass* 5 (2010): 91–101.

Swerdlow, N. M. *The Babylonian Theory of the Planets*. Princeton, NJ: Princeton University Press, 1998.

Tambiah, Stanley Jeharaya. *Magic, Science, Religion, and the Scope of Rationality*. Cambridge: Cambridge University Press, 1990.

Taylor, Daniel M. *Explanation and Meaning: An Introduction to Philosophy*. Cambridge: Cambridge University Press, 1970.

Thomsen, Marie-Louise, and Frederick H. Cryer. *Witchcraft and Magic in Europe: Biblical and Pagan Societies*. Athlone History of Witchcraft and Magic in Europe 1. London: Athlone Press, 2001.

Thorndike, Lynn. *A History of Magic and Experimental Science*. Vol. 1. New York: Columbia University Press, 1923.

———. "The True Place of Astrology in the History of Science." *Isis* 46 (1955): 273–78.

Tinney, Steve. "Tablets of Schools and Scholars: A Portrait of the Old Babylonian Corpus." In *The Oxford Handbook of Cuneiform Culture*, edited by Karen Radner and Eleanor Robson, 577–96. Oxford: Oxford University Press, 2011.

Turner, Terence. "The Crisis of Late Structuralism: Perspectivism and Animism; Rethinking Culture, Nature, Spirit, and Bodiliness." *Tipití: Journal of the Society for the Anthropology of Lowland South America* 7 (2009): 3–42.

Tversky, Amos, and Daniel Kahneman. "Judgment under Uncertainty: Heuristics and Biases." *Science* 185 (1974): 1124–31.

Van de Mieroop, Marc. *Philosophy before the Greeks: The Pursuit of Truth in Ancient Babylonia*. Princeton, NJ: Princeton University Press, 2016.

Van Dijk, J. J. *LUGAL UD ME-LÁM-bi NIR-GÁL: Le récit épique et didactique des Travaux de Ninurta, du Déluge et de la nouvelle Création*. Leiden: Brill, 1983.

Van Fraassen, Bas. *The Empirical Stance*. New Haven, CT: Yale University Press, 2002.

———. "The False Hopes of Traditional Epistemology." *Philosophy and Phenomenology Research* 60 (2000): 253–80.

Veldhuis, Niek. *The Cow of Sîn*. Library of Oriental Texts 2. Groningen: Styx, 1991.

———. "Divination: Theory and Use." In *If a Man Builds a Joyful House: Assyriological Studies in Honor of Erle Verdun Leichty*, edited by A. Guinan et al., 487–97. Leiden: Brill, 2006.

———. *History of the Cuneiform Lexical Tradition*. Guides to the Mesopotamian Textual Record 6. Münster: Ugarit Verlag, 2014.

———. "How to Classify Pigs: Old Babylonian and Middle Babylonian Lexical Texts." In *De la domestication au tabou: Le cas des suidés dans le Proche-Orient ancien*, edited by Cécile Michel and Brigitte Lion, 25–29. Travaux de la Maison René-Ginouvès 1. Paris: Éditions de Boccard, 2006.

———. "The Poetry of Magic." In *Mesopotamian Magic: Textual, Historical, and Interpretative Perspec-*

tives, edited by T. Abusch and K. van der Toorn, 35–48. Studies in Ancient Magic and Divination 1. Groningen: Styx, 1999.

———. *Religion, Literature, and Scholarship: The Sumerian Composition "Nanše and the Birds."* Cuneiform Monographs 22. Leiden: Brill and Styx, 2004.

———. "The Theory of Knowledge and the Practice of Celestial Divination." In *Divination and Interpretation of Signs in the Ancient World*, edited by Amar Annus, 77–91. Oriental Institute Seminars 6. Chicago: Oriental Institute, 2010.

Verderame, Lorenzo. "*Enūma Anu Enlil* Tablets 1–13." In *Under One Sky: Astronomy and Mathematics in the Ancient Near East*, edited by John M. Steele and Annette Imhausen, 447–57. Alter Orient und Altes Testament 297. Münster: Ugarit Verlag, 2002.

———. "The Halo of the Moon." In *Divination in the Ancient Near East: A Workshop on Divination Conducted during the 54th Rencontre Assyriologique Internationale, Würzburg, 2008*, edited by Jeanette Fincke, 91–104. Winona Lake, IN: Eisenbrauns, 2014.

———. *La Tavole I–VI della serie astrologica Enuma Anu Enlil.* NISABA 2. Messina: Di.Sc.A.M, 2002.

Versnel, H. S. "Some Reflections on the Relationship Magic-Religion." *Numen* 38 (1991): 177–97.

Virolleaud, Charles. *L'astrologie chaldéenne: Le livre intitulé "enuma (Anu)ilu Bêl," 2 suppléments.* Paris: P. Geuthner, 1908–10.

Volk, Katharina. 2010. "Aratus." In *A Companion to Hellenistic Literature*, edited by James J. Clauss and Martine Cuypers, 199–210. Oxford: Wiley-Blackwell, 2010.

Von Staden, Heinrich. "Galen's Daimon: Reflections on 'Irrational' and 'Rational.'" In *Rationnel et irrationnel dans la médicine ancienne et médiévale: Aspects historiques, scientifiques et culturels*, edited by Nicoletta Palmieri, 15–43. Saint-Étienne: Publications de l'Université, 2003.

Walker, C. B. F. "Babylonian Observations of Saturn during the Reign of Kandalanu." In *Ancient Astronomy and Celestial Divination*, edited by Noel. M. Swerdlow, 61–76. Cambridge, MA: MIT Press, 1999.

———. "Notes on the Venus Tablet of Ammiṣaduqa." *Journal of Cuneiform Studies* 36 (1984): 64–65.

Wapnish, Paula Claire. "Animal Names and Animal Classifications in Mesopotamia: An Interdisciplinary Approach Based on Folk Taxonomy." PhD diss., Columbia University, 1984.

Wasserman, N. *Style and Form in Old Babylonian Literary Texts*. Cuneiform Monographs 27. Leiden: Brill and Styx, 2002.

Watson, Rita, and Wayne Horowitz. *Writing Science before the Greeks: A Naturalistic Analysis of the Babylonian Astronomical Treatise MUL.APIN*. Leiden: Brill, 2011.

Wee, John Z. "Discovery of the Zodiac Man in Cuneiform." *Journal of Cuneiform Studies* 67 (2015): 217–33.

Weill-Parot, Nicholas. "Astrology, Astral Influences, and Occult Properties in the Thirteenth and Fourteenth Centuries." *Traditio* 65 (2010): 201–30.

Wengrow, David. "The Intellectual Adventure of Henri Frankfort." *American Journal of Archaeology* 103 (1999): 597–613.

Westbrook, Raymond. "Judges in the Cuneiform Sources." In *Law from the Tigris to the Tiber: The Writings of Raymond Westbrook*. Vol. 2, *Cuneiform and Biblical Sources*, edited by B. Wells and R. Magdalene, 197–210. Winona Lake, IN: Eisenbrauns, 2009.

———. "Law Codes and Omen Series: Practical Application." In *Law from the Tigris to the Tiber: The Writings of Raymond Westbrook*. Vol. 1, *The Shared Tradition*, edited by B. Wells and R. Magdalene, 9–20. Winona Lake, IN: Eisenbrauns, 2009.

Westenholz, Joan Goodnick. "The Clergy of Nippur." In *Nippur at the Centennial*, edited by Maria deJong Ellis, 297–310. Occasional Publications of the Samuel Noah Kramer Fund 14. Philadelphia: University Museum, 1992.

Wiggermann, F. A. M. "Ring und Stab (Ring and Rod)." In *Reallexikon der Assyriologie und vorderasiatischen Archäologie*, vol. 11, 414–21. Berlin: De Gruyter, 2008.

Wiggermann, F. A. M., and W. Van Binsbergen. "Magic in History, a Theoretical Perspective, and Its Application to Ancient Mesopotamia." In *Mesopotamian Magic: Textual, Historical, and Interpre-*

tative Perspectives, edited by T. Abusch and K. Van der Toorn, 1–34. Ancient Magic and Divination 1. Groningen: Styx, 1999.

Wilcke, Claus. *Early Ancient Near Eastern Law: A History of Its Beginnings; The Early Dynastic and Sargonic Periods*. Bayerische Akademie der Wissenschaften Philosophisch-Historische Klasse. Sitzungsberichte Jahrgang 2003, Heft 2. Munich: Verlag der Bayerischen Akademie der Wissenschaften, 2003.

Williams, Clemency. "Signs from the Sky, Signs from the Earth: The Diviner's Manual Revisited." In *Under One Sky: Astronomy and Mathematics in the Ancient Near East*, edited by John M. Steele and Annette Imhausen, 473–85. Münster: Ugarit Verlag, 2002.

Williams, Raymond. *Key Words: A Vocabulary of Culture and Society*. New York: Oxford University Press, 1976, 1983 rev. ed.

———. *Problems in Materialism and Culture*. London: Verso, 1980.

Wilson, Robert A., Matthew J. Barker, and Ingo Brigandt. "When Traditional Essentialism Fails: Biological Natural Kinds." *Philosophical Topics* 25 (2007): 189–215.

Winitzer, A. "Writing and Mesopotamian Divination: The Case of Alternative Interpretation," *Journal of Cuneiform Studies* 63 (2011): 77–94.

Wiseman, D. J., and J. A. Black. *Literary Texts from the Temple of Nabû*. London: British School of Archaeology in Iraq, 1996.

Wittgenstein, Ludwig. *On Certainty*. Edited by G. E. M. Anscombe and G. H. von Wright. Translated by Denis Paul and G. E. M. Anscombe. New York: Harper Torch Books, 1972.

Yoffee, Norman. "Context and Authority in Early Mesopotamian Law." In *State Formation and Political Legitimacy*, edited by Ronald Cohen and Judith D. Toland, 95–113. Political Anthropology 6. New Brunswick, NJ: Transaction Books, 1988.

Zgoll, Annette. *Traum und Welterleben im antiken Mesopotamien: Traumtheorie und Traumpraxis im 3.–1. Jahrtausend v. Chr. als Horizont einer Kulturgeschishte des Träumens*. Alter Orient und Altes Testament 333. Münster: Ugarit-Verlag, 2006.

Zgoll, Annette, and Kai Lämmerhirt. "Lachen und Weinen im antiken Mesopotamien." In *Überraschendes Lachen, gefordertes Weinen*, edited by August Nitsche, Justin Stagl, and Dieter R. Bauer, 449–83. Vienna: Böhlau Verlag, 2009.

Zhmud, Leonid. "On the Concept of 'Mythical Thinking.'" *Hyperboreus* 1/2 (1994–95): 155–69.

Zilsel, Edgar. "The Genesis of the Concept of Physical Law." *Philosophical Review* 51 (1942): 245–79.

Zwart, Hub. *Understanding Nature: Case Studies in Comparative Epistemology*. Berlin: Springer Verlag, 2008.

Index